Practical Systems Analysis

Practical Systems Analysis

Anthony Chandor
John Graham
Robin Williamson

Rupert Hart-Davis Educational Publications
3 Upper James Street Golden Square London W1

SBN: 298 17868 0

Printed in Great Britain by
Willmer Brothers Limited
Birkenhead

Contents

INTRODUCTION   17

PART 1   THE ROLE OF THE SYSTEMS ANALYST   19

1.1   THE SYSTEMS ANALYST   21
Qualities and abilities – contemplative – single-minded – receptive – energetic – communication skills – accuracy – responsibility.

1.2   THE DATA PROCESSING FUNCTION   24
Place in an organisation – responsibility for data processing – steering committee – structure within the department – functions to be fulfilled.

1.3   STAGES OF SYSTEMS ANALYSIS   27
Tasks involved – grouping of functions – four main stages.

PART 2   INVESTIGATION AND ANALYSIS   35

2.1   ESTABLISHING SYSTEMS OBJECTIVES   37

2.1.1   Feasibility studies   37
Defining the problem – objects of study – project assignment.

2.1.2   Identifying areas for systems study   38
Preliminary assignment brief – standard format – scope.

2.1.3   Identifying management requirements for information   39
Conflicting constraints – uncertainty of real need – the base for further development.

2.1.4   Assessment of benefits and costs   40
Current costs – likely economies – assessment of benefits – detailed costs.

2.1.5   Formal definition of purpose and objectives   43
Instruction to systems analyst – systems project assignment brief – advantages of formal procedure.

2.1.6   Establishing project teams   45
Dangers of loss of contact – involvement – guide for project teams.

2.1.7   Summary of section 2.1   46

2.2   INVESTIGATION AND ANALYSIS OF EXISTING PROCEDURES   47

2.2.1   Introduction   47

PART 2 (continued)

2.2.2   Methods of fact finding   47
Reading – charts, procedure manuals, files – questioning – observing.

2.2.3   Interviewing techniques   50
Promoting confidence – preparation for interviews – structure of the interview – two main tasks – seven hints for interviewing.

2.2.4   Methods of fact recording   54
Systems file – directory section – document section – form descriptions – file descriptions – record descriptions – interview section – chart section – block diagram – vertical section chart – horizontal flowchart – computer procedures flowchart – suggestions section.

2.3   ANALYSIS OF RECORDED DATA   72
2.3.1   Analysis as a separate stage   72
2.3.2   Interpretation   74
2.3.3   Analysis   75
The time for opinion – system review – alternative solutions.

PART 3   DATA PROCESSING EQUIPMENT   77
3   INTRODUCTION   79
3.1   PROCESSORS   82

3.1.1   Information patterns   82
General – basic characteristics – internal and external formats parity checks.

3.1.2   Control logic   86
Registers – control unit – arithmetic unit – method of operation.

3.1.3   Core storage in relation to central processors   88
General – single-level storage.

3.1.4   Compatibility   89
Types of compatibility – methods of achieving compatibility.

3.1.5   Data operations   90
General – arithmetical operations – logical operations – George Boole – floating point and fixed point representation.

3.1.6   Timesharing and multiprocessing   93
Peripheral transfers – multiprogramming – priorities – security.

3.2   PERIPHERAL UNITS   94

3.2.1   Interface   95
General – transfer modes.

3.2.2   Card reader   95
General – card codes – operation of card reader – characteristics – error checking.

3.2.3   Card punch   98
General – card codes – operation of card punch – characteristics – error checking.

3.2.4	Paper tape reader	100
	General – paper tape codes – modes of reading – characteristics – error checking.	
3.2.5	Paper tape punch	102
	General – paper tape codes – modes of punching – characteristics – error checking.	
3.2.6	Printer	103
	General – types of line printer – operation – control loop – error checking.	
3.2.7	Typewriter	105
	General – types of typewriter – method of operation.	
3.2.8	Visual display unit	106
	General – types of visual display units – method of operation.	
3.2.9	Graph plotter	108
	General – types of graph plotter – method of operation.	
3.2.10	Character recognition equipment	109
	General – types of character recognition equipment – method of operation – MICR – OCR – mark reading.	
3.2.11	Communication equipment	113
	General – modulation – line systems – message switching on-line and off-line working – error detection units – Post – Office transmission facilities.	
3.3	STORAGE	115
3.3.1	Core storage	116
	General – magnetic cores – operation of core storage.	
3.3.2	Magnetic tape	117
	General – operation of magnetic tape unit – characteristics – error checking.	
3.3.3	Magnetic drums	120
	General – operation of magnetic drum – organisation of data on a drum – error checking.	
3.3.4	Magnetic disc storage	121
	General – operation of magnetic disc – data organisation – error checking.	
3.3.5	Magnetic card files	123
	General – magnetic card – magazine – retrieval unit – data organisation – error checking.	
3.4	PROGRAMMING	128
3.4.1	Flowcharting	128
3.4.2	Decision tables	129
3.4.3	Coding	135
	Instructions – basic principles – segmentation and overlays – communication – documentation.	
3.4.4	Languages	138
	Low level – high level – commercial – scientific – general languages – generators – compilers and assemblers – simulators.	

PART 3 (continued)

3.4.5   Diagnostics   141
Error types – checking – compiler messages – program testing –
monitor prints – trace routines – store prints.

3.4.6   Remote testing   145
Open and closed shop – on-line testing.

3.4.7   Programming techniques   146
General – modification and loops – subroutines – switches –
table look-up – buffering.

3.5   UTILITY SOFTWARE   151

3.5.1   Housekeeping packages   152
General – basic peripheral control software – data editing
software – storage device control – interrogation and display
unit control.

3.5.2   Generators   153
General principles – report generators – sort/merge
generators.

3.5.3   Sorts   154
Principles of sorting – sort keys – sort techniques (magnetic
tape) – sort techniques (other devices).

3.5.4   File control   157
Principles – labelling – scratching – copying – printing –
merging – maintenance – loading.

3.5.5   Data management software   159
General – data management operations – systems applications.

3.5.6   Program handling software   160
General – loading – run-time amendment – dumping.

3.5.7   Program maintenance   161
General – program libraries – documentation.

3.5.8   Dump and restart   163
General – dumping – restarting.

3.6   OPERATING SYSTEMS   164

3.6.1   General principles   164
General – foreground and background working –
hardware considerations.

3.6.2   Command languages   165
Job description – job set-up – job running – software
utilisation – operating system control.

3.6.3   On-line and off-line working   167
General – the document concept.

3.6.4   File storage   168
Types of file – file creation – file storage utilisation –
file security.

3.6.5   Job scheduling   171
Principles of scheduling – job concept.

3.6.6   Conversational mode   172
         General – types of conversational mode – function of
         operating system in conversational mode.
3.6.7   Multi-access   173
         General – operating systems and multi-access.
3.6.8   Systems design for operating systems   174
         Input and output functions – file structures – program
         specification.

  3.7   DATA PREPARATION   178
3.7.1   Punched cards   178
         Introduction – 80 column card – punching codes – card layout
         punching cards – programming a card punch – gangpunching –
         speeds of card punching – verifying punched cards.
3.7.2   Paper tape   188
         Introduction – characteristics of paper tape – error detection –
         verification methods – systems considerations.
3.7.3   Mark sensing   191
         General – method of operation – characteristics of mark
         sensing systems.
3.7.4   Keyboard to magnetic tape   192
         General – description of unit – error checking – systems
         advantages.
3.7.5   Source documents direct to computer   193
         Reference to 3.2.10 character recognition – review.

  3.8   PUNCHED CARD EQUIPMENT   194
3.8.1   Sorter   194
3.8.2   Reproducer and gangpunch   197
3.8.3   Interpreter   198
3.8.4   Electronic calculators   198
3.8.5   Collator   199
3.8.6   Tabulator and summary punch   200

PART 4   DESIGN OF COMPUTER PROCEDURES   205
         INTRODUCTION   207

  4.1   ESTABLISHING THE OUTPUT REQUIREMENTS   207
4.1.1   Importance of a clear objective   207
         Output requirements – timing of data – volume and sequence
         of data – approval of output design.
4.1.2   Definition of output formats   209
         Print layout charts – dummy reports – print schedules –
         field limitations.
4.1.3   Approval of output formats   211
         Approval levels – supporting data – appearance of output –
         multiple copies.

9

*A

PART 4 (continued)

4.1.4   Choice of output medium   214
Turn-around documents – magnetic tape output – on-line output.

4.1.5   Principles of exception reporting   215
*Need for exception reporting – examples – need for flexibility –
programming for flexibility.*

4.1.6   Error reporting conditions   217
*Requirement for error reporting – acceptance of error –
guide lines for error reporting.*

4.1.7   Appraisal of output effectiveness   218

4.2   ANALYSING DATA COLLECTION REQUIREMENTS   219

4.2.1   The impact of output requirements on data collection   219
*Classification of data elements – reporting frequency –
transaction sources.*

4.2.2   Developing data collection procedures   220
Data characteristics – choice of input medium – volumes.

4.2.3   Source documents and data preparation   222
*Source documents – design of source documents – routing of
source documents – batch totals – batch sizes – invalid data.*

4.2.4   Summary of data collection principles   226

4.3   SYSTEMS CONTROLS   226

4.3.1   Standard hardware and software controls   226
*Hardware controls – file processing software controls – write
permit or write inhibit rings – program lockout – monitoring live
operations – check digits – transcription errors – transposition
errors – shift errors – modulo 11 check.*

4.3.2   Control accounts   229
*Use of control accounts — control stages — maintenance of
control procedures.*

4.3.3   File control and file security   232
*General – physical control of files – grandfather, father, son –
dumping of disc files – release of files.*

4.3.4   Systems controls and auditors   234
*Role of the audit department – initial approval methods –
verifying control procedures – verification of a live system.*

4.4   PROCESSING METHODS FOR MAGNETIC TAPE   236

4.4.1   Characteristic structure of magnetic tape systems   236
*Serial medium – batch processing – file relationships –
basic run structures.*

4.4.2   Input transcription programs   240
*Comparative speed of input – batch totals – check sums –
validation – example of logical validation – input editing –
input processing.*

4.4.3   Sorting and merging   244
*Sequencing considerations – record type coding – minimising
sort operations – editing while sorting – key changes on a main
file – sorting multi-reel files.*

10

4.4.4 File updating methods   250
Types of master data – distinctions between updating and
maintenance – the nature of basic updating runs – processing
operations in a simple updating run – dealing with end of
file conditions – key changes during updating.

4.4.5 File maintenance   260
What is file maintenance – user participation – techniques of
file maintenance — maintenance of values for fixed fields –
deletion of records – maintenance of correct sequence and
structure – need for maintenance.

4.4.6 Error reporting   263
Error conditions – dealing with errors – re-entry of corrections –
errors and control totals – maintenance of an error file.

4.4.7 Reporting from magnetic tape files   268
Creation of sub-files – flexibility – economy of print out.

4.4.8 Multiprogramming with magnetic tape files   269
Design standards – peripheral-limited and processor-limited
runs – mixing run types – scheduling runs – peripheral utilisation.

4.4.9 Dump and restarts using magnetic tape   272
Dump routines – restart routines – use of dumps and restarts.

4.5 THE DESIGN OF A MAGNETIC TAPE SYSTEM   275

4.5.1 Magnetic tape files – structure and relationships   275
General aims – scheduling for multiprogramming – input/output
balance – frequency of file processing.

4.5.2 Record design for magnetic tape file   277
General influences on record design – minimising input/output
time – hardware/software implications – fixed length and
variable length records.

4.5.3 Block organisation for magnetic tape files   281
General influences on file organisation – main memory utilisa-
tion – input/output balance – variable length blocks.

4.5.4 Timing and evaluation of magnetic tape   287
Purpose of timing estimates – general timing method – peripheral
time – total elapsed peripheral time – input/output channels –
data conditions – processor time – overall running time for a
program – peripheral times – paper tape files – card files – printer
output files – magnetic tape files.

4.6 DIRECT ACCESS PROCESSING   296

4.6.1 Direct access file storage principles   296
Use of direct access – modes of processing – the seek area
concept – a bucket, the unit of data retrieval – basic principles
of file organisation – bucket packing density.

4.6.2 File types and processing modes   301
Serial files – sequential files – random files.

4.6.3 Indexing methods for sequential files   303
Self indexing – partial indexing – packing density and bucket
overflow – summary.

11

PART 4 (continued)

4.6.4   Overflow of sequential files   306
         Reasons for overflow – chaining – tagging – file reorganisation–
         analysing overflow.

4.6.5   Timing characteristics of sequential files   309
         Processing methods; sequentially, selective-sequentially,
         random – timing estimates – effects of overflow – file processing
         times – dealing with existing overflow records – dealing with
         new overflow records – summary.

4.6.6   Random files and address generation   314
         Address generation principles – overflow in random files.

4.6.7   Inverted files   316
         Descriptive record structure – problems of information
         retrieval – inverted file structure – information retrieval from
         inverted files – hardware and software considerations.

PART 5   SYSTEMS IMPLEMENTATION   321

5.1   INVOLVEMENT IN IMPLEMENTATION   323
5.1.1   User participation   323
         General – implementation stages.
5.1.2   Project teams   323
         General – types of project – special considerations.

5.2   EDUCATION FOR IMPLEMENTATION   325
5.2.1   Planning education   325
         General programme – staff categories – time scales.
5.2.2   Types of training   326
         General – full-time courses – part-time courses – week-end
         courses – lectures – on the job – simulation – scheduling.

5.3   TESTING AND ACCEPTANCE   330
5.3.1   Test data   330
         Importance of testing – program test data – procedure test data.
5.3.2   Program acceptance   332
         Modifications to programs – testing against expected results –
         acceptance procedure.
5.3.3   Procedure acceptance   332
         Major acceptance points – input documents – input controls –
         validation procedures – output formats – system timings.

5.4   CHANGEOVER   334
5.4.1   Changeover planning   334
         Scheduling of changeover – continuity – timing.
5.4.2   Pilot schemes and parallel running   335
         Need for smooth transition – pilot schemes – parallel running.

5.5   FILE CREATION AND CONVERSION   336
5.5.1   Initial file creation: data collection   336
         Parallel data collection – phased file creation – dummy file
         creation – special exercise – coding.

5.5.2   Converting existing files   339
         Record structures – simulated updating – use of software –
         partial file creation.

PART 6   DOCUMENTATION STANDARDS   341

6.1   WHY DOCUMENTATION IS NECESSARY   343
         Long lead times – communication – standardisation.

6.2   THE ASSIGNMENT BRIEF   344
         Two stages – initiation – approval.

6.3   THE INVESTIGATION FILE   345
         Purpose – check list of contents.

6.4   WORKING FILES   346
         Purpose – contents – eight sections.

6.5   FEASIBILITY REPORTS   348
         Purpose – contents – twelve sections – summary.

6.6   THE SYSTEMS DEFINITION   353
         Purpose – contents – nine sections.

6.7   COMPUTER OPERATIONS GUIDE   359
         Purpose – check list.

6.8   OPERATING INSTRUCTIONS   360
         Purpose – sample instructions.

PART 7   SYSTEMS MAINTENANCE   363

7.1   INTRODUCTION   365

7.2   TYPES OF CHANGE   365
         General – equipment changes – organisation – user attitudes.

7.3   TYPES OF FLEXIBILITY   368
         General – data – program – procedure – documentation.

7.4   SYSTEMS REVIEW   369

         BIBLIOGRAPHICAL NOTE   371

         INDEX (key words in context)   371

13

Introduction

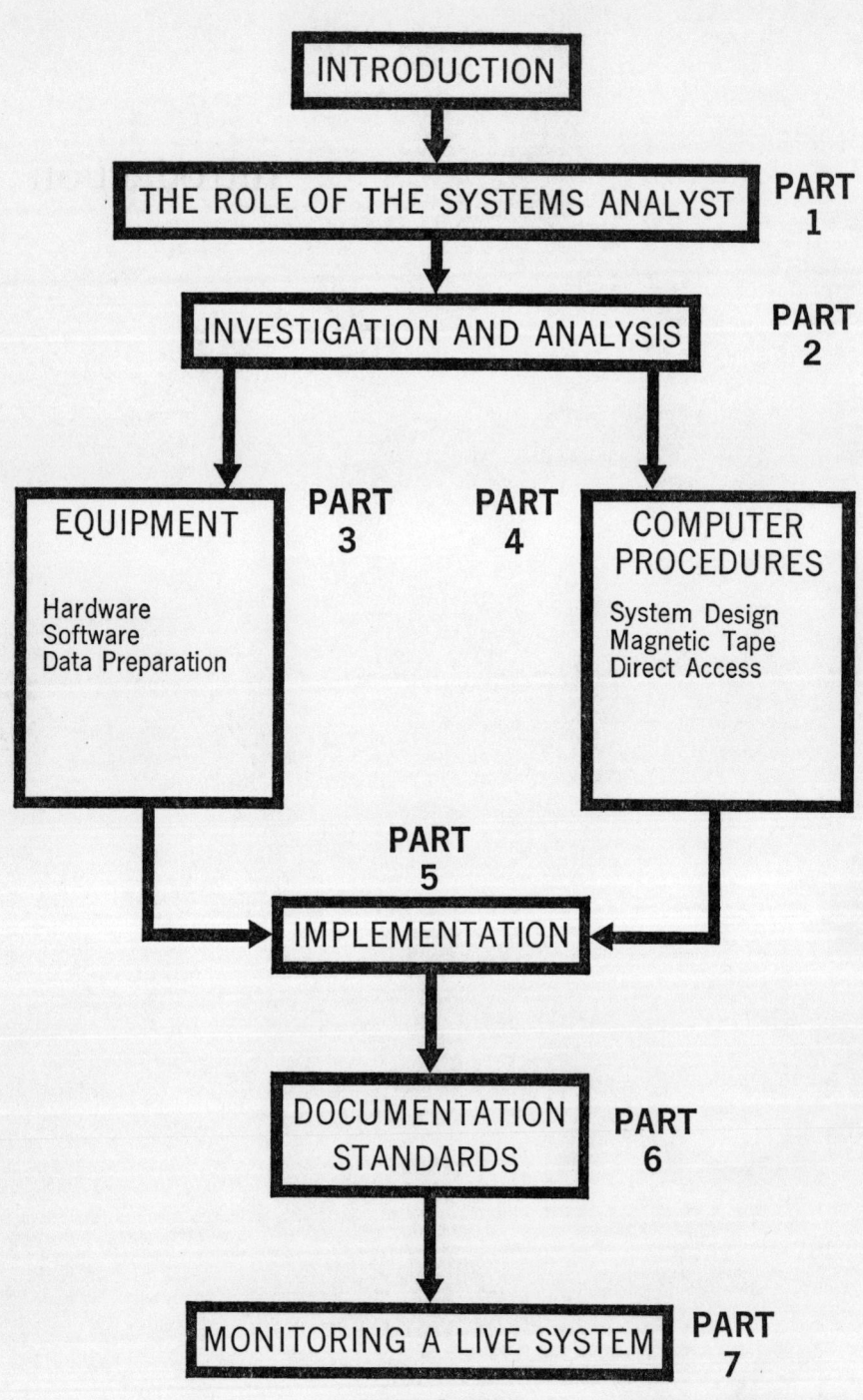

1. Outline flowchart

INTRODUCTION

The systems analyst's job is to make sure that systems work in the best possible way, and it is not hard to see that anyone carrying out such a widely-defined task must be endowed with many skills and be familiar with many different techniques. During the course of this book we shall be examining these techniques in considerable detail, and the Contents pages provide a check list of all the things a systems analyst needs to know before he tries to design a major system. It will be seen that this check list is by no means confined to purely technical matters such as the mysteries of block organisation and direct access timing: given time and resources, even a bad systems analyst will eventually persuade a machine to provide satisfactory output from satisfactory input, but most systems are far more dependent on people than on machines and it is just as much the systems analyst's task to make sure that *people* are fulfilling the proper functions in his system as to make sure that the *machine* is doing so. In designing this book we have therefore given due weight to the many systems considerations involving people so that potential systems analysts who read this book will know how to implement systems for *users* rather than merely for machines.

The second main factor has been the realisation that it is just as important to know what is likely to go wrong as it is to know what happens when everything goes right. We have therefore made sure that solutions to problems are, throughout the book, given in the context of likely errors and have stressed the assumption that every systems analyst should make at all times: 'I am probably wrong'. The emphasis is on the practical problems involved in the design and implementation of a system.

Thirdly, we have accepted that it is not always possible to group subjects in a way that will be satisfactory both for learning and for later reference, and have therefore tried to make sure that each small section of the book can be useful as a reference document as well as being easy to read in continuous narrative. Sometimes the grouping of these 'modules' may need to be altered when a particular subject is under review for reference, and we have adopted a systems technique known as 'flowcharting' to solve this problem: the main flowchart of the contents pages (figure 1) shows the major blocks into which the book is divided while the lists in each block show the level of detail discussed. Each main section is preceded by a further flowchart which will help in establishing suitable routes from one module to another, and the form of the index, where key words are shown in context, will also allow these routes to be easily found.

PART 1
The Role of the Systems Analyst

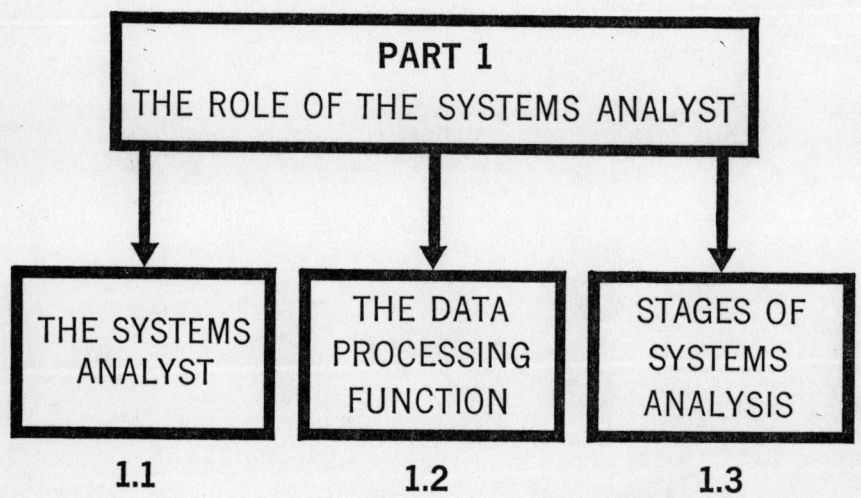

PART 1
THE ROLE OF THE SYSTEMS ANALYST

THE SYSTEMS ANALYST

1.1

THE DATA PROCESSING FUNCTION

1.2

STAGES OF SYSTEMS ANALYSIS

1.3

1.1 THE SYSTEMS ANALYST

First, the name: it is regularly pointed out that systems analysts do very much more than analyse, and various attempts have been made to gain acceptance for such phrases as 'systems designer'. It is probably true to say, though, that the name 'systems analyst' is so well established that, even though the job is known to include systems design, procedure design, and responsibility for implementation, its component parts are no longer considered separately and most people are happy enough to accept that the phrase covers all aspects involved in making sure that a system works. This still leaves some argument about whether the man is a 'systems analyst' or a 'system analyst', but it's not an argument which rouses great passion; in either case the analyst has a very considerable responsibility and must look on himself as a master of all trades—he may be sure that any failure will be blamed on his ignorance of some aspect of the system for which he is generally responsible.

Systems analysts are not, of course, all alike, but there are various qualities and abilities which all should possess in some degree. For example, one could say that if a systems analyst is contemplative and single-minded, yet energetic and receptive, and at the same time technically able, then he is well equipped for the task. Let us look at each of these in turn, and in so doing build up a picture of why each attribute is required.

Contemplative He must be able to examine quietly and thoroughly, and to continue examining however acceptable or accepted something may be. Many systems have taken as their cornerstone some statement which has appeared to be true and which everyone has supported as true but which, in certain circumstances, has proved to be disastrously untrue, and many a systems analyst has regretted the fact that an 'absolutely invariably' was not cross-questioned hard enough to be modified to an 'almost always'. An ability to question everything is therefore very important, but it should be made clear that this is not necessarily questioning out loud. Few things are more annoying than having everything you say questioned, particularly with a refrain of 'Are you sure?' and in section 2.2.3 on Interviewing more is said on this subject. But whether a question is asked out loud or not, the systems analyst should be posing the question and reviewing each new fact in the light of those he has already gathered.

Single-minded It is very unusual for the requirements of a system to remain unmodified from the time they are first analysed to the time the system is implemented, and in an ordinary commercial environment, where an original requirement is often likely to be altered as a result of organisational change, the systems analyst sometimes seems alone in his determination to see a system implemented: others may be anxious to see a system *exist* but are only too ready to alter its input, its output, its surroundings

21

and even its purpose. Many types of change can be catered for without difficulty, but the systems analyst needs to be always aware of the ease with which changing organisational structures can affect the implementation of a system, and must sometimes therefore exert considerable pressures to protect the investment already made in the system he has designed. It may, for example, be better to modify a system *after* rather than during implementation, and persuading others of this may take more than a little singleness of purpose.

Receptive   Being single-minded, however, does not mean that a systems analyst can afford to shut out the views of others once he has made up his own mind. Later on in this book, section 2.1.6, strong emphasis is placed on the fact that, while the systems analyst is *responsible* for the design of a system, he should in fact never attempt to impose on others a system he has designed in a vacuum. Receptiveness to other people's ideas is essential, combined with an ability to elicit ideas and build a tentative half-suggestion into a firm plan. Very often the ability to understand and accept other people's ideas is not enough to ensure that such ideas are forthcoming, and there must therefore be combined with receptiveness a quality of sympathy and enthusiasm—the half-idea needs encouragement and an atmosphere of acceptance and understanding.

Energetic   There are many occasions when a systems analyst has to be almost frenetic at times, when the volume of work and the apparently insurmountable difficulties involved in getting a thousand things done at once seem to ensure that the system is doomed to total failure. No system occurs naturally: it needs constant harrying and guidance, and its effective implementation may depend in the end entirely on the remorseless energy of one man. At such times the contemplative and receptive systems analyst gives way to the determined fury who will be checking long test-prints, calming anxious users, pressing for more computer time, helping with a parallel run and generally doing his utmost to ensure that if the dozens of activities laid down in the timetable are not completed on schedule, it won't be for lack of effort on his part.

In addition to these general moods there are some specific qualities without which it is hard to be a capable systems analyst. These include the qualities and skills associated with communication, with accuracy, and with responsibility, and are discussed below.

Communication   Systems analysis begins and ends with communication: in order to establish a problem it is necessary to listen to and understand those most closely connected with it, and in order to implement a solution it is necessary to instruct and advise a variety of people, from programmers and operators to clerks and line management. Good systems are not designed in ivory towers and the ability to extract information in conversation and to pass on information in writing is very necessary. Clear and

22

concise writing is perhaps more important than anything—a good system badly interpreted is, after all, a bad system.

Accuracy There is no room for broad generalisations, guesses, or approximations in systems analysis, and a very high standard of accuracy has to be maintained at all times. Carelessness can be heavily penalised in terms of expensive computer time, specialist time and, most important, wasted time by user departments, and the ability to work accurately under pressure is one that must be acquired as quickly as possible during training.

Responsibility A very considerable weight of responsibility is given to a systems analyst and he must be mature enough to appreciate this and ensure that the power at his disposal in terms of manpower and computer time is not frittered away. Without a sense of responsibility towards the users of the system and those who will be operating it, a systems analyst can do a great deal of harm and the necessary maturity of outlook is a requirement which ranks very high in the list of essential qualities.

It is unlikely that anyone equipped with all the attributes described above would have any difficulty getting on with people, and it is probably not necessary to include the ability 'to be accepted'. It should by now go without saying that any degree of bumptiousness, arrogance, or assumed superiority is poison to a system since there are few jobs where it is easier for other people to help one to a disaster: an arrogant systems analyst is remarkably vulnerable when people are not anxious to help.

In the following pages we shall see more clearly why the qualities outlined above are so important in a systems analyst. Before we go on to discuss the different functions which will require these qualities, however, we should consider the general environment in which most systems analysts will be working, and section 1.2.2 describes the standard pattern for a data processing department. In discussing this general background we shall need to be aware that the systems analyst will have ultimate responsibility for certain main activities. These activities are each described fully later on in this book, but a short statement about them here will help us to realise the logic behind the organisation described in the next section. First, the systems analyst is responsible for investigating and recording the requirements and success of an existing system; then he must analyse this in the light of new requirements; he must then, if necessary, organise the design and documentation of a new system which will meet the new requirements; and finally he must ensure that the new system is implemented and efficiently maintained.

1.2 THE DATA PROCESSING FUNCTION

Its place in the organisation
In the early 1950s when data processing departments were responsible only for unit record equipment (see 3.8) there was very little problem about the place of the department in an organisation: quite regularly the output was required by the Chief Accountant and it was perfectly proper that he should be ultimately responsible for the information produced and the deadlines which had to be met. Between the later tabulators and the earlier computers, however, an uneasy period led to some unfortunate developments: as the calculating equipment grew more and more sophisticated and yet still relied exclusively on equipment which dealt with one card for one transaction, so there seemed no reason for the department to be moved out of the control of the Chief Accountant. The equipment manufacturers, excited by the new possibilities of their machines, tried to persuade the directors of other commercial enterprises that the data processing department should be placed directly under the managing director. This did not often happen, however, and it took some years of using a computer before the senior management of a company came to realise that if the computer was directly controlled by any one director, the other directors' requirements were likely to suffer when placed in competition with the requirements of the director controlling the machine. In addition to this it was not easy for systems analysts to exert influence themselves in areas of the company outside, for example, the Chief Accountant's domain if they were preceded by mutters of, 'Here comes the man from the Chief Accountant's office!' Both these facts were found to disturb the efficient creation and implementation of a computer system, and the procedure described below is now often adopted as the best answer to a difficult problem.

Responsibility for data processing
The data processing department, which has responsibilities to all parts of the company, must report to all parts of the company. In practice it is undesirable for the department to report to the managing director himself, and therefore a regular committee known as the steering committee is appointed. This committee includes as members certain senior members of the company such as the Chief Accountant, the Production Director, and the Marketing Director, and is responsible to the managing director for formulating data processing policy and ensuring that it is implemented. This is easier said than done, and the main difficulty usually lies in the probability that, unless a considerable amount of work is devoted to the subject, the steering committee will be working in almost entire ignorance of what they are trying to do. The fact is that a measure of detailed knowledge of existing systems is required before decisions are taken which will affect them, and it is hard for this level of detail to be obtained at

infrequent meetings of a steering committee. It is not realistic to expect the members to study and absorb the explanatory documents produced by the systems analysts, and these are anyway produced at the *end* of systems design rather than during the design stage when the proposals are most likely to be affected by policy decisions. What is necessary is some sort of regular updating method which will keep the steering committee fully informed in general of what is going on in the various systems they are 'steering' and yet relieve them of the very arduous and not always productive task of reading through a number of quite complicated technical documents. When the system is complete, of course, the systems definition (section 6.6) will contain a description of the system in English (as opposed to computer jargon) but until this is ready it is a rewarding practice for the steering committee to insist that at each of their regular meetings one or other of the senior systems analysts describes the system he is currently working on, and makes the members of the committee aware of those factors which are affecting his progress. This procedure has several advantages:

1. The steering committee is *regularly* advised of problems, and does not suddenly become aware of difficulties.
2. The fact that a systems analyst has to report from time to time to senior members of the company is a considerable spur to progress.
3. The contact maintained between steering committee and the system itself (through the systems analyst) can be enormously valuable when organisational changes are being considered.

Structure within the department
We now have, then, an organisation serving all parts of the company and reporting directly to a body representing all parts of the company. Before defining the specific structure of the department it may be useful to consider the number of activities which will have to be carried out by such a department. Quite often a department's structure is built around the particular qualities of the available staff, and therefore, in deciding how to group people, it is as well for the manager to know what functions they must fulfil. This is particularly so in a newly-formed department, when there may well be only three or four people among whom the manager can distribute the many tasks the department will be expected to carry out.

As a first grouping, some major headings of activities are shown in figure 3, organised by the approximate levels of seniority required to carry them out. For example, it can be assumed that only the most senior level in the department will be responsible for overall control of the department's budget, and this activity is therefore shown at line management level (level 1). Six levels are given only and, while it is quite possible for one person to carry out more than one of the activities shown, and those activities to be themselves on different levels, it should be clear that all

26

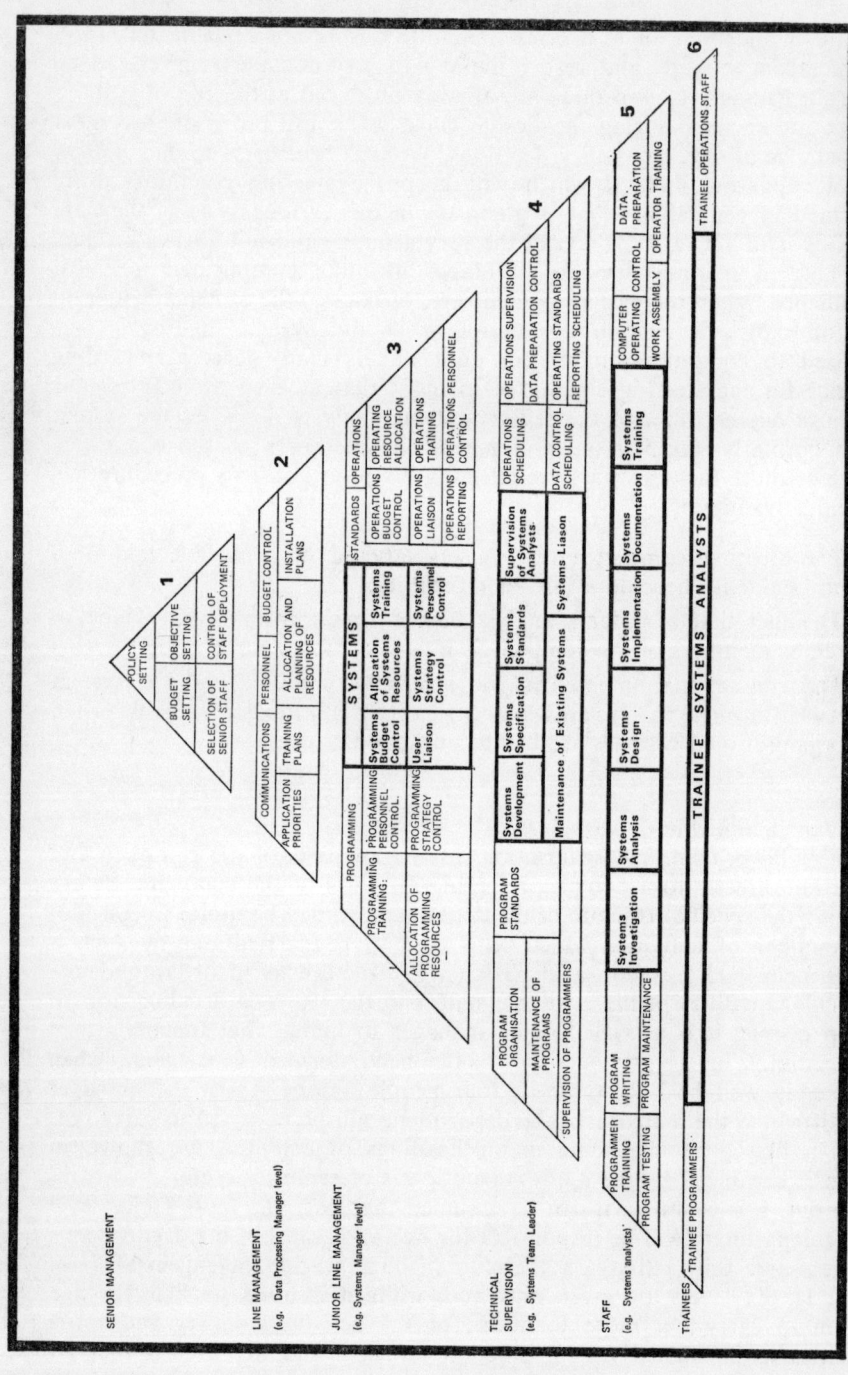

3. Data processing activities. Each major function is shown at an appropriate level of responsibility. The functions are shown in greater detail in figures 4a, 4b, 4c, and 4d.

the activities will have to be covered in some way if the department is to function successfully.

But this functional chart does no more than indicate the broad levels of activity; the charts in figures 4a, 4b and 4c show under the same main headings, those functions which need to be carried out at some time by someone within the department. Many of these activities are the functions of ordinary management or office procedure, and many others are related only to programming or operating, but the bulk are specifically duties of a systems analyst. These are covered in detail in the course of this book. The charts may at first sight appear rather formidable, but they do no more than cover those functions and events which must be catered for in any data processing department.

1.3 STAGES OF SYSTEMS ANALYSIS

So far we have seen what sort of person the systems analyst is and where he fits into the department. We have also looked briefly at the kind of function he is expected to handle, and we shall now consider how these functions are grouped into separate stages. Each of these stages must be aimed at obtaining the maximum result for the minimum expenditure, and it will soon become apparent that a large part of the systems analyst's task is in ensuring that a system is created *efficiently*—given enough time and resources anyone can design a system using a computer, but unless the whole operation is carefully planned and executed a very considerable amount of money can be wasted.

To begin with, clear objectives must be set—a vague realisation that computerisation could improve some aspect of a company's business can too easily become a headlong rush into a long period of systems design continually extended to cope with newly understood requirements. In any computer system the requirements can change, just as a company's objectives can change, and it is right that a system should be prepared to cater for changing requirements; what is dangerous, however, is that the requirements are not carefully enough stated at the beginning, so that each addition is a change of course rather than an extension.

Objective setting, therefore, includes the functions of

Making feasibility studies (section 2.2.1)
Identifying areas for systems study (section 2.2.2)
Identifying management requirements for information (section 2.2.3)
Assessment of benefits and costs (section 2.2.4)
Formal definition of purpose and objectives (section 2.2.5)

When the objectives have been defined it will be necessary to make a detailed investigation into existing methods to see how they are achieving

27

the current objectives and where they need altering if at all. There are certain recognised techniques for this sort of investigation and these are explained in section 2.2. Once again, this stage must be the subject of very tight discipline and a constant awareness of the objectives—it is sometimes hard to decide where one system finishes and another begins and the systems analyst must make sure that he restricts himself to an analysis only of the system concerned with the objectives he has been set.

The investigation stage covers these functions

Fact finding	(section 2.2.2)
Interviewing	(section 2.2.3)
Fact recording	(section 2.2.4)

If an analysis of the existing system has shown that it falls short of the objectives, it will be necessary to design a new procedure and, since this job is the most naturally creative of the systems analyst, it must also be the one most strictly controlled by disciplines. A large part of the book is devoted to the discipline of the design activity, which include

Knowledge of equipment	— Part 3
Computer procedures	— Part 4

Finally, the newly designed system must be implemented and maintained, and once it is running efficiently its performance must be regularly monitored to ensure that changing requirements do not erode its efficiency. These aspects are dealt with in Parts 5 and 7 of this book. Throughout these activities standards of documentation must be maintained, and these are discussed in Part 6.

POLICY SETTING
1 Set company policies
2 Set data processing policy
3 Set policy on career structures

BUDGET SETTING
1 Set budgets to achieve plans
2 Review expenditure against budgets

OBJECTIVE SETTING
1 Set objectives
2 Set priorities of objectives

SELECTION OF SENIOR STAFF
1 Select and appoint senior staff

CONTROL OF STAFF DEPLOYMENT
1 Control organization changes

SENIOR MANAGEMENT FUNCTIONS (RELATED TO D.P. ACTIVITY)

COMMUNICATIONS
1 Co-ordinate the D.P. department
2 Liaise with other line managers
3 Set internal lines of communication
4 Establish lines of communication at all levels with computer manufacturer
5 Arrange and maintain stand-by arrangements
6 Arrange education in appropriate areas on D.P. subjects
7 Arrange education on D.P. applications being implemented
8 Communicate company policy changes to staff
9 Decide policy on standards
10 Control visits to computer installations
11 Inform staff of current plans
12 Disseminate within the company information about D.P. activity
13 Publicise D.P. activity outside the company where appropriate

PERSONNEL
1 Review the department's organisation
2 Review the specifications
3 Review the evaluations
4 Review and control career paths of staff
5 Advertise for new staff
6 Select staff
7 Control maintenance of staff records
8 Maintain records of sickness, holidays and training
9 Administer appropriate company benefits
10 Review expenses, overtime, etc.
11 Maintain internal security
12 Review terms of employment
13 Approve salary changes

BUDGET CONTROL
1 Review forecast work load
2 Review manpower resources
3 Review machine resources
4 Review space resources
5 Review supplies resources
6 Estimate budget requirements
7 Arrange recovery of expenditure where appropriate
8 Control expenditure against budget
9 Arrange budget divergence where necessary

APPLICATION PRIORITIES
1 Plan long-term development
2 Initiate feasibility studies
3 Review feasibility studies
4 Authorise development of new applications
5 Assign development and implementation priorities after appropriate discussion
6 Review against objectives all completed work
7 Keep informed of hardware and software developments

TRAINING PLANS
1 Review plans for D.P. training
2 Arrange training of managers of D.P. departments
3 Arrange technical and managerial succession plans
4 Arrange membership of technical societies, etc.
5 Co-ordinate attendance at conferences, etc.
6 Review training plans for non-D.P. skills (industry training, accountancy)
7 Review induction training
8 Arrange D.P. training of user department staff

ALLOCATION AND PLANNING OF RESOURCES
1 Plan manpower resources including trainee departments
2 Set performance standards
3 Select and order D.P. equipment
4 Control orders of D.P. accessories, furniture and stationery
5 Order ancillary equipment
6 Attend progress meetings with computer manufacturers
7 Attend internal progress meetings
8 Review fire precautions
9 Review insurance arrangements
10 Review legal requirements

INSTALLATION PLANS
1 Plan accommodation for staff and equipment
2 Arrange for maintenance and cleaning
3 Arrange security control
4 Review legal requirements of installation plan
5 Plan transport of equipment
6 Plan transport of data where necessary

4A. Data processing activities. Senior Management and Line Management.
(See also Figure 3 and Figures 4B, 4C and 4D)

PROGRAMMING TRAINING	PROGRAMMING PERSONNEL CONTROL	ALLOCATION AND PROGRAMMING RESOURCES AND BUDGET CONTROL	PROGRAMMING STRATEGY SELECTION
1 Arrange training in equipment and software 2 Arrange training in standards 3 Arrange training in operating requirements 4 Arrange training in the department's activity 5 Arrange non-D.P. training 6 Arrange technical and managerial succession 7 Supervise on the job training	1 Assist in staff selection 2 Carry out induction of new staff 3 Review salaries 4 Recommend promotions 5 Approve absences (holidays, training, etc.)	1 Forecast work load 2 Forecast manpower requirements 3 Forecast space requirements 4 Forecast supplies requirement 5 Forecast machine requirements 6 Recommend programming budget 7 Control expenditure against budget 8 Arrange budget divergence where necessary 9 Allocate programming resources and review constantly	1 Liaise with systems staff on future software/hardware requirements 2 Liaise with operating staff on current programming strategy 3 Review programming strategy 4 Review standards maintenance

ALLOCATION OF SYSTEMS RESOURCES	USER LIAISON	SYSTEMS STRATEGY CONTROL	SYSTEMS PERSONNEL CONTROL
1 Agree detailed schedules for systems work 2 Assist in planning application priorities 3 Control progress against schedules within budget 4 Arrange succession for all development jobs to ensure continuity 5 Review machine loads for testing, parallel running and conversion	1 Review audit participation 2 Review user participation 3 Approve systems changes 4 Authorise start of live running 5 Review existing systems regularly	1 Maintain awareness of new systems techniques 2 Review systems standards maintenance 3 Liaise with programming and operating staff on future requirements 4 Review systems strategy	1 Assist in staff selection 2 Carry out induction of new systems staff 3 Review salaries and recommend changes 4 Recommend promotions, ensuring suitable succession 5 Approve absences (holidays, training, etc.)

SYSTEMS BUDGET CONTROL	SYSTEMS TRAINING
1 Forecast work load 2 Forecast manpower requirements 3 Forecast space requirements 4 Forecast supplies requirement 5 Forecast machine requirements (development and new production work) 6 Recommend systems budget 7 Control expenditure against budget 8 Arrange budget divergence where necessary	1 Arrange training in equipment and software 2 Arrange training in behavioural sciences 3 Arrange training in standards 4 Arrange training in programme requirements 5 Arrange training in operating requirements 6 Arrange training in other systems activity 7 Arrange training in the department's activity 8 Arrange training in company and industry affairs 9 Arrange training in non-D.P. but related subjects 10 Arrange technical and managerial succession 11 Supervise on the job training 12 Arrange D.P. training for all user departments 13 Arrange application training for all user departments

4B.   Junior Line Management Functions.

STANDARDS CONTROL

1 Ensure that suitable standards are agreed for all departmental procedures
2 Enforce such standards

OPERATIONS BUDGET CONTROL

1 Forecast work load
2 Forecast man-power requirement
3 Forecast space requirements
4 Forecast supplies requirements
5 Firecast machine requirements
6 Recommend operations budget
7 Control expenditure against budget
8 Arrange budget divergence where necessary

OPERATIONS RESOURCE ALLOCATION

1 Agree detailed work schedules
2 Control progress of all operations functions
3 Assign operators to shifts
4 Provide appropriate cover for illness, etc.
5 Review peak operating loads regularly

OPERATIONS LIAISON

1 Review stand-by arrangements
2 Liaise with auditors
3 Agree operating instructions in system documentation
4 Review operating implications of proposed systems strategy
5 Review operating implications of new equipment and software

OPERATIONS TRAINING

1 Arrange training in computer equipment and software
2 Arrange training in punching and ancillary machines
3 Arrange training in standards
4 Arrange data control training
5 Arrange training in the department's activity
6 Arrange non-D.P. training where appropriate
7 Arrange technical and managerial succession
8 Supervise on the job training

OPERATIONS REPORTING

1 Review operations statistical data
2 Review scheduling
3 Review library operations
4 Review maintenance policy
5 Review disciplines and methods

OPERATIONS PERSONNEL CONTROL

1 Assist in staff selection
2 Carry out induction of new staff
3 Instil discipline
4 Review salaries
5 Recommend promotions
6 Approve absences (holidays, training, etc.)

4B.—*continued.*

PROGRAM ORGANISATION

1. Advise programmers throughout programming
2. Structure appropriate modules from specifications
3. Co-ordinate testing of modules
4. Co-ordinate assembly or compilation of modules
5. Co-ordinate testing and writing of programs
6. Co-ordinate pilot run testing
7. Liaise with systems analysts on hardware and software requirements
8. Agree program specifications
9. Arrange specifications for standard routines
10. Arrange systems test on stand-by machines
11. Liaise with operations staff
12. Liaise with auditors

PROGRAM STANDARDS

1. Write programming standards
2. Evaluate changes to standards
3. Enforce programming standards

MAINTENANCE OF PROGRAMS

1. Review program operation
2. Evaluate programming aspects of proposed changes to systems
3. Review existing programs in the light of enhanced software capabilities

SUPERVISION OF PROGRAMMING

1. Assist in preparation of time and manpower schedules
2. Advise and train juniors
3. Maintain documentation of standard subroutines
4. Maintain library of codes
5. Maintain data element index
6. Maintain library of program names and module names
7. Maintain register of console messages
8. Maintain program specification library
9. Maintain register of printer forms, loops, etc.
10. Report on progress against schedules

SYSTEM DEVELOPMENT

1. Plan initial feasibility studies
2. Carry out feasibility studies
3. Prepare assignment briefs

SYSTEM SPECIFICATION

1. Assess differing systems proposals
2. Select appropriate system solution
3. Arrange approval and acceptance of specification

SUPERVISION OF SYSTEMS ANALYSTS

1. Assist in preparation of time and manpower schedules
2. Advise and train juniors
3. Maintain documentation of feasibility studies
4. Maintain documentation of specifications
5. Maintain library of system definitions
6. Report on progress against schedules
7. Make recommendations for promotion and salary reviews

SYSTEMS LIAISON

1. Propose and agree terms of reference for project teams
2. Propose training for users where appropriate
3. Agree implementation plan
4. Maintain audit liaison
5. Maintain programming liaison
6. Maintain operation liaison

MAINTAINING EXISTING SYSTEMS

1. Review operation
2. Review performance
3. Assess proposed changes
4. Recommend resource allocation to cater for changes

OPERATIONS SCHEDULING

1. Plan daily schedule
2. Plan analyses for machine and operating faults
3. Plan machine time usage analyses

OPERATIONS SUPERVISION

1. Assist in preparation of time and manpower schedules
2. Advise and train juniors
3. Maintain equipment inventory
4. Maintain supplies inventory
5. Maintain documentation libraries
6. Report on progress against schedules

SYSTEM STANDARDS

1. Write systems standards
2. Evaluate changes to standards
3. Enforce systems standards

DATA PREPARATION CONTROL

1. Supervise data preparation
2. Schedule daily work
3. Exercise quality control
4. Review through-put and error rate

DATA CONTROL SCHEDULING

1. Supervise data control
2. Supervise reception, despatch and transport
3. Supervise data security
4. Control and organise re-runs where necessary

OPERATING STANDARDS

1. Write operating standards
2. Evaluate changes to standards
3. Enforce operating standards

REPORTING SCHEDULING

1. Review output of all reports
2. Maintain quality control on reports
3. Arrange immediate reaction to report queries

4C. Technical Supervision Functions.

PROGRAMMER TRAINING

1 Software training
2 Associated technical training
3 Company training
4 Review of new developments

PROGRAM WRITING

1 Carry out programming functions under supervision
2 Advise juniors and trainees
3 Maintain appropriate liaison with systems and operating staff

PROGRAM TESTING AND DOCUMENTATION

1 Carry out testing and documentation functions
2 Correct and re-document new errors in production programs
3 Maintain software documentation
4 Participate in approved tests

PROGRAM STANDARDS

1 Abide by existing standards
2 Propose additions to standards where appropriate

SYSTEMS TRAINING
SYSTEMS INVESTIGATION
SYSTEMS ANALYSIS
SYSTEMS DESIGN
SYSTEMS IMPLEMENTATION
SYSTEMS DOCUMENTATION

COMPUTER OPERATING

1 Operate console. peripheral units, auxiliary machines
2 Maintain machine room logs
3 Exercise initial quality control

DATA CONTROL

1 Control receipt of data
2 Control despatch of output
3 Co-ordinate control totals, etc.

DATA PREPARATION

1 Carry out data preparation
2 Carry out verification and control

WORK ASSEMBLY

1 Control supplies (e.g. special stationery)
2 Ensure that data, programs, etc., are prepared for operating
3 Maintain production schedules
4 Design quality control

OPERATOR TRAINING

1 Hardware and software training
2 Training in current systems
3 Company training
4 Review of new developments

4D. Staff Functions.

PART 2
Investigation and Analysis

```
┌─────────────────────────────────────────────┐
│  PART 2 :  INVESTIGATION AND ANALYSIS         │
└─────────────────────────────────────────────┘
                        │
                        ▼
┌─────────────────────────────────────────────┐
│  2.1   ESTABLISHING SYSTEMS OBJECTIVES        │
└─────────────────────────────────────────────┘
                        │
                        ▼
┌─────────────────────────────────────────────┐
│  2.2   INVESTIGATION AND ANALYSIS             │
│        OF EXISTING PROCEDURES                 │
└─────────────────────────────────────────────┘
                        │
                        ▼
┌─────────────────────────────────────────────┐
│  2.3   ANALYSIS OF RECORDED DATA              │
└─────────────────────────────────────────────┘
```

2.1 ESTABLISHING SYSTEMS OBJECTIVES

2.1.1 Feasibility studies

Defining the problem

At the beginning of a major project it is quite possible for management to have a reasonably clear idea of what the objective appears to be without really having any understanding of how much money is liable to be spent in machine time, systems analysis time, programming time, clerical time, and—by no means the least expensive—effort from all parts of the company in getting a new system off the ground. It is therefore sound business practice for a detailed estimate of costs and time scales to be made before the go-ahead is given to anything but the most trivial of computer projects. Sometimes it will be found early on that a computer is not really necessary for the successful solution of a problem, and sometimes it will be found that a very much more exact definition of the problem must be provided before any estimate can be made of the costs involved in solving it.

The latter—the definition of a problem—is one of the greatest difficulties facing a systems analyst—and a properly organised initial study can be the foundation of a successful project. Such an initial study is usually called a feasibility study, although this means that it is sometimes confused with the 'feasibility study' made by representatives of computer manufacturers or consultants in an attempt to determine whether a company requires a computer or not. The study should be designed to answer the following questions about any potential application.

1. Where can the boundaries of this application be drawn—what other related work is likely to be affected?
2. What is the current cost of processing each item? Does it seem probable that this cost would be reduced if the work was processed by computer?
3. What are the volumes involved—data, likely processing time, frequency of processing?
4. Are existing procedures and methods satisfactory, and can they be continued as at present if no further systems work is done?
5. If the application is not currently carried out by some means, what benefits are now expected from doing it, whether on a computer or not?

Once answered, these questions can be the basis of a formal assignment brief (section 2.1.5) for the systems analyst, but it will be readily appreciated that any request for a systems study received from a line manager is unlikely to be as closely defined as necessary and a considerable amount of time and money can be wasted when an analyst starts working to an incomplete brief.

The remaining parts of this section are concerned with the problem of ensuring that a satisfactory project assignment is achieved before any major resources are committed to a systems project. The usual pattern will be that any preliminary project assignment brief is prepared, a feasibility study is made and then, if the project is approved, a final project assignment brief institutes the full scale systems study, design and implementation.

2.1.2 Identifying areas for systems study

Preliminary assignment brief

The need for a systems study can arise as a result of (a) a predetermined major plan in which various applications are to be implemented in a sequence designed and approved by a steering committee; or (b) a sudden request from a line manager asking for a particular need to be satisfied. In either case, however august the steering committee, or whatever urgency may be demanded by the line manager, a clearly defined project brief must be raised and agreed by all those concerned. It could be, for example, that the steering committee has considered that the company's very large inventory holding may not be justified and that the current methods used for forecasting demand should be replaced by a computer system. This apparently innocent directive can be the start of a very major investment: perhaps in order to measure past demand on which to base demand forecasts, it will be necessary to change the existing recording methods completely—perhaps the order-taking process will also have to be changed and even the order-reporting may have to undergo change as a result. An alteration to the suppliers' ordering pattern may then be necessary and perhaps the entire inventory system may be physically altered. If the steering committee compared the cost of this operation with, say, the cost of a new machine tool, they would realise the foolhardiness of lightly undertaking so considerable an investment. The purchase of a machine tool may be the result of months of detailed cost comparison and measurement of effectiveness. Ideally, the steering committee will be working against the background of a predetermined integrated system and will be concerned either with additions to a data base construction or with the introduction of certain applications using data already held on the data base. Even so, a close watch must be kept on the boundaries of the selected applications and a formal, written statement of these should be provided to the analyst who will be conducting the feasibility study. In less ideal circumstances, which may well begin as an informal request to 'get something going' the written statement becomes even more important. This applies to any project, however small, and it is advisable if a preliminary assignment brief in a standard format is used, including the following headings:

Project Title
Purpose of the project
Scope of the project, indicating the existing procedures to be covered and
   the departments involved.
Name of person initiating project

Names of people available for advice and assistance

Time limitations, such as 'system must start at beginning of a tax year'

Other information relevant to the project, such as existing reports, descriptions of current clerical procedures, etc.

Time Scale, indicating a date by which the feasibility study must be completed and the final assignment brief, if appropriate, must be prepared. (The final assignment brief will in its turn, show the date by which the project should be implemented.)

Completion of the sections 'purpose' and 'scope' should ensure that a preliminary identification of the systems area has been achieved, even though some reappraisal of this may be necessary after the feasibility study. For an operational system, such as payroll, an assignment brief containing paragraphs under these headings will have gone a long way to establishing the requirements; for information systems, designed to produce specific reports, it is likely that further action will have to be taken to ensure that the system produces what is required. This subject is considered below.

2.1.3 Identifying management requirements for information

Conflicting constraints

Any system designed to produce information for management (such as order reporting) rather than information for an operation (such as invoicing) will be subject to two major constraints, each conflicting. The first is the assumption that information and the communication of information are fundamental to the management of an organisation and allow the various parts of an organisation to be co-ordinated into a productive whole. This assumption demands that each manager receives information that is accurate, up-to-date, consistent, and easy to understand. Because the firm is growing and becoming more and more complex, so more and more factors are of relevance, and greater and greater accuracy is required; one parameter, therefore, is that *the manager needs as much accurate information as he can get.* The other parameter is almost the reverse: in general, *managers receive far more data than they can possibly absorb,* and are at any time liable to be swamped with too much information. One of the main reasons for this is that the designers of systems start by asking managers what information they require, and at any one time a manager is likely to ask for anything he can think of. This is almost certainly more than he needs, although it will include some data he *might* need, and the systems analyst is likely to cap this request with suggestions of other clever variations of raw data which he will be able to produce. In the end, the output of the system has been designed as a monster catalogue of almost everything a manager might want to know but probably won't have to know, and the information he really requires will be drowned in a rising tide of information which is available but unwanted.

It is therefore important that any statement of management information requirements begins with an analysis of what the information is *for*—what decisions will have to be taken. Output reports should always be considered in the light of the question 'What shall I do with this report when I receive it?' If the answer is 'Keep it until I need to look something up', then every item on the report should be considered with the greatest care—it is likely to be not only unused but unusable. Section 4.1.1 deals in detail with the definition of output reports and establishes the importance of making sure at the beginning that they are right. If the thinking in setting the requirement in the assignment brief has not been clear, then the eventual system will be on a very shaky base: management will change the information requirement more and more rapidly, and output will be produced which will satisfy none of the decision-making needs.

Stages of system approval

Some systems projects are easy to identify and proposals can be made quickly and evaluated without too many formal stages of approval. However, major projects are likely to have long lead times and their development and operational costs have to be carefully assessed and reviewed as the work proceeds. With such projects there are three stages at which approval should be established. These are:

1. When the initial terms of reference are formulated in the preliminary assignment brief.
2. When an opening investigation has been conducted and possible solutions to the problem have been put forward in a feasibility report.
3. When the detailed system design has been completed to guidelines laid down in the final assignment brief, and the systems definition has been produced to final approval.

This process is represented as a flowchart in figure 6.

2.1.4   Assessment of benefits and costs

Current costs

In the preliminary project assignment brief, prepared before any detailed feasibility study, it is unlikely that there will be any clear indication of costs involved and potential savings, although other desired benefits will already have been identified by this stage. A considerable part of the effort given to the feasibility study must therefore be allocated to examinations of current and proposed costs for the following reasons:

1. To ensure that the proposed expenditure on a full systems study is justified as opposed to continuing with existing methods.
2. To provide a basis against which performance can be monitored.

During the feasibility study it will be possible to measure quite accurately the costs of the present system. The operating budgets of the line managers involved will show their annual expenditure under such main heads as plant and equipment, accommodation, staff, and stationery and

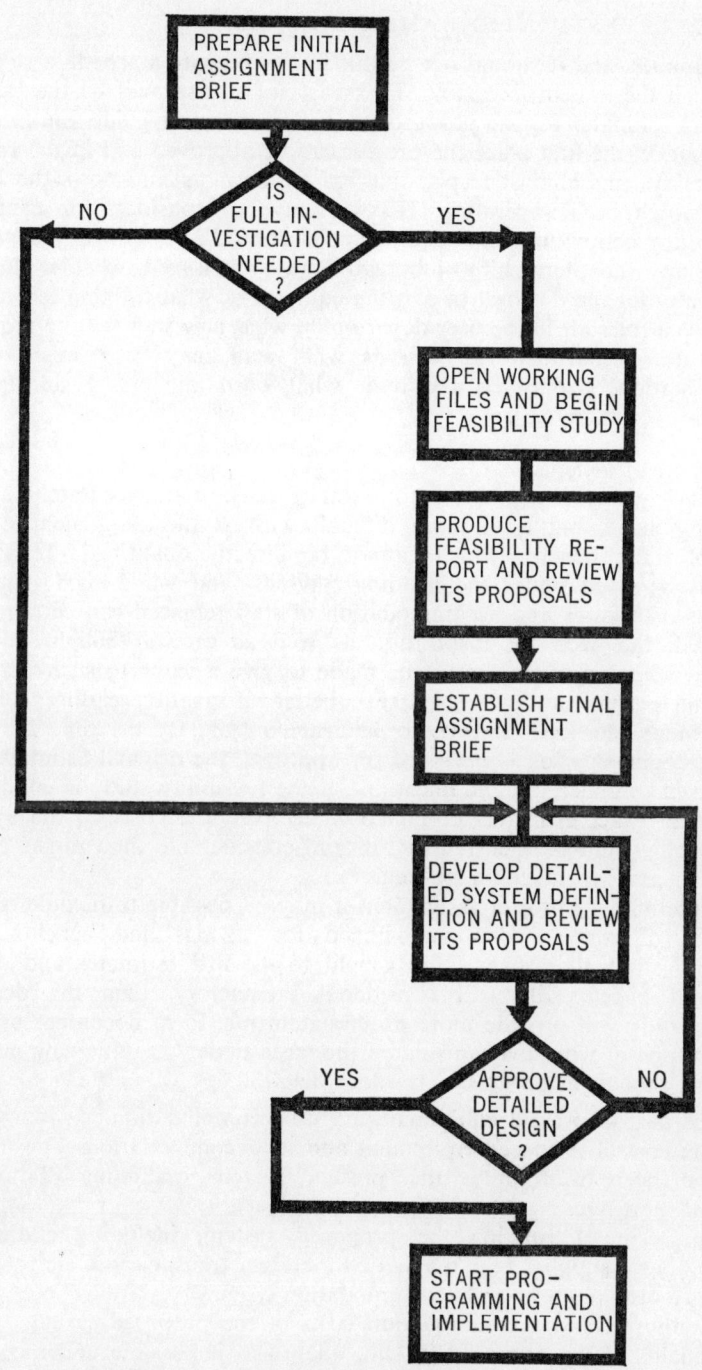

6. Stages of system approval

*B

other supplies, and it should not be difficult to prepare a schedule showing how much the system is costing. Less easy, of course, will be the task of providing estimates for the proposed system, but a clear guide-line must be set so that, in the first place the project can be approved and in the second place the systems analyst responsible for the new system knows the limits within which he is operating. He will have to consider, for example, whether any communications equipment (section 3.2.12) will be economic, whether any short-term hire or bureau time is envisaged, whether there is any penalty for the disposal of existing equipment, what training fees might be involved for staff in the user department, what new staff will be required by the user department as well as what staff may be released, what special stationery may be required, what costs are likely during file conversion (section 6.5).

Assessment of benefits
It will not always be possible to estimate these costs accurately at the feasibility stage, but even more difficult will be the assessment of the value of those benefits which cannot be directly quantified. These are generally grouped under the heading 'savings' and while such items as the costs of salaries and accommodation of staff replaced can certainly be quantified, the need for inspiration as well as careful thought become apparent when an attempt must be made to give a value to such items as 'more efficient utilisation of resources', 'better information leading to better management decisions' and 'more efficient control'. By the time the final systems documentation is prepared for approval, the original estimates are likely to have undergone modification—if the feasibility study is conducted to produce exact estimates of all costs involved the costs of the system study itself would be incurred in full and considerable time might elapse before an assessment could be achieved.

In a preliminary project assignment it may be possible to include overall estimates of current costs and hoped for savings and benefits. The feasibility study then gives more weight to the first estimates and, if the costs and benefits are still considered satisfactory, then the detailed systems study will provide more exact statements. Each document used to obtain approval will have considered the costs under the following general headings, though in progressively more detail.

1. The cost of present equipment and its accommodation.
2. The cost of proposed equipment and its accommodation.
3. The cost of running the present system, including stationery, transport, etc.
4. The cost of running the proposed system, including stationery, computer supplies (e.g. magnetic tape) data transmission, etc.
5. Current manpower and accommodation costs.
6. Manpower and accommodation costs of the proposed system.
7. Benefits of the proposed system, quantified if possible under control, efficiency and direct savings.

8. Cost of file conversion to proposed system.
9. Cost of feasibility study. (This will be replaced in the final assignment brief by 'Cost of study and implementation.)
10. Estimates of possible systems maintenance costs.

2.1.5 Formal definition of purpose and objectives

Instruction to systems analyst
The preliminary project assignment brief, prepared by the steering committee or a line manager requiring systems assistance in his area of operation, included a date by which the feasibility study was to be completed. Let us assume that as a result of the study a report on the feasibility or otherwise has now been prepared and a decision needs to be taken by the steering committee on whether resources should be allocated to the project under review. If the committee decides that the project should be instigated, it is important that a formal document is prepared which can act as the instruction for the systems analyst given the task. The feasibility report itself may be enough, since it will certainly include all the relevant information, but it is probably better if all project assignment briefs follow the same format, and it is just as well if they follow the same format as the original preliminary brief.

At this stage, when a thorough systems design phase is about to be launched, the steering committee will want to make sure that all those who are going to be associated with the project are fully aware of its aims and the overall timescale involved. The final assignment brief will be used as a means of disseminating this information, and, as this will be sent to many people who may not readily understand the implications of a major systems study, it is helpful to have a standardised document which concentrates on covering effectively but economically all these items which need to be brought to the attention of potential users of the system. 'Users' in this case will include auditors and others who may be heavily but indirectly involved. It is for this reason that many companies use a pre-printed form on the lines of that shown in figure 7—most of what is said on the form can as well be indicated on an ordinary piece of paper, but the formality given to the presentation is helpful and the form has the added advantage of making sure that not too many words are used in the sections for 'purpose' and 'scope'.

Once the project assignment brief has been approved and distributed, it is a formal indication that senior management of the company has reviewed the costs and advantages of a new system and requires it to be implemented. The task of designing and implementing the system has been passed to the appropriate authorities, who now have a clearly defined brief and particular targets to aim at. Other interested parties in the company are aware of the project at an early stage and can take steps to keep themselves further informed if necessary. It will be accepted by all that certain modifications to either the aims or the method of achieving them may have to be made, but such changes, like the assignment itself, will have to be made formally. The care and thought that will have been expended in making the feasibility study and producing the report will have been by no means wasted: it sometimes appears that the objectives are quite clear and that any time spent in delineating them will be time wasted, but it is still not unusual to hear of the wreckages of systems which were undertaken without enough thought on the objectives or the costs of achieving them.

SYSTEMS PROJECT ASSIGNMENT BRIEF	No.

PROJECT NAME

PURPOSE OF PROJECT

SCOPE OF PROJECT

TIMING LIMITATIONS

TIMESCALE

AVAILABLE REPORTS

SPECIAL ADVISERS

REPORTING DATES

INITIATOR

APPROVAL

DISTRIBUTION

7. Assignment brief format

2.1.6   Establishing project teams

Dangers of loss of contact

In some systems organisations it is always the same staff who carry out feasibility studies—they are 'feasibility experts' as it were—and the systems analyst who will eventually undertake the task of designing and implementing a new system may not be the one responsible for the feasibility study. If this is so, one of his first jobs will be to forge links with the eventual users of his system, as a great deal will clearly depend on his ability to interpret them and ensure that they are thoroughly satisfied with the new system. To begin with, contact will be quite straightforward, since it will be necessary for the analyst to conduct interviews for fact-gathering and at the beginning of the design period he will be a familiar figure to the potential users of the system. It is from here on that he is likely to lose contact, and as soon as this happens he will be running into two distinct dangers: the first is that, when the new system is ready for explanation the users will feel that it is one designed in a vacuum by someone who does not really understand their job and is therefore a system which is being foisted on them somewhat against their will; and the second is that many minor points, unnoticed in the investigation stage, still remain uncovered in the final system.

Involvement

These dangers can both be avoided by the formation of appropriate systems project teams consisting of members of user departments and the systems analyst conducting the design. Such project teams must be based on certain very clear rules if they are not to achieve the fate of old soldiers and last for ever, and it will very often require a considerable effort on the part of the systems analyst not to convince himself that a part of the work could not better be done by one man working on his own than by a group of people. Once the project team principle is adopted in a company, however, it has been shown to pay very great dividends in both the creation of a good system and the control of the effort involved. Basic rules for the conduct of a project team can be laid down as follows:

1. Each project team will be set up with a clear definition of what its task is to be and how long it is expected to take. The timescale will not exceed six weeks for any one project team. (This means, of course, that each major task is broken down into a number of smaller ones, each capable of being completed in six weeks).

2. The members of a project team will include those members of the user departments who are best able to deal with the detail matter concerning the team. This will often mean that comparatively junior staff are members of a team, which has the twofold benefit of making sure that current situations are discussed in the light of what actually happens rather than what a manager thinks is happening, and

also that a pride of ownership for the new system as it is developed is fostered among junior staff in the user department.

3. Formal minutes of each meeting are kept, and these are designed to form part of a report produced as the final result of the project team. Naturally systems analysts on the team will ensure that, as far as possible, these reports can form the basis of the eventual systems documentation.

4. The terms of reference for each project team should include an indication of how much time is to be spent by the members of the team in meetings, etc. For example, while the systems analyst on a project team would expect to be working full time in collating the necessary facts and papers, the other members might have to give up two half-days a week. A clear statement to this effect means that there should be no difficulty when a clerk has to leave his desk to attend a meeting of the project team.

This somewhat formal approach to the business of designing a new system offers several advantages in addition to those already stated. To begin with, it allows the fact-gathering and discussion processes to take place in an environment other than the busy office of the user, and the systems analyst, instead of being perched uncomfortably at the edge of someone's desk, fitting his questions in between telephone calls and other interruptions, can conduct the proceedings in the quiet of an office arranged especially for the project team.

For general reporting purposes, too, the project team becomes a useful tool: it is likely that the first project team of a series will do no more than set the major steps which must be undertaken to design and implement a system, with an overall timetable for these steps. The next team, perhaps, will be set up to implement the first of these steps according to the timetable, and both the systems manager and the manager of the line department have a series of target dates before them which must be completed before the system is operational.

The tight control exercised by the project team over all the activities which will be undertaken in getting the system under way does a great deal to ensure that minor points are not overlooked: they will be continually brought to the attention of the members of the team, rather than dealt with at one discussion between one systems analyst and one member of the user department.

Finally, the members of the user department will realise that they have more than the usual interest in making the system a successful one: their names are on the project team reports which have gone towards the design of the system, and if the system is successful it will be to *their* credit and not merely that of a systems analyst.

2.1.7 Summary of Section 2.1

In this section we have seen how a systems investigation is placed under way, how a feasibility study is made as a result of a preliminary assignment brief, how this becomes a final assignment brief including a statement on expected benefits and costs, and how the systems analyst invokes the aid of the members of the user department in reaching the objectives of the assignment brief. In the next section we shall review the methods of establishing how an existing system works as a preliminary to designing a new system.

2.2 INVESTIGATION AND ANALYSIS OF EXISTING PROCEDURES

2.2.1 Introduction

A significant proportion of any computer program prepared for a commercial data processing activity will be concerned with events which will probably *not* happen but which *may* happen: in other words, the systems analyst needs to be aware not only of the regular pattern of events but of the unlikely exception. For most systems, it is fairly straightforward to design a method of coping with the ordinary day-by-day events, and manual systems are usually designed only for the usual, relying on human intervention and ingenuity to deal with the unusual. Computer systems, however, do not take kindly to unforeseen events, and must be prepared to recognise the unusual and take appropriate action. This means that a systems analyst designing a new system must be thoroughly aware of all the exceptions to the general run of the mill activity, but in designing a new system which is going to replace or extend an existing system it is not easy to envisage the likely exceptions since there is no past history or experience to draw on. Where there is already an existing system, however, no opportunity should be lost to examine the current procedures in detail and before the start of the design of a new system the systems analyst should be thoroughly steeped in all the procedures on which the existing system is founded.

In the remaining parts of this section we shall consider different methods of establishing facts about existing systems, with particular reference to the techniques of interviewing, and a survey of the usual methods of recording the facts once they have been established.

2.2.2 Methods of fact finding

Three methods
Facts can be established by any of three methods—reading, questioning and observing—and are best confirmed by a combination of all three. One of the main difficulties about fact collecting, however, is knowing when

you have collected enough and in this respect the value of a properly prepared assignment brief can be most apparent. It is helpful, too, to work within the boundaries of a detailed schedule, and if this has been prepared by a project team aware of some of the intricacies of the procedures to be investigated, then a realistic schedule can be of very great assistance in making sure that a project team covers the ground thoroughly and yet does not become too involved in one aspect of the work.

Before the three methods of fact collecting are considered in detail, it should be stressed that the process is by no means merely the collection of existing facts: far more important is the task of questioning *apparent* facts in the light of what is actually happening. *Why* does this document always take four days to pass from A to B? What do people *really* use the 'general remarks' column for? Is it still necessary for the eight-part set to be in eight different colours? And so on and so on—any fact presented by anyone should always be the subject of question, although the techniques employed in questioning need not be as blunt as this implies!

Reading

At an initial project team meeting to investigate an existing system, it is almost certain that one of the members of the team will point with pride to a pile of documents relating to the system. Such documents are likely to include an organisation chart and may also include a procedures manual of some kind. With luck there may also be job specifications for members of the department under investigation, and the combination of three such documents will form a good basis for an investigating project team. They will not be enough, however; for almost inevitably such documents will to some extent be out of date. Even if the head of the department is convinced that the documents represent a true statement of what is going on in his department, the chances are that they are not entirely accurate as procedures have a way of altering themselves to suit individuals or to cover slightly changing requirements. Organisation charts and procedure manuals should, therefore, be taken as a basis for questioning only, and not accepted at their apparent value.

Other existing written material is unlikely to be of very much use—time spent browsing through old files and memoranda is likely to be time wasted and should be avoided although it is frequently suggested by line managers as a way for the systems analyst to find out what is going on. Such a suggestion can be turned to advantage, however, if the systems analyst asks for copies of letters *from now on* to be shown to him: this will allow him to note anything interesting or unusual and ask questions about it when it is still fresh and can provide a useful way of learning about procedures.

Under 'reading' we can also include the technique of analysing answers to specially prepared questionnaires. This can be helpful when two or three particular, well-defined facts need to be established from a large number of people (for example, the average time for the journey to and from work) and a well-prepared questionnaire can be the best method of fact-finding in such circumstances. There are difficulties, however, which should be carefully considered. To begin with, the questions asked must be very precise and not capable of misinterpretation. The question 'Do you have a difficult journey to work?', for example, is clearly too vague to produce a helpful reply, while 'Does your journey to work take more than half an hour on average?' is more precise. If a questionnaire is distributed which gives any grounds for argument or even extended discussion not only will it result in incorrect answers, but its existence may be the cause of a lack of confidence in the systems analyst and any system he may later design.

This lack of confidence can also be caused when insufficient thought is given to the possible reception of the questionnaire. Will people be suspicious of it? What is it for? Who does it come from? Who else is being asked to fill it in? These questions, if unanswered, can cause many problems and are best avoided by attaching to the questionnaire a carefully planned covering letter explaining the background to the questionnaire and what results are hoped for from it. Except for those questionnaires requiring limited answers from large numbers of questions, however, this approach is not often successful. All too often the questionnaire will be put on one side and left unanswered and the value of the few answers received will not then be great. Even if, by constant nagging, a satisfactory reply rate is achieved, the work involved in sorting and assessing replies to anything but the most simple series of questions is unlikely to be justified.

Questioning
One of the major lessons underlined in section 2.2.3 (interviewing techniques) is that all through an interview one question must be used to support another, so each answer confirms or contradicts information given previously. This is the secret of all successful fact-finding by questioning —every apparent 'fact' is always open to question. In order to carry out questioning with this in mind, of course, it is important to be quite clear— before the questioning begins—on a general plan which it is expected the questioning will take. Questions should as far as possible be asked with some definite intention: the man who goes out with the hope that general answers to broad questions will establish all he needs to know will usually waste a good deal more time than he gains.

Formal questioning is discussed in section 2.2.3 but it is worth while remembering that answers to questions can be elicited as the result of informal discussion as well as by direct questioning and that a systems analyst should miss no opportunity of arming himself with facts he will find so necessary when he comes to analyse a system and design one to replace it.

Observing
Once a systems analyst has some grasp of the functioning of the system

he is investigating, it will often be helpful if he can arrange to spend some time in the main office where the work is carried out. The offer of a desk to sit at is frequently made by line managers concerned, and, while it is not always helpful at the beginning of an investigation, it should be readily accepted at the stage when a regular observation of the processes under investigation can be intelligently observed. If this offer is accepted too early there is the danger that the systems analyst will watch the procedures without fully understanding them and, by the time he is ready to notice irregularities, will be too familiar with the routines to notice that such irregularities exist. On the other hand, if he comes to the office with a good understanding of the main routines and their purpose he will be well placed to spot anything unusual and to ask positive questions about it.

2.2.3   Interviewing techniques

Promoting confidence
We have already seen that the best systems on paper may turn out to be the worst in practice and that the reason for this is that people are the unpredictable element on which every system depends. Very often the problems are due to five aspects of the way in which the system is viewed—

> *ignorance,*
> *disbelief* (because of experience of bad systems)
> *lack of time* for discussion
> *suspicion* about the way in which jobs will be affected
> *fear* for security

In section 2.1.6 we have discussed the project team method of involving users in systems design, and a well-established routine for project teams goes a long way to solving these problems. It is usually in an individual interview, however, that the systems analyst has the best opportunity of promoting a potential user's confidence in his ability, and this confidence is a very important factor in the successful design and introduction of a new system. Very great care should therefore be taken in planning initial interviews, and it should be recognised that the primary object is as much to promote an atmosphere of trust as to establish facts about procedures.

Preparation for interviews
Interviews should be arranged well in advance (including the obtaining of a supervisor's consent for members of his staff to be interviewed) and the person to be interviewed should be given a clear statement of the ground to be covered and how long the interviewed will take. Time before an interview can be well spent in reading as many documents as are available on the subject of the interview, but knowledge gained by this means, should on no account be elaborately displayed at the interview—an air of already knowing all the answers can be very damaging indeed. As far as the

preparation of a room for interviewing is concerned there are several general rules: interviews should, if possible, take place away from the desk of the person being interviewed, and therefore away from the usual pressures around the desk such as people looking in for answers to urgent problems and telephone conversations which 'won't take a minute' but destroy the thread of an interview. If the interview is to take place in the interviewer's office, then, of course, he must make sure that he himself is not going to be interrupted: all telephone calls must be transferred elsewhere and it must be made quite clear to all those likely to interrupt that an interview is taking place and interruptions are unacceptable. Interviews should be person-to-person if possible—it is a constraint for either interviewer or interviewee to be talking to more than one person—and should take place in a single room rather than one with other people busy in the background. The person being interviewed should be at least as comfortable as the person conducting the interview and should have as easy access to ash-trays etc. as the interviewer. Some care should be taken about the positioning of the interviewee's chair: it should not be immediately opposite the interviewer separated by the long width of a desk, but preferably to one side of the desk so that there is no obvious physical barrier between the two. The chair should also be sited in such a position that the interviewee will not be troubled by the glare of sun through a window or too distracting a view.

Finally it is often helpful to prepare a draft outline of some general questions which will require answering: these may be useful in filling any awkward pause, and the act of preparation will itself help to get some ideas clear in the mind of the interviewer.

Structure of the interview
In the same way that one of the requirements of an interview between prospective employer and employee is for the employer to explain the nature of the job the employee will be required to do, so in a systems enquiry interview it will be just as important for the analyst to explain *why* he wants to find out as *what* he wants to find out. The interview must therefore be planned to ensure that, while the interviewee does the major part of the talking, the systems analyst has an opportunity to explain his part in the operation.

The first few minutes of an interview will set the tone for the rest, and must be used to establish an air of friendliness and interest. Without such an atmosphere, the interviewee is unlikely to speak freely and any suspicion he starts with will be nourished throughout the interview. Some attempt must be made to find a common area of discussion—a familiar part of the country, a hobby, or even the easiest route to the interview room. Such preliminary conversation, as well as setting an atmosphere of rapport, allows both people to get used to the sound of the other's voice and to get comfortably settled.

It has been said that after this point, the really successful interview

51

consists of talk by the interviewee and a series of encouraging murmurs from the interviewer. This would in fact lead to an uncontrolled exposition by the interviewee which would be by no means as valuable as a series of lengthy answers to some well-directed questions, but might well be preferable to a number of 'Yes' or 'No' answers to a long list of prepared questions.

A general framework for an interview might look like this: some minutes establishing rapport and covering such administrative activities as ensuring that the interviewee's name has been spelt correctly and that his telephone number is known. The systems analyst should then explain carefully the background to the systems project, making sure that any misunderstandings about its purpose are removed and that the interviewee has a clear understanding of the area to which the investigation is limited and the part the interview is to play in the investigation. It is usually a good idea to emphasise that the interview does not mean in any way that the present activities of the interviewee are being criticised. By the time this has been done, most suspicions should have been allayed, and the time will be right to start asking the main questions which will lead to any accurate statement by the interviewee of what his job entails or whatever the subject of the interview may be. The systems analyst should use his previous knowledge of the system he is investigating to ask the right sort of questions to begin with, but he should *not assume anything:* even if he is sure he knows some simple procedure he cannot accept it until it has been described in detail by the interviewee, and even then the facts should be checked later on in the interview. Sometimes the re-asking of questions already answered can be a difficult thing to do, particularly when the interviewee has made it clear that he is a busy man with little time to spare, and it will be useful here if the analyst has taken notes which he can ask the interviewee to check at the end of the interview. In the next section we discuss various methods of recording the results of an interview, and it will be seen that the methods which allow the results to be recorded during the interview are the most suitable for checking at the end of the interview.

Two Main Tasks

While the interviewee is explaining his job the interviewer has two main tasks: one is to make sure that he keeps to the subject and gives his description in more or less the right order; and the second is to make sure that each new fact builds up an overall picture, one new fact serving to check another. One of the main difficulties will be that clerical staff are likely to rattle off form numbers, names of colleagues, initials of other departments, etc. without feeling it necessary to define or explain them, and yet frequent interruptions will upset the flow of the description. Here again some advance knowledge of the system gleaned from reading reports will help the analyst keep up without having to interrupt.

The analyst should not be afraid of being complimentary at any suitable point, and in particular when the interviewee has just presented his idea for a way of improving current procedures. The idea may be quite impracticable, but cold water should *not* be poured on it—no harm will be done by making it clear that you are at least impressed by the determination of the interviewee to change things for the better, and a sympathetic reception given to a poor bad idea may lead to the production of a good one later.

During the course of the interview the analyst should keep one eye on the clock: the interviewee has been given an idea of how long the interview is expected to take, and if possible that time should be kept to, even if this means a second interview later. Naturally this should be avoided if possible, and the analyst should therefore try to make sure that the procedure which is the subject of the interview has been covered, leaving enough time for the analyst to describe it back in his own words. This should make sure that inevitable misunderstandings are ironed out, and will give the interviewee an opportunity to see how quickly the analyst has understood the current procedures and, incidentally, how well he himself must have explained them. Finally the analyst should ask some general question like 'Well, do you think we've covered everything or is there anything else I ought to have asked you?' This allows the interviewee to bring up any point which may be troubling him, and if there is no such point then he will feel that he has been given every opportunity to make one. The interview should therefore end with the interviewee feeling that he has covered the subject of the interview clearly and that a capable systems analyst has listened carefully and with interest to what he has had to say, with the interview reaching a close at the expected time.

Seven hints for interviewing

1. The success of the interview may depend on the interviewer's ability to be interested in the interviewee and the subject of the interview: if the analyst's honesty and interest are apparent the interview is likely to be successful.
2. In general the interviewee must do the talking, but the analyst must do the leading.
3. The analyst should not try to be anything but himself; it may be tempting to pretend a knowledge of the local cricket team if the interviewee is an enthusiast, but this can be disastrous when ignorance is inevitably exposed.
4. Avoid leading questions like 'Do you send this form to Contracts Section?' Frame the question as 'What happens to the form then?'
5. Avoid questions beginning 'Do' and 'Did'; aim at short questions and long answers.
6. Use one question to confirm the answer to another; do not constantly ask for confirmation of one answer, but do not assume the answer was correct.

7. If you are taking notes *never* do so just after the interviewee has admitted an inadequacy, and *always* do so when he suggests an improvement.

2.2.4 Methods of fact recording

In a forty-minute interview it is possible to fill many pages with closely-written detailed notes, and however well organised the initial plan for the interview may have been, the notes are almost certain to be a little haphazard. This may not seem very significant at the time, when the interviewer has formed a clear general impression of the work being carried out by the person interviewed, but when the interviewer returns to his notes (after perhaps some weeks of further interviewing) and attempts to analyse the system they describe, lack of order may make them almost incomprehensible and of very little use. There is, of course, some value in merely recording, since the act of recording itself often helps to fix facts in the memory; we are all familiar with the practice at school of listening while writing, with the pen a phrase or two behind the speaker. But when a multitude of facts and figures are building up a complete picture, it is important that each fact should be both capable of isolation and at the same time related to its context. Facts are recorded so that they can be of use in the next stage of systems design, and a method must be found of ensuring that the facts are recorded not only correctly but usefully.

Systems File
It is obvious that all notes, documents, etc., collected during the course of an investigation should be kept in an orderly way, but much self-discipline is required to ensure that original commendable plans are not dissipated under a welter of apparently unrelated paper-work. Some balance should therefore be drawn between a detailed file structure which may prove impossible to maintain and an 'open-drawer' technique which will merely result in a confused mass of papers. The followed broad sections for a file have been found useful in practice:

Directory section containing organisation charts, names and telephone numbers of people connected with the system.

Document section containing copies of all forms used in the system, together with notes of regular letters and telephone messages which play their part in the system.

Interview section containing notes of all interviews.

Chart section containing all attempts to illustrate the system graphically.

Suggestions section containing all notes for improvements to the system and indications of those areas which seem readily capable of improvement.

Directory Section
The drawing of organisation charts is a regular game in most commercial

54

groups, and the task of keeping up to date with names and job titles is not usually an easy one. The systems analyst needs to be well informed on this aspect of the system he is analysing, however, and it is important for him to keep as accurate as possible a chart of the organisation with which he is dealing. Apart from the usual family tree type of chart, he should also have a detailed structure showing in broad outline the activities carried out by each section and the numbers of staff involved. The drawing of such a chart is a good way of obtaining as quickly as possible a general picture of the way in which the main functions of the organisation are carried out and the sort of effort required for the activity. Figure 8 illustrates a function chart, and figures 9 and 10 show how the basic chart has been easily extended to give some basic facts and figures about the cost of the operation.

As well as containing background information of this kind, the Directory Section should also be used for records of names and telephone numbers of people whom the analyst may have to consult during the course of his investigation. (This section is often extremely useful *after* the investigation is closed as it provides a good list of contacts for anyone wishing to re-consider a point in the system or re-design some aspect because of changing circumstances.)

Terms of reference for each project team involved in the investigation should also be filed in the Directory Section: a series of single-sheet terms of reference can provide a very useful record of the course taken by the investigation, and can show later users exactly the level of detail considered without requiring them to study each section of the file.

Finally, the Directory Section should include an up-to-date index to topics covered in other parts of the file. The maintenance of such an index is often a chore, but can pay considerable dividends in ensuring that the file is as orderly—and therefore as useful—as possible.

Document Section

This section should provide a complete record of all documents used in the system. It is rare for all forms in a system to be the same size and it can be an awkward business sorting through a number of differently shaped forms to establish, for example, whether a certain item of information is included on a particular form. Because of this, a useful general practice is to describe all forms on a standard size of paper under standard headings. These standard form descriptions can all be kept together and will be more easily handled and used for reference than the forms themselves. Completed examples of the forms (if possible, completed with actual data as used in the system) should be filed to support the form descriptions. This is because—as anyone who has ever filled in a form will know—the form headings can seem straightforward and self-explanatory until the form actually has to be completed, and it is often only by examining a completed example that one can see exactly what is required.

55

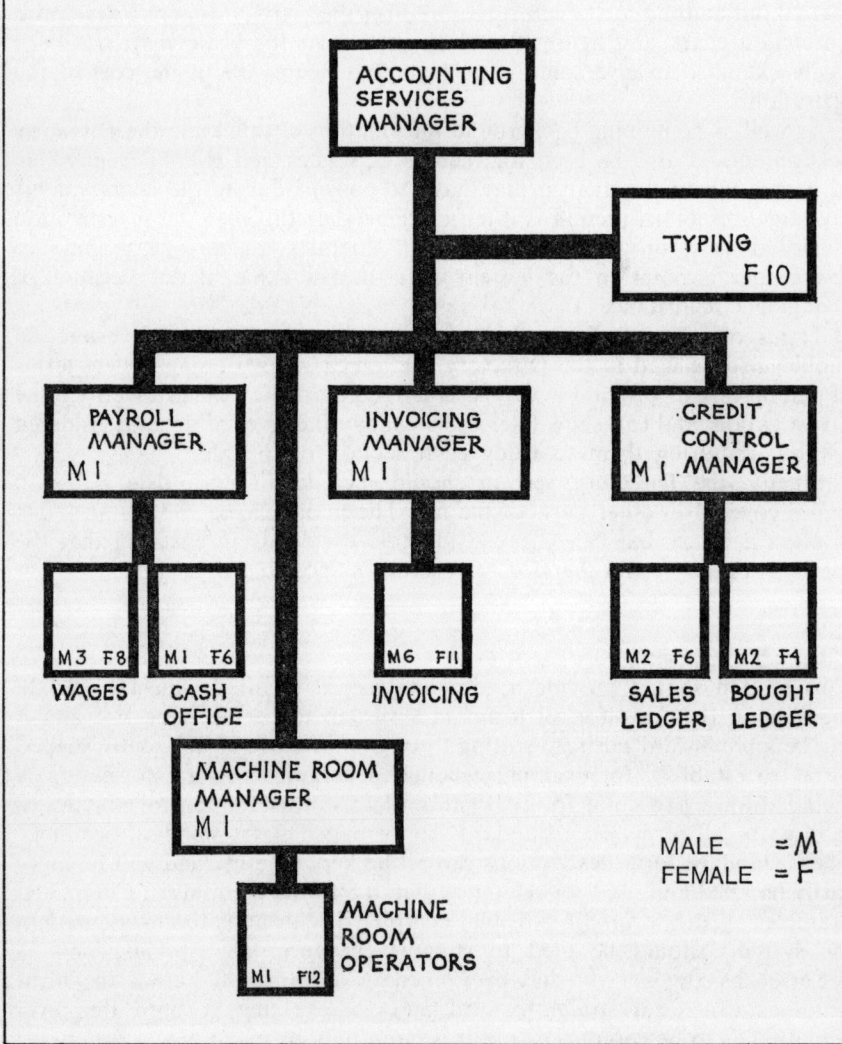

FUNCTION CHART

DEPT : ACCOUNTING SERVICES
DATE : 9·11·69
ANALYST : J. HODSON

8. Function chart

ESTABLISHMENT TABLE

DEPT : ACCOUNTING SERVICES
DATE : 9.11.69
ANALYST : J. HODSON

SECTION	MANAGER	SUPERVISOR	STAFF
OFFICE MANAGER	1		
PAYROLL	1	2	16
TYPING		1	9
INVOICING	1	2	15
CREDIT CONTROL	1	2	12
MACHINE ROOM	1	2	11
TOTALS	5	9	63

9. Establishment table

COST TABLE				
DEPT: ACCOUNTING SERVICES DATE: 9·11·69 PREPARED BY: S.J.MURRAY				
SECTION	SALARY PENSIONS ETC.	OVER – –HEADS BILL	EQUIP– – MENT AND SUPPLIES	TOTAL
PAYROLL	22,000	1,600	1,450	25,050
TYPING	10,050	900	500	11,450
INVOICING	20,150	1,100	700	21,950
CREDIT CONTROL	18,100	1,050	600	19,750
MACHINE ROOM	9,100	6,050	2,060	17,210
TOTAL	79,400	10,700	5,310	95,410

10. Cost table

Form descriptions: Examples of standard form descriptions are shown in figures 11 and 12. It will be seen that the descriptions are not prepared on pre-printed stationery, but follow the same pattern in the way the facts are presented. This pattern is accentuated because after each heading the analyst has ruled a heavy line, and this should make it easy for him to track down information later when he comes to analyse the system. The pattern followed in the examples is one which has gained general acceptance and is fairly representative: it does not very much matter what method is adopted so long as it is standard within an organisation, and it is better to follow any standard method than to follow no standard while deciding which might be the best.

FORM SECTION A   *Heading information* including the title of the project, the name of the systems analyst and the date. The top right hand corner should show boldly the name or number by which the form is known and a cross reference to the place in the Document Section where a copy of the form itself is filed.

FORM SECTION B   *Form description* including a brief statement of the main purposes of the form. This section should also include such facts about the form as its size, colour, and type of paper. Alternative names for the form should also be noted, and particular care should be taken to establish any 'local' names for forms in early discussions.

FORM SECTION C   *Distribution.* This section should show how the document is originated, the location at which it originates, and details of the routes it then follows, with a different route for each copy if necessary. Some indication should also be given of what action is taken at each location visited by the form.

FORM SECTION D   *Volumes,* showing the average number of copies of the form generated or handled during a given time, with a note drawing attention to any seasonal variations. This section should also show the frequency with which the forms are met, i.e. if they arrive in batches twice a day, or are raised throughout the day for immediate action.

FORM SECTION E   *Data.* A detailed description of each type of entry on the form must be made here, showing all the fields of data which can be entered and a statement of their size. This should show whether the size is fixed (e.g. a 6-figure customer number) or variable (e.g. a customer's name) and should show either the actual field length for fixed fields or the minimum, average and maximum field lengths for variable fields. This section should also include an indication of whether each field is used regularly, only occasionally, or invariably. The source of each item of data should also appear in this section.

FORM SECTION F   *Related documents.* All documents known to be affected by the information contained on the document under review should be listed here, and a note should also be made of all files and records which will need to be updated as a result of such information.

FORM SECTION G   *Remarks.* Any further notes should be made here. For

FORM DESCRIPTION

A   HEADING INFORMATION
PROJECT TITLE: STOCK RECORDS
DATE: 6·11·69
ANALYST : J. HODSON

B   FORM DESCRIPTION

<u>NAME</u> FACTORY RECEIPT FORM NO. 5.
<u>PURPOSE</u> SENT FROM WAREHOUSE TO INDICATE
RECEIPT OF GOODS.   USED:- EXISTING STOCK
UPDATING SYSTEM.   <u>SIZE</u> A4 BLUE.
USUALLY KNOWN AS "RECEIPT FORM" OR "A BLUE."

C   DISTRIBUTION

ORIGINATED MANUALLY IN FACTORY DESPATCH SENT
TO STOCK RECORDS SECTION (MR. PHELPS) USED TO
UPDATE STOCK FILE

D   VOLUMES

AVERAGE - 8 FORMS DAILY (RECEIPT OF 80 PRODUCTS)
FORMS ARRIVE THROUGHOUT DAY
NO SEASONAL FLUCTUATION

E   DATA

DATE 6 DIGITS (FIXED). DATE OF DELIVERY 6 DIGITS (FIXED).
PRODUCT NO. PER ITEM 5 DIGITS (FIXED). DESCRIPTION PER
ITEM 65 CHARACTERS (MAX), 30 CHARACTERS (AVERAGE).
QUALITY PER ITEM 6 DIGITS (MAX) 3 DIGITS (AVERAGE) 1 DIGIT (MIN)
ALL FIELDS ARE ALWAYS USED

F   RELATED DOCUMENTS
ALL STOCKS FILE RECORDS (PAGES 18,19, IN DOCUMENT DESCRIPTION)

G   REMARKS

11. Form description

FORM DESCRIPTION

A HEADING PROJECT TITLE: STOCK RECORDS DATE: 7·11·69 ANALYST: S.J.MURRAY	FORM NO. 183 REQUISITION FILED PAGE 34 DOCUMENT SECT.

B FORM DESCRIPTION

<u>NAME</u> FACTORY REQUISITION FORM NO.183
<u>PURPOSE</u> SENT TO FACTORY TO WAREHOUSE TO RE-ORDER
ORIGINATED DURING STOCK FILE UPDATING   SIZE A5
(TWO COPIES ARE MADE) COLOUR WHITE
USUALLY KNOWN AS "WHITE REQUISITION"

C DISTRIBUTION

<u>TOP COPY</u> - TO FACTORY (MR.JOHNSON) TO ORIGINATE
PRODUCTION   <u>2ND COPY</u> - RETAINED BY STOCK CONTROLLER
TO MATCH AGAINST LATER RECEIPT

D VOLUMES

AVERAGE OF 8 DOCUMENTS DAILY (CALLING FOR 80
PRODUCTS) NO SEASONAL FLUCTUATIONS

E DATA

FORM NO. 4 DIGITS (FIXED) — DATE 6 DIGITS (FIXED)
PRODUCT NO. 6 DIGITS (FIXED)(SOURCE - STOCK CARD)
DESCRIPTION PER ITEM 65 CHARACTER (MAX)
                  30 CHARACTER (AVERAGE)
QUANTITY - 6 DIGITS (MAX) 3 DIGITS (AVERAGE) 1 DIGIT (MIN)

F RELATED DOCUMENTS

STOCK FILE RECORDS (PAGES 18,19 IN DOCUMENT DESCRIPTION)

G REMARKS

CURRENTLY A BAD PUNCHING DOCUMENT FOR M/C ROOM

12. Form description (see also fig. 11)

example, suggestions for changes in the format, or an indication that the information contained on the form is usually passed verbally.

File descriptions: The Document Section should also contain accurate descriptions of existing files of information and a similar technique for describing them is in general use. In the same way that documents were described under general headings in a uniform pattern, so the files are described under the general headings shown below. An example of their use is shown in figure 13.

FILE SECTION A *Heading information* including the title of the project, the name of the systems analyst and the date. The top right hand corner should show boldly the name of the file.

FILE SECTION B *File description,* giving a brief statement of the type of file (e.g. punched card), the use of the file, and the organisation responsible for the data contained in it.

FILE SECTION C *Data.* This section should include a list of all the different types of record contained in the file. This should show the sizes of each type of record and an indication of the relative size of the record group to the whole file. Each record type should be supported by a more detailed description showing the composition and size, as shown under Record Description below.

FILE SECTION D *Maintenance.* This section should contain details of the sequence in which the file is held (e.g. Salesman within Territory) and a statement of the frequency with which the file is updated.

FILE SECTION E *Volumes.* All relevant facts about the volumes in the file should be established and recorded here. Such facts will include the size of the file, the number of records it contains, the average number of transactions resulting in amendments to the file in a given period, and estimates of the maximum size of the file for any seasonal variations.

FILE SECTION F *Remarks.* Any special considerations can be noted here.

Record description Record descriptions, kept in File Section C, are detailed descriptions of each record type, and are used to support the general picture shown in a list of different types of record. An example of a record description list is shown in figure 14, and it will be seen that each element of data in the record is described accurately and is completed by a statement of maximum and average sizes. It is important to remember that these figures may well be used as a basis for deciding file lengths in the systems design stage, and great care should be taken in establishing these figures, particularly that for the maximum number of characters.

Document section—summary The document section consists of two main parts, one for forms and one for files. Forms and files are both described in a standard way which allows ready access to the information later, and ensures as far as possible that important information does not remain uncollected.

FILE DESCRIPTION	STOCK
A HEADING PROJECT TITLE: STOCK UPDATING DATE: 3·11·69 ANALYST: J. HODSON	FILE

B FILE DESCRIPTION

CARD FILE REPRESENTING ALL CURRENT
STOCKS. MAINTAINED CLERICALLY BY STOCK
CONTROL DEPT.

C DATA

ONE HEADER CARD IS KEPT FOR EACH PRODUCT,
FOLLOWED BY A STOCK CARD. EACH RECORD IS
MAXIMUM 13 CHARACTERS. SEE RECORD
DESCRIPTION 4. NO OTHER RECORD IS ON THE FILE

D MAINTENANCE

HELD IN NUMERIC SEQUENCE OF PRODUCT CODE
UPDATED DAILY

E VOLUMES

830 PRODUCTS, TWO CARDS PER PRODUCT.
AVERAGE OF EIGHT AMENDMENTS DAILY

F REMARKS

CYCLICAL STOCK CHECK AGAINST CARDS.
ALL STOCK IS CHECKED ONCE YEARLY.

13. File description

RECORD DESCRIPTION

STOCK FILE RECORD

PROJECT TITLE : STOCK UPDATING
DATE : 3·11·69
ANALYST : J. HODKIN

STOCK FILE RECORD (SEE STOCK FILE DESCRIPTION. P.12 IN DOCUMENT SECTION.)

FIELD NAME	TYPE	SIZE MAXIMUM	AVERAGE	USE	REMARKS
PRODUCT No.	99999	6	6	100%	
PRODUCT CODE	9999	4	4	100%	
PRODUCT SIZE	9	1	1	100%	
PACK SIZE	9	1	1	100%	
PRODUCT NAME	ALPHA	65	30	30%	
RE-ORDER LEVEL	999999	6	4	100%	
MAXIMUM LEVEL	999999	7	4	10%	
DATE OF TRANSACTION	39·19·99	6	6	100%	
DESPATCHES	999999	6	3	100%	
RECEIPTS	999999	6	4	100%	
STOCK	9999999	7	4	100%	
REQUISITION DATE	39·19·99	6	6	5%	
REQUISITION QTY.	999999	6	4	5%	
REQUISITION No.	9999	4	4	5%	
		131	81		

14. Record description

Interview section Rough notes made at interviews should be converted to more formal reports as soon as possible. This is primarily to ensure that hastily jotted glosses do not lose their significance (the act of converting the gloss to a considered statement will make sure that any necessary action is taken as a result of it) but also they make easy reference possible later on. One of the best ways to help here is to break the continuous narrative of the notes down into a series of paragraphs each with a heading which will be meaningful later on, for example, 'Action on Receipt of Form 63', 'Distribution Problems', etc. Much of the matter contained in the interview section will be repeated in chart form and be held in the next section, and it will therefore be useful for cross-references to this section to be made, particularly so that a study of the chart can be supported if necessary by reference to the detailed narrative in the interview section.

Chart section In any attempt to represent a system graphically, a systems analyst should be quite clear why he is trying to do it. There is a tradition that systems can be well represented in the form of a number of diagrams, and very often this is so; this should not, however, be used as a reason for a determined effort to cover yards of paper with numerous small, differently-shaped boxes, because unless these can actually help in the understanding of a system and recording of it for future use, then it will be a remarkable waste of time. The main fault with most charting activities in systems work is that the relevant actions cannot be immediately detected by looking at the chart and it is possible to work out what is happening only by a painstaking reference through each detailed step. If the number of steps followed is so many that the follower has forgotten how a particular sequence began by the time he has reached the end, then the chart will not have fulfilled its function of informing—only that of recording. It is important therefore that the analyst should not seize paper and pencil and start a feverish action-after-action diagram. He should instead spend some time in considering the main points of the system that he wishes to make clear, and should then plan the best method of grouping the activities of the system so that these main points are thrown into relief. Sometimes this will best be done after all by a straight narrative description, with a number of short sentences each describing a step in the system, with each sentence numbered and starting on a new line.

Block diagram: Very often the bald statement can be given greater power if it is represented in block diagram form, each step in the system being drawn in an appropriately sized box. This technique allows the analyst to show how various steps interlink, and how each step may be repeated as a result of certain later actions. An example of the block diagram technique is shown in figure 15 and it will be seen that great care has been taken to follow the main rules of block diagrams:

> Few Words
> Much Space
> Many Cross references

c

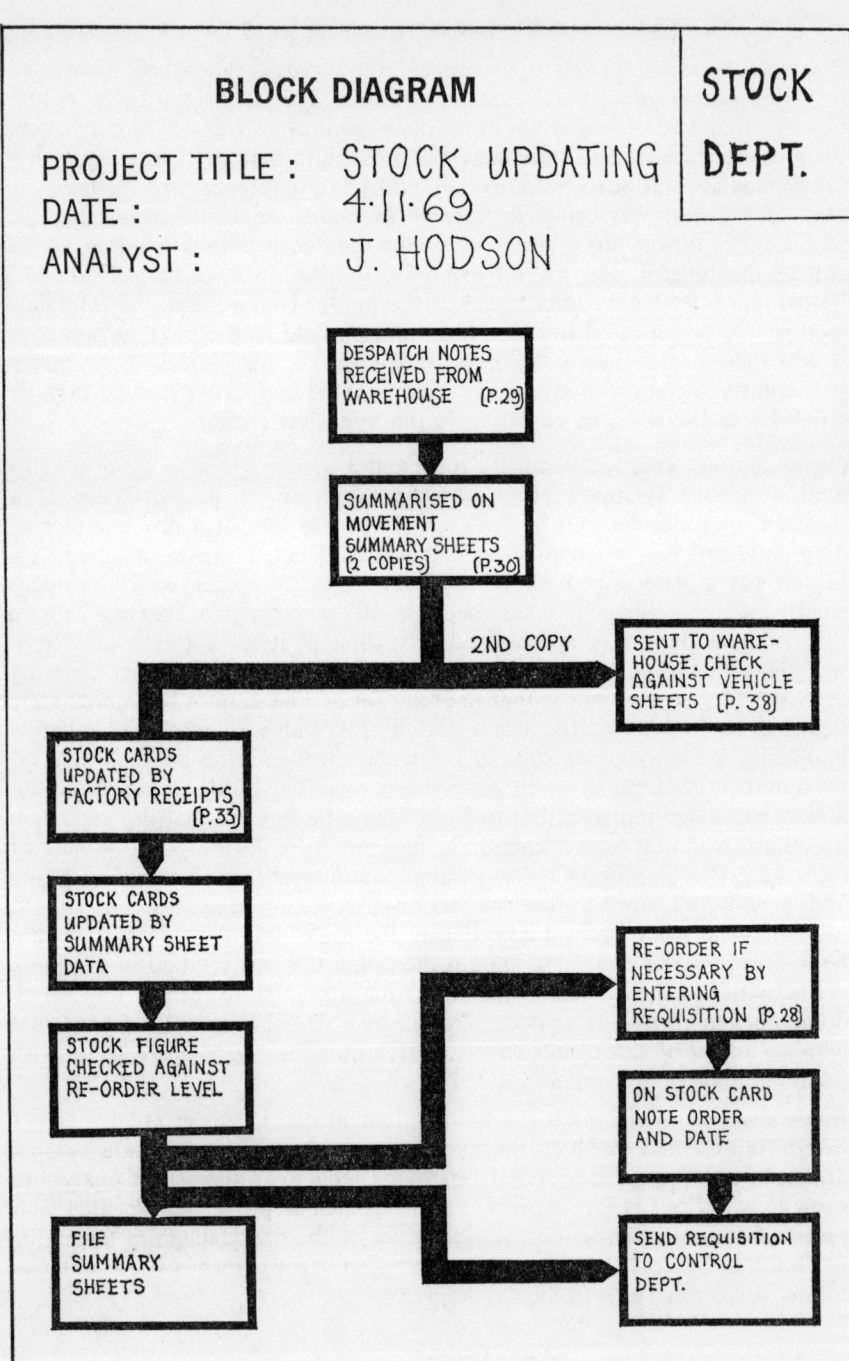

BLOCK DIAGRAM

STOCK DEPT.

PROJECT TITLE : STOCK UPDATING
DATE : 4·11·69
ANALYST : J. HODSON

DESPATCH NOTES RECEIVED FROM WAREHOUSE (P.29)

SUMMARISED ON MOVEMENT SUMMARY SHEETS (2 COPIES) (P.30)

2ND COPY

SENT TO WAREHOUSE. CHECK AGAINST VEHICLE SHEETS (P. 38)

STOCK CARDS UPDATED BY FACTORY RECEIPTS (P.33)

STOCK CARDS UPDATED BY SUMMARY SHEET DATA

RE-ORDER IF NECESSARY BY ENTERING REQUISITION (P.28)

STOCK FIGURE CHECKED AGAINST RE-ORDER LEVEL

ON STOCK CARD NOTE ORDER AND DATE

FILE SUMMARY SHEETS

SEND REQUISITION TO CONTROL DEPT.

15. Block diagram

66

The requirement for space is two-fold: it helps to make the diagram visually attractive (and therefore more easily understood) and it allows changes to be made when further areas of the system are uncovered later in the investigation. Because of this liability to change, it is desirable for all diagrams representing the system to be drawn in pencil.

Vertical sectional chart: The block diagram represents each step in the system in chronological order, and is used when it is required to record each action of the system and show how it interacts with other activities; the vertical sectional chart follows the same general pattern of plotting movements, but shows them in streams or channels of actions carried out by separate departments or individuals. This technique is used when different people are required to check out the part they play in what may be a complex inter-active system, and it is felt that, if they are presented with a conventional block diagram they will have to look at activities carried out by others and, in the mass of detail, may overlook an incorrectly recorded action or not notice that one has been omitted altogether. The vertical organisation allows each participant in the system to follow his own actions in one stream, and yet allows the whole system to be plotted as for a block diagram. An example of the vertical sectional chart is shown in figure 16. The chart can, of course, be drawn horizontally, but is known as a 'vertical' chart in order to avoid confusion with the horizontal chart described below.

Horizontal flowchart: Sometimes a system can best be described solely in terms of the documents used in the system, and it is then easier if each document is considered separately in the system, from its inception to the time when it is destroyed. A useful technique for illustrating the life of each document is the horizontal flowcharting method, which uses a limited number of special symbols to indicate the operations involving each document, and allows each document's life to be plotted on a separate line. As will be seen from the example in figure 17 the relationship between documents in a system can be shown without difficulty, and the fact that each line must begin with an 'origin' symbol and end with a 'destroy' symbol means that each document is likely to be thoroughly plotted.

This technique has the advantage of being readily understood by anyone who may have to confirm that the system in fact operates as represented. The charts themselves are sometimes unwieldy and it is not always possible to detect likely areas for improvement from an analysis of the chart, but as a technique for ensuring that all details of movements of documents are plotted the horizontal flowchart is extremely useful.

Computer Procedures Flowcharts: Existing or proposed computer procedures may be illustrated by means of a computer procedure flowchart. Standard symbols are used to represent files of data held on different types of device, and also processes through which the data passes. A single symbol in a procedure flowchart is used for each complete program,

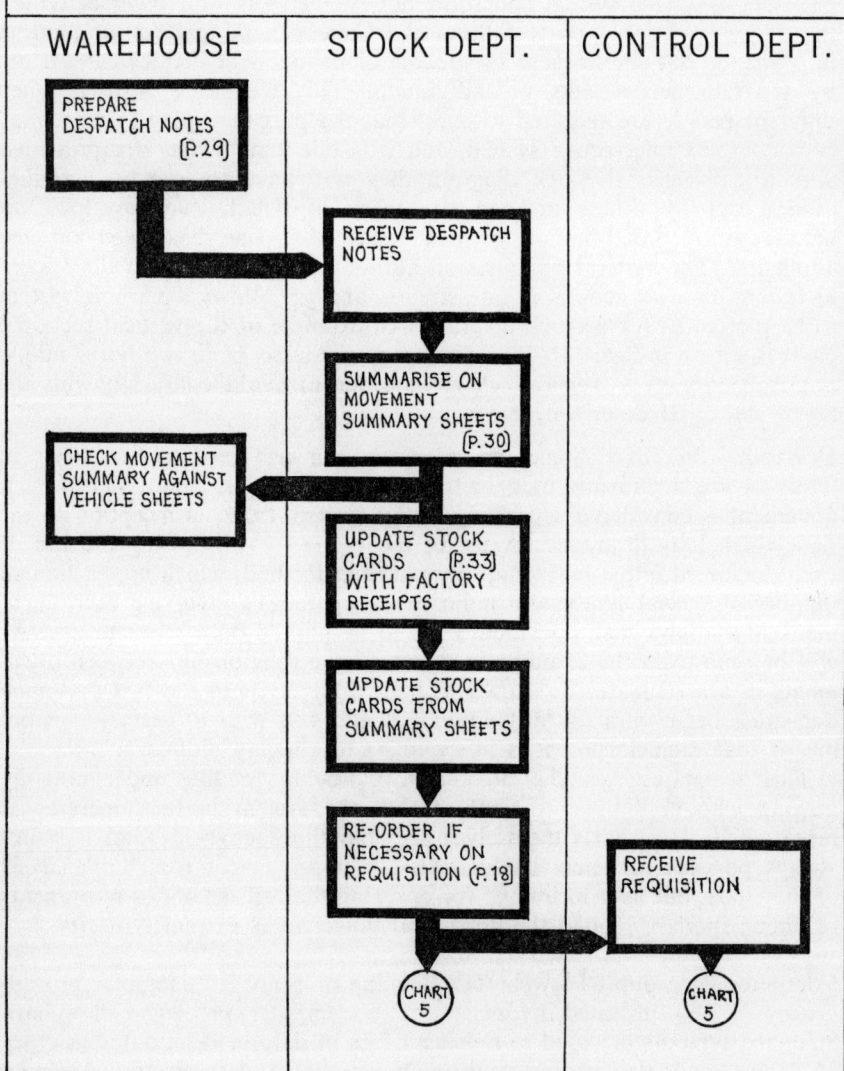

VERTICAL SECTIONAL CHART

PROJECT TITLE : STOCK UPDATING
DATE : 6·11·69
ANALYST : J. HODSON

STOCK
UPDATE
CHART 4

WAREHOUSE	STOCK DEPT.	CONTROL DEPT.

PREPARE
DESPATCH NOTES
(P. 29)

RECEIVE DESPATCH
NOTES

SUMMARISE ON
MOVEMENT
SUMMARY SHEETS
(P. 30)

CHECK MOVEMENT
SUMMARY AGAINST
VEHICLE SHEETS

UPDATE STOCK
CARDS (P. 33)
WITH FACTORY
RECEIPTS

UPDATE STOCK
CARDS FROM
SUMMARY SHEETS

RE-ORDER IF
NECESSARY ON
REQUISITION (P. 18)

RECEIVE
REQUISITION

CHART
5

CHART
5

16. Vertical sectional chart

68

and a brief narrative text explains the function of the process. Standard symbols are also used for outputs from each run. A set of standard symbols is shown in figure 18 and a sample flowchart is shown in figure 19. Where necessary, a flowchart can be accompanied by text describing the various runs. A flowchart should always aim at a clear and unambiguous representation of the interrelationships of the various programs and associated input and outputs.

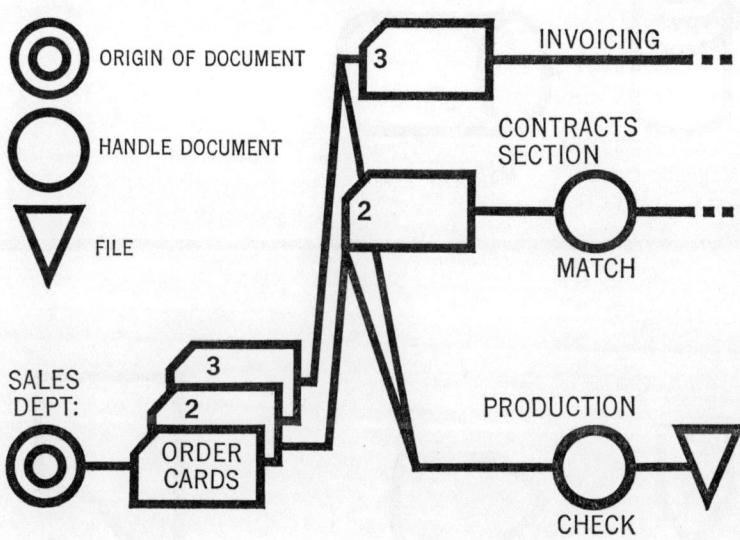

17. Horizontal flowchart

Suggestions section All through the investigation stage of a system, various ideas for improvement will either be made to the analyst by those he is interviewing or will occur to him as he prepares his charts and descriptions of the system. These ideas, however trivial they may appear, must be carefully nurtured as they are the raw material for the design of a new system. They must therefore be recorded in such a way that they can be of use in the design stage, and this means that they must have room to grow and have adequate cross-references.

Each suggestion or idea should be written out, however sketchily, on a separate piece of paper clearly marked with a cross-reference to that part of the main chart or description which deals with the area of possible improvement. Essentially all such notes of ideas should be brief, as there is

69

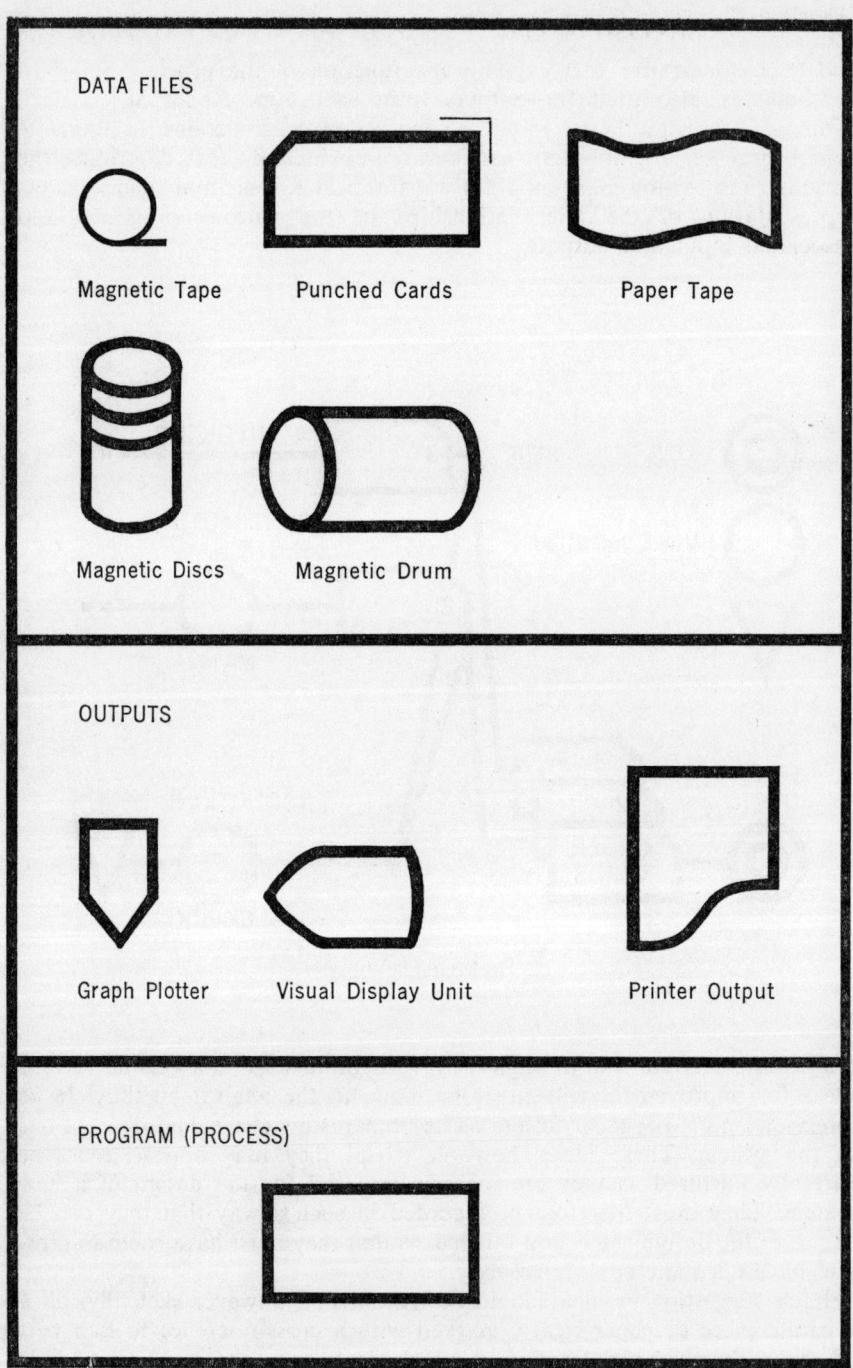

DATA FILES

Magnetic Tape   Punched Cards   Paper Tape

Magnetic Discs   Magnetic Drum

OUTPUTS

Graph Plotter   Visual Display Unit   Printer Output

PROGRAM (PROCESS)

18. Computer procedure flowchart symbols

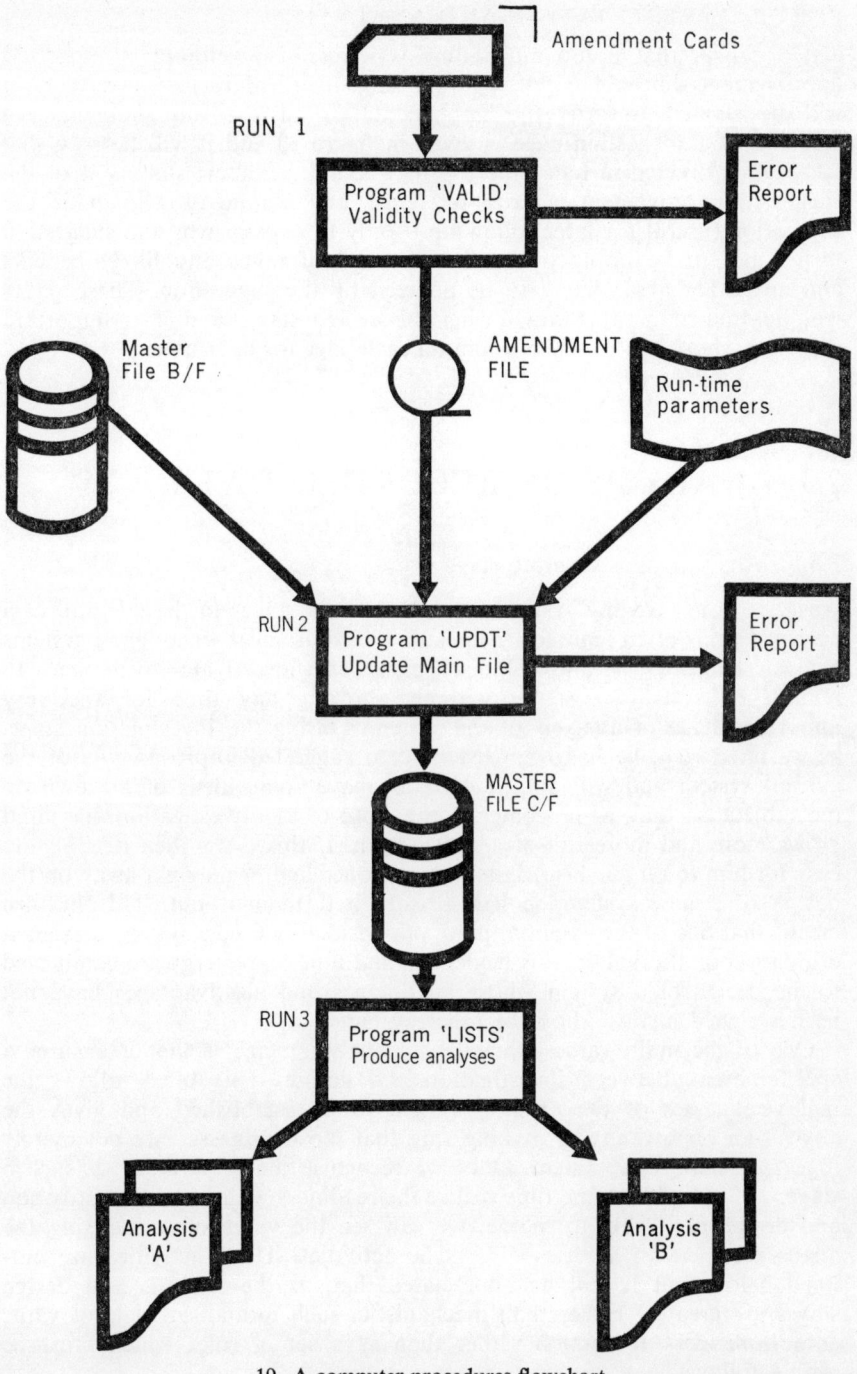

Amendment Cards

RUN 1

Program 'VALID'
Validity Checks

Error Report

Master File B/F

AMENDMENT FILE

Run-time parameters

RUN 2

Program 'UPDT'
Update Main File

Error Report

MASTER FILE C/F

RUN 3

Program 'LISTS'
Produce analyses

Analysis 'A'

Analysis 'B'

19. A computer procedures flowchart

a real danger that if too much time is spent on developing a suggested improvement during the investigation stage then all further investigation will be slanted towards the implementation of the improvement. An example of a suggestion page is given in figure 20 and it will be seen that the page is divided in half—the top half contains a short statement of the suggested improvement, a cross-reference, and a note of who made the suggestion (useful for later follow-up if only to explain why the suggestion turned out to be impractical); the second half notes any likely benefits and any other areas likely to be affected by the suggestion. These pages will be studied many times during the design stage, and it is important that there should be plenty of room on each page for later notes.

2.3 ANALYSIS OF RECORDED DATA

2.3.1 Analysis as a separate stage

Analysis of an existing system, one of the main stages in the creation of a new system, is often ignored by the man called, ironically enough, a systems analyst. Too often he allows himself as systems investigator to merge with himself as systems designer without allowing any time for the very important stage of analysis of the system. During the investigation stage, as we have seen, he has been listening to suggested improvements of the current system and will almost certainly have some ideas of his own on the subject, so that, as he comes to the close of the investigation, his mind grows more and more full of a new system. If this is so, then it is all too easy for him to pursue his original idea at once and begin right away on the design of a new system he has already half thought out. This in turn means that one of the essential parts of considering a new systems design— evaluation of alternatives—is neglected, and time and energy are committed to the design of a system whose advantages and disadvantages have not been weighed against those of other systems.

One of the main values, then, of an analysis stage is that it ensures a specific break between investigation and design. This break allows the real weaknesses of the existing system to be established and gives the analyst an opportunity of making sure that the weaknesses are not merely repeated in the new system. Once we recognise the importance of such a stage, and yet at the same time realise the readiness with which investigation and design are likely to merge, we can see the value of formalising the analysis stage into a series of specific activities. These activities are outlined below, but it will be appreciated that, as the analysis and design stages are creative rather than mechanistic, such formalisation is of value as a *framework* for action rather than as a set of rules which must be rigidly followed.

72

```
┌─────────────────────────────────────┬──────────────┐
│ PROJECT TITLE  STOCK UPDATING        │ STOCK        │
│ DATE    2·11·69                      │ DEPT:        │
│ ANALYST    S.J. MURRAY               │ SUGGESTION   │
│                                      └──────────────┤
│                                                     │
├─────────────────────────────────────────────────────┤
```

SUGGESTION: J. BATES, OF ORDER SECTION:
WHY DON'T WE MAKE <u>THREE</u> COPIES OF WHITE
REQUISITIONS (P.36) AND SEND THE EXTRA ONE
TO WAREHOUSE AS A CHASE DOCUMENT?

CHECK ON WAREHOUSE — THIS MIGHT SAVE
LAST-MINUTE PREPARATION OF RECEIPT
SCHEDULES (P.37)

20. A suggestions page

2.3.2 Interpretation

When an investigation has been completed, the analyst should find himself with a well-organised file containing the details of the system; he should also have a very clear picture himself of how the system works and where some of its weaknesses lie. It is very often difficult to see from this just how well a system is achieving its objectives, and, since this is the purpose of interpretation, it will be useful for him to recast the facts and figures of the system into some shape that allows them to be looked at in a new light. For example, a flowchart of the system can be examined in the light of the data elements appearing on each document—tables should be drawn up showing all the data elements used in the system and which documents are used to record them. A further table can be drawn showing the numbers of movements from one place to another made by each form, and how many alterations or additions have to be made to the form as a result of these movements. Next, comparative statements can be made showing how long information and instructions take to reach their recipients—are these adequate or is there some hidden loss to the organisation as a result of the time taken for information to travel? These areas for analysis can be summarised as,

1. Data elements
2. Movements
3. Information travel times.

It will be seen that these activities are a re-drawing of existing information, gathering data already in the file into a new order. This fulfils two purposes:

(a) The resulting data is more easily accessible than when it is dispersed in the systems file.

(b) The re-grouping of existing facts will almost certainly have highlighted some aspect that had remained hidden while the investigation was mirroring the existing system.

The interpretation activity can sometimes seem an unnecessary and tedious drudge, and a systems analyst will often feel that his knowledge of the system is good enough to indicate that no benefit will accrue from a mechanical process of re-drawing tables of facts already familiar to him. It has been proved again and again, however, that a simple re-statement in a new form can reveal areas of redundant paperwork, opportunities for combination, and possibilities of reduction—areas that had never been noticed while the system remained in the context of something that 'is done this way, has been done this way and probably always will be done this way'.

74

2.3.3 Analysis

The time for opinions

So far the analyst has been dealing with facts, the way the current system operates, and a re-organisation of the facts embedded within it. Now comes the time for opinion. He must consider critically the operation of the system, not in the light of any ideas he has for improving it but in the light of the overall objectives. In other words he must look at each aspect of the system thinking 'Does it do what it sets out to do as economically as possible?' rather than 'Is it as good as my own idea for improving it?' This is, of course, because the current system will always appear to be second-best to something not yet thought out and possibly full of unforeseen snags, and a constant measuring of the two systems will convince the analyst that his first idea is in fact the final answer. This is likely to be very far from the truth, and, in order to arrive at a proper assessment of the current system unclouded by comparisons with a projected system, the analyst must confine himself to a relentless probing of the existing system on its own.

System review

Later, when the design of a new system is complete, it, too, should be subjected to as searching a probe as that undergone by the system it supersedes. In fact the analyst will find himself continually questioning his own work from now on, and the questions outlined below apply as well to the new or developing system as to the old one. Those systems being designed to meet a new need (and not therefore replacing an existing system) will particularly benefit from the technique of probing since there is no initial yardstick against which they can be measured. It is usually helpful to group questions in the following way:

1. Are the outputs of the system satisfactory? If not, how can they be improved?
2. Is data being collected satisfactorily? Are changes to input permissible? What improvements could be made?
3. What obvious weaknesses in the system must be avoided in the new design?
4. What existing resources (e.g. Telex, office machinery) could profitably be used in the system as it is?
5. Are existing files maintained in the best sequence? Is there any obvious duplication of file maintenance?
6. Is there an opportunity to improve form design within the system?
7. Are there any legal requirements which must be borne in mind? (other than auditing requirements which should be catered for under the project team approach).
8. Which aspects of the clerical activity in the system seem the least attractive to those carrying them out? (These aspects are likely to be

75

those where the most mistakes are bred, and therefore the most sensitive for a computer system.)

9. Finally, a question for the analyst to ask himself throughout the analysis and design stages: 'If it were my own money being spent, where would I be worried about waste?'

Alternative solutions

The answers to these questions will form the basis for a system designed to improve the existing system and meet any new objectives. Once they have been answered the creative activity can begin, and a large part of the rest of this book is concerned with systems techniques for the design and implementation of new systems. One point to remember here is that it should be the intention at the outset of design to prepare at least two outline proposals for the solution of the problems posed: a comparison between two proposals (rather than between a proposal and an existing system whose faults are already known) is the surest way of indicating any likely flaws in each of the two proposals. For there will be flaws: even though all the right questions have been asked about the proposals, it is almost certain that some aspect of the new system has not been sufficiently thought out in the outline stage, and a comparison with another proposal is one of the best ways of bringing this to light.

The next part of this book provides a review of the equipment at the disposal of the systems analyst, both hardware and software, and this review is followed by a detailed consideration of the techniques of systems design. For the design to be successful, a thorough and careful investigation and analysis must have preceded it.

PART 3
Data Processing Equipment

PART 3A HARDWARE

INTRODUCTION

PROCESSORS   3.1

Information Patterns
Control Logic
Core Storage
Compatibility
Data Operations
Timesharing

PERIPHERAL UNITS   3.2

Interface
Card Reader
Card Punch
Paper Tape Reader
Paper Tape Punch
Printer
Typewriter
Visual Display Unit
Graph Plotter
Character Recognition Equipment
Communications Equipment

STORAGE   3.3

Core Storage
Magnetic Tape
Magnetic Drum
Magnetic Disc
Magnetic Card File

TO
PART 3B

3 INTRODUCTION

This part of the book describes the tools at the disposal of the systems analyst. It is clearly important for him to have a clear and sound knowledge of the basic equipment he is to use, but this knowledge will not be that of the computer salesman, the engineer, or the software writer: all of whom require a very detailed knowledge of the precise matters of the equipment and the way everything works. The analyst must know what the equipment does, what may be achieved by using different types of equipment, and what characteristics will affect his system design.

He need not know every detail of how the equipment works: The average motorist is able to drive a car, and knows enough about how it works to take care of simple maintenance: oil change, radiator checks, tyre checks. He does not need to understand the more esoteric techniques of piston-boring, or know how to fit a new big-end. The purpose of this part of the book is to give the analyst the information he needs to 'drive' a computer, with enough detail to enable him to have a broad understanding of the nature of the equipment he is using.

The tools available to the analyst may be broadly categorised into three main groups:

> Hardware
> Software
> Data preparation equipment

These groups are described in outline below, and the various types of equipment falling into these categories are described in more detail in further sections.

Hardware
Hardware comprises all the visible equipment which makes up a computer installation, and may be divided into three main categories:

> Processors
> Basic peripherals units
> Storage devices

The function of the *processor* is to carry out all the operations on basic data which are defined by a series of instructions, known as a *program,* also held within the processor. Thus the processor is the essential part of a computer system, since it is by means of the processor that the requirements of the user, as defined by the analyst and converted into instructions by the programmer, are finally carried out. A processor can be imagined as a giant chessboard: the pieces contain information, each piece occupying a square in the board. Each square or location in the processor has a unique identifying number, known as the location's 'address'. Within the processor there is a control unit which shuffles pieces from location to location, modifying the pattern of information according to the program of instructions. Some locations contain instructions, and the control unit can look at each in turn to see what to do next: others contain data on which operations are to take place, for example a gross wage to be converted to net pay after tax. Processors are described in more detail in section 3.1.

22. A computer room

Basic peripheral units are needed to supply the processor with the data on which operations are to take place, and, once these operations are completed, to produce them in a form suitable for the user, or for re-entry to other programs for further processing. The processor manipulates data in the form of patterns of magnetic charge: data must thus be converted into this form from an 'external' format. This external data will be on punched cards or paper (described in section 3.7.1 and 3.7.2) or in the form of machine readable print or marks on ordinary paper (section 3.7.3) or possibly in some other directly acceptable format. Once the data has been processed it must be passed out of the processor to the user, in the form of printed paper, or drawings and graphs or as visual images: output devices perform this function. All these units are known as peripheral units since they are devices separate from but under the control of the processor. They are described in section 3.2.

During the course of processing, it may be necessary to store large quantities of data for further processing or for reference. For example in a payroll, details of all employees must be stored, and modified each week with such variable details as actual hours worked. *Storage devices* enable information to be held in a form which can be accessed rapidly for processing when required. These devices are described in section 3.3.

Software

In order to use computer hardware to produce required results, it is necessary to prepare detailed instructions, which are then stored in the computer's processor and performed in sequence on data to generate the desired output. Software in its most general sense is a term used to describe all instruction sets, or *programs* which are used together with appropriate hardware to produce the required results from a computer system. The expression 'Software' is sometimes restricted to apply solely to general-purpose programs prepared by manufacturers or other specialists for doing standard operations, contrasted with more specialised programs written by the user performing particular operations applicable solely to his own requirements. In this book we describe three categories of software.

Programming
Utility Software
Operating Systems

Within any system it will usually be necessary to devise some operations which require special-purpose rather than general programs. A computer's processor has available a set of operations or instructions, known as its instruction set or repertoire: any program is basically a combination of some of these instructions which, when performed in sequence, ultimately perform the overall task required. *Programming* is the art of creating such a set of instructions, given an overall description of the required task. Section 3.4 describes the various stages in preparing such a program, from the initial planning of the logical steps involved to the testing of the resulting program to see that it in fact works. Although an analyst may well not himself be involved in the actual detail of program writing, he will certainly be required to specify program requirements and must thus be aware of the problems and techniques of programming.

Many operations carried out on a computer, however, are of a routine nature. Standard programs known as *utility software* are frequently available for carrying out these operations, and considerable savings of time and money may be made by using such programs whenever possible. Section 3.5 describes some of the common items of utility software generally available.

As computers increase in power and complexity, so the problem of utilising their power economically and efficiently becomes more complex. *Operating systems,* described in section 3.6 put the computer's own power to use to solve such problems by means of highly sophisticated programs which themselves organise and control the operation of other programs and hardware within the system.

Data preparation
Before data can enter a computer system to undergo processing it must generally be converted into a form which can be read by a peripheral device and converted into the signals which can be manipulated within the computer processor.

Data preparation equipment is used to transfer data from the format in which it enters a system, for example, order forms, despatch notes, onto cards, paper tape, magnetic tape or other forms from which data can be transferred to the computer for processing. Section 3.7 describes various common forms of data preparation equipment, discusses the nature of punched cards and paper tape as input media, and considers methods of data capture which avoid the stage of transcribing date from a document to an input medium.

Since computer systems are frequently used to replace existing punched card systems, various types of equipment used specifically with punched cards are described in a separate section on *Punched card equipment* (Section 3.9).

3.1 PROCESSORS

This section describes the features common to the central processor of a computer system. A computer is, logically, a sophisticated tool which converts a raw material, basic data, into a finished product, information. It thus needs to be fed its raw material, perform the operations which convert this into information, and produce the finished result. The part played by the processor in this basic cycle of events is the key part of processing the raw material according to a fixed pattern of instructions to produce the final requirement. The processor must thus be able to carry out operations, a process described in section 3.1.1 (control logic); it must have somewhere to hold instructions and the basic data on which it is operating—section 3.1.2 (core storage). Compatibility, as described in section 3.1.3, explains how different machines may perform the same operations and use the same instruction programs. The way data is converted into information by operations performed on information patterns is described in sections 3.1.4 and 3.1.5. Section 3.1.6 describes how processors can perform several different operations at the same time.

3.1.1 Information patterns

General
The processor's core storage, or memory, consists of thousands of core units. Each core can represent the state of one binary digit, or bit, which has the value 0 or 1. Information in memory can be represented by the particular arrangement or pattern of a number of bits. Thus the organisa-

tion of groups of bits in store is of fundamental importance to computer design. Since a single bit conveys very little information on its own, the basic unit of data in store consists of a group of bits. This group of bits is treated as a unit for all purposes of processing. A group of bits may vary in size, and the information conveyed by the bits may vary in type. For example, a string of 24 bits may represent a number in binary notation, four alpha-numeric characters, or the coded form of a computer instruction.

Considerable variation between processors exists as to the methods by which bit patterns are used to convey information of different types: the analyst must be aware of the particular characteristics of the machine in so far as they affect methods of file design and data handling.

Basic characteristics
In all machines, data is handled in basic groups of bits. These basic groups or units vary in size and significance between different types of computer, and, within any particular range, the size and significance of the unit may vary according to the type of information carried.

The basic distinction between types of digital computer is that between word machines and character machines. In a word machine, numeric data is represented in binary form by the bit pattern of the word, and arithmetic operations are carried out directly on the number in its binary form. In a character machine, a grouping of bits represents a character, either a letter of the alphabet, a special symbol or a numeral. Arithmetic is thus performed not on a binary number, but by the serial processing of each numeral in the number.

A word may be either a fixed number of bits, or may itself be sub-divided. The most common sub-division of words is into sections known as *bytes*. A byte may represent an alpha-numeric character, two decimal digits, or part of a pure binary field. Numeric data can thus be held in this system either in binary form or as binary coded decimal, in which each decimal digit in a number is represented by its binary form. In machines in which the word rather than the byte is the smallest accessible unit, alpha-numeric characters may be represented as bit-patterns within the word, up to the maximum number of distinct character patterns which can be contained in the word.

Program instructions are also represented by bit patterns, either as numbers of characters or by the arrangement of bits within a word or byte. Depending on the type of logic employed by the machine, an instruction must contain an *operator* which specifies the action to be carried out, and one or more *addresses* specifying the location of data to be operated on, also known as operands. Some instructions also contain a *modifier*. This is either an absolute quantity, or the address of a location containing a quantity, which is automatically added to one or more of the operands each time the instruction is obeyed. The value of the modifier itself may be changed by program or may be increased or decreased automatically.

83

Modification as a programming technique is discussed more fully in section 3.4.7.

Internal and external formats

In character machines, data stored within a processor or other storage device may be considered as a 'mirror image' of data as input from a basic input device. For example, the number '1234' will be punched or printed as four separate numeric characters, and on input will be converted to the four separate bit patterns representing the characters 1, 2, 3 and 4. If this number is multiplied by 2, the result will likewise be stored as '2468' and output as four characters. In the case of word machines in which arithmetic operations are carried out on numbers held in binary form, the character form of a number must first be converted to binary. This conversion usually takes place the first time such a number enters the processor. Numbers on which arithmetic is to be performed are then held in binary format through all intermediate stages until immediately before output, when they are again converted to character form. Such conversion from *external* to *internal* format is generally performed by means of conversion routines supplied by manufacturers which can be simply incorporated into the users' programs.

When designing file structures the analyst must specify the format in which numeric data is to be held on files. In word and byte machines, the analyst must choose between conversion of data to internal formats where appropriate, thus involving the use of conversion routines on input and output, and retaining data in character form. Since arithmetic is much faster on binary formats, fields which will be used in arithmetic are almost invariably converted to binary format. Fields which are merely for information or reference such as file keys are retained in character form. However, since the character form of a large number generally requires more bits than the converted form, in cases where storage capacity is at a premium such non-arithmetic data may be converted.

Parity checks

Parity checks are effected by adding up the individual bits that make up each word or character to see whether an odd or even number of bits are present. A computer system that uses even parity checking would expect to find an even number of '1' bits in each word or character, whereas an odd parity check would look for an odd number of '1' bits. In the following explanation we have assumed an even parity system. When a unit of information (i.e. word or character) is transferred from one location to another, a parity bit is added to make the number of '1' bits equal an even number.

Character pattern	Added parity bit
011010	1
111100	0
101010	1
100010	0

Arabic Numerals	Octal	Binary
0	0	0
1	1	1
2	2	10
3	3	11
4	4	100
5	5	101
6	6	110
7	7	111
8	10	1000
9	11	1001
10 symbols	8 symbols	2 symbols

23. Table showing different number systems

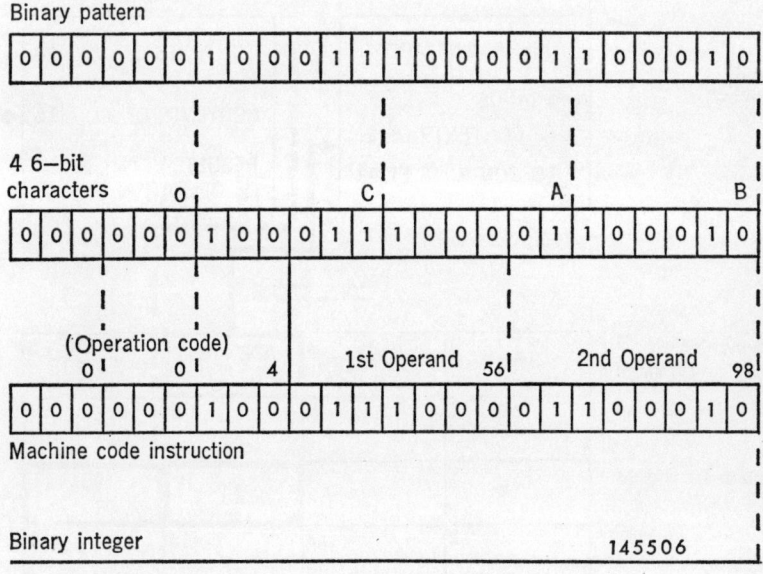

24. The same binary pattern representing three types of information

85

When the data is received and stored at its new location a check is made to see that there is an even number of '1' bits present; when this check is satisfied the value of the parity bit has no significance in any future arithmetic or logical operations that are performed.

3.1.2 Control logic

Introduction

Within a processor, the control unit performs two basic functions: it accesses and interprets program instructions, and performs the logical and arithmetical operations specified by the instructions, where these are directly performed by hardware. The *control unit* performs the function of accessing instructions; the *arithmetic unit* performs logical and arithmetical operations. Both units use *registers* for storage of instructions and data. Control logic storage is part of high speed or immediate access store: either at the same level as the main storage of the processor (usually core store) or a specially high speed storage provided solely for the control logic. Figure 25 shows the function of control logic diagramatically.

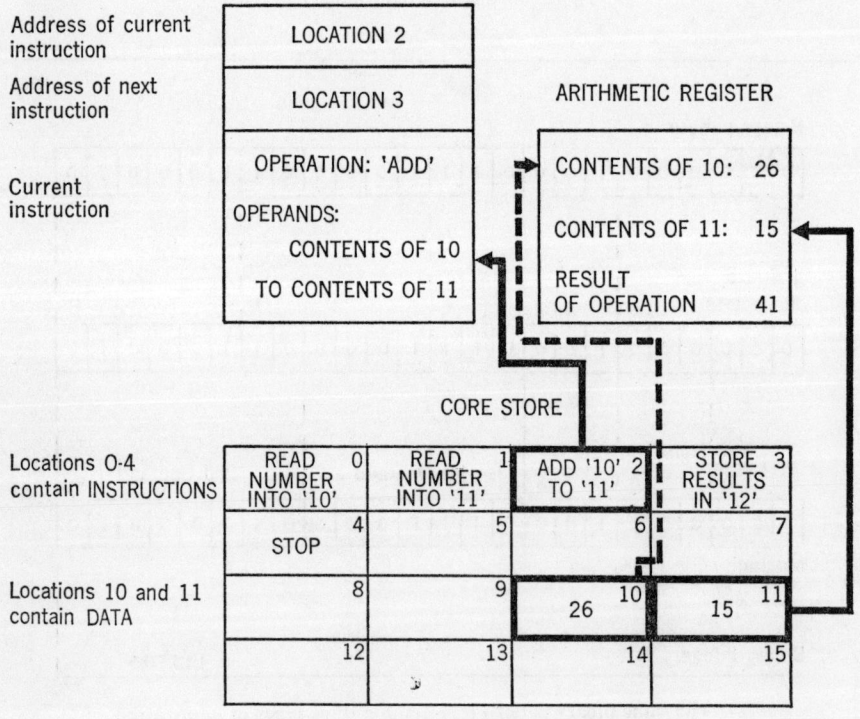

25. Function of control logic

Registers
Registers are locations of high speed or ultra high speed storage used for the temporary storage of instructions and data during the operation of the control and arithmetic unit. The number and function of the registers depend on the logic design of the control unit, and on such factors as the instruction format of the processor system. Registers are also sometimes referred to as accumulators, particularly registers used in the arithmetic unit. Registers generally have special hardware circuitry associated with them to enable the required arithmetic and logical operations to be performed.

Control unit
The function of the control unit is to access each instruction in turn, and obey the operation contained in the instruction. Normally instructions are accessed consecutively as they are located in store. Branch instructions are the exception: they cause the control unit to 'change direction' and start accessing instructions at a different part of store. Such instructions are thus also known as 'change of control' instructions. Registers in the control unit are used to hold the addresses of the next instruction to be performed and also may contain the operation code of the instruction being performed and addresses of storage locations to be operated on by the instruction. In the case of peripheral unit instructions the control unit activates the device and monitors the control signals to and from the device, such as interrupt signals and signals denoting the operating state of the device. Contents of control registers provide useful information about the progress of a program, and in some cases may be printed out as part of diagnostic routines (see section 3.4.5). Registers are also used to hold constants to be added to addresses held in relative form in order to make them absolute, i.e. to refer to actual locations in store. This technique is used to enable programs to be relocated to different parts of storage without disrupting the program's operation. Addresses in the program are held relative to the first location of the program, regardless of where the program is placed in store. This technique is used particularly with multiprogramming systems (see section 3.1.6).

Arithmetic unit
The arithmetic unit contains the hardware circuitry required for the manipulation of data in core store. This can include the normal arithmetic operations of adding, subtracting, multiplying and dividing, and such logical operations as shifting, logical 'and' and 'or' operations, negating. These are further described in section 3.1.5. In addition the unit will perform comparisons between elements of storage, and set switches or signals as a result of these operations. Registers are required to hold intermediate results, and to enable special hardware to be associated with specific storage locations: it would obviously be wasteful to connect every

87

location in store with the circuitry required to perform every arithmetical or logical operation.

Method of operation

Control logic units differ widely in methods of operation, depending on the type of computer, the extent of the operations in the instruction set and the organisation of data. However in most processors the pattern of events follows a general cycle which can be described in outline. The instruction address to be accessed is extracted from a control unit register, and the contents of the address are transferred to the control unit. The unit then decomposes the instruction into its consistent parts, basically operation code and one or more storage location addresses. If the instruction is an arithmetic or logical operation, the contents of addresses to be operated on (either data or other program instructions) are transferred to registers in the arithmetic unit. The operation is performed in the arithmetic unit, and the results of the operation placed in further registers. Depending on the instruction format, the result may be placed back into the required area of store in one cycle of operations, or it may be left to the next or subsequent cycles to do this. If the instruction involves a branch or 'change of control', the control unit replaces the address of the next instruction to be performed with the address specified in the branch: otherwise the next address is incremented by one, and the cycle of events is repeated. In the case of peripheral unit instructions the control unit activates the control unit of the peripheral, which may be located either in the central processor or via standard interface (see section 3.2.1) to a control unit within the peripheral itself.

In some cases, arithmetic operations and control of peripherals is performed by software rather than hardware. Such software is held permanently in store in the form of small programs. The arithmetic unit is replaced by a branch to the appropriate routine (which itself may use the arithmetic unit for its own functioning). This technique is used with executive or supervisory routines or operating systems (see section 3.6).

3.1.3   Core storage

General

The general characteristics of core storage are described more fully in section 3.3.1. In this section core storage is discussed in relation to its use in the central processor. Storage associated with the processor is used to hold programs and that part of the data available to the system which is being processed by the program. In a commercial environment it is unlikely that all the data being processed by the program will be held on the same storage medium as the program operating on it: backing stores of one sort or another are available for holding large quantities of data. Sufficient data for action by the relevant program is read into the processor's store when required: it is processed in this store and then the results of the processing are output either onto another backing store or an output peripheral. Core store is used for programs and the data on which they operate because of its fast access time, required because of the relatively large amount of manipulation of items of data and program between core store and the control unit and arithmetic unit described in section 3.1.1. In some configurations alternative forms of storage device with fast access times, although not as fast as core store, are used to hold some of the data on which a program is working. In particular, magnetic drums are used for this purpose. Large programs which cannot fit into core store in one segment may be divided into sections known as *overlays* (see section 3.4.3). Overlays may also be held on backing store such as magnetic tape or a direct access device, and are called into core store when required.

Single level storage
An important feature of the use of core store to hold programs and data is that, as far as access is concerned, no distinction is made as to which elements of the core store hold data and which hold programs. Program and data are held as a continuous area of store, and data locations and program locations are addressed in exactly the same way. This means that arithmetical and logical operations may be performed equally on data or on program locations. This is a vital factor in the art of programming, for by manipulating individual instructions, or by using store locations to hold the results of arithmetic or logic, program modification can be performed. Modification is described in more detail in section 3.4.7. It is the fact that core storage is used as a single level store, treating data and program in the same way, that enables this flexibility to be achieved.

3.1.4 Compatibility

General
Compatibility is a term used to describe an important feature of later generations of computers: the ability to run programs on different configurations and different sizes and speeds of processor without modification. In earlier generations of computers, processors differed not only in size of core store available, but also in control logic associated with individual processors. This meant that programs written for one processor might not be operable on a different processor designed by the same manufacturer, because its control logic would not accept some of the program's instructions. Another problem was that of the addressing of locations in store: this might differ from processor to processor, again causing difficulties in running programs in different processors. A further problem causing difficulties in running programs on different processors was that of internal speed, or cycle times. In some early processors instruction timing was a highly critical programming problem, and the programming solution for one speed of processor might well prove unworkable at a different cycle time.

89

In a compatible range of processors some of these difficulties are eliminated. A compatible range of processors is thus defined as a series of processors which differ in size and speed, but within which programs are interchangeable.

Types of compatibility
Processors differ in size of core storage and in cycle times, and differ also in their control logic, i.e. whether certain instructions are performed by hardware or software. A range of processors may be said to be upwards compatible if the limiting factor on running programs on different processors is solely that of size of core store. In this sense, any program which will 'fit' into the core store of a processor will run satisfactorily on that processor. The technique of overlaying programs, described more fully in section 3.4.3 often allows programs which are too big for a smaller core store to be divided into sections, each of which is read into core when required. On a larger processor the whole program might be loaded at once.

Another form of compatibility is achieved by the use of a standard inter-face between processor and peripheral devices, described in section 3.2.1. This allows programs to be written to operate with any type of peripheral unit, since data reaches the program in the same format regardless of unit. This not only applies to different models of the same type of device, but may even allow the program to run unchanged with different types of device, e.g. the program may be fully compatible with any type of direct access storage device.

Methods of achieving compatibility
For a program to run on a given processor, two factors must be satisfied: the storage capacity of the processor must be large enough to hold either the whole program or the largest overlay into which the program is divided, and the control logic must be capable of accepting all the operation codes and addresses in the program. It is in satisfying this second factor that different techniques may be used. Instructions may be carried out either by *hardware*, i.e. special circuitry, or by special *software* or small programs designed to perform the required operation. In some ranges of processor, instructions performed by hardware on some processors may be performed by software on others. If the range is fully compatible, the user and programmer need not know which method is being used: the only difference will be that programs making extensive use of instructions performed by software will tend to take longer than when the same instructions are performed by hardware.

3.1.5   Data operations
General
The information patterns in the processor which represent different forms of data, may be subjected to different types of operation. These operations

90

on data alter the information patterns according to the fixed rules of arithmetical and logical operations.

Arithmetical operations
The basic arithmetical operation performed in a processor is addition. The contents of two locations in store may be added together, and the result put into a third location: in special cases, any two or even all three of these locations may be the same location. Arithmetic operations may be performed directly between any store locations, or may be performed via a special location sometimes known as an accumulator or register. Operations may be performed entirely by hardware circuitry, or may use software routines to simulate the operation. All arithmetic operations may be reduced to the basic addition operation coupled with the operation of negating a value. Thus subtraction is adding a negated value. Multiplication and division are effectively repeated additions and subtractions. In some processors some or all of these operations are performed by hardware: in others, the slower but cheaper method of using software is used. For scientific and mathematical work, additional arithmetic operations such as taking of square roots and exponentiation are frequently required, and special scientific processors may provide hardware circuitry rather than software for these. In addition, floating point arithmetic is used extensively in scientific applications, and floating point hardware may be provided on scientific processors. Floating point arithmetic is described more fully below.

Logical operations
Logical operations in terms of computer terminology are operations which are performed according to the rules of Boolean algebra: logical in this context should not be confused with the everyday use of the expression. Boolean algebra is a system of notation devised by the mathematician George Boole (d.1864) to express formal logical relationships in the same way that mathematical algebra is used to express mathematical relationships. The significance of formal logic to computers is in the fact that formal logic deals with expressions which may have one or two values, 'true' and 'false'. Since these values may be expressed as '0' and '1' and the 'rules' of logic are purely formal ways of generating a value of '0's and '1's, the rules of logic may be used in operating on the binary patterns used to represent information in a computer system. Although the logical operations are given such names as 'logical and', 'logical or', these are simply formal ways of talking about the pattern generating rules of logic and have no 'meaning' in the sense of ordinary language. A logical operation may be thought of as a gate through which two signals try to pass simultaneously: the gate only allows one signal to exit, and the particular rule defines the value of the output signal given the values of the two input signals, each signal having one of two values, '0' and '1'. The 'logical and' and 'logical or' operations are shown in figure 26.

91

Input signals		Output signal for 'logical and'	Output signal for 'logical or'
1	1	1	1
1	0	0	1
0	1	0	1
0	0	0	0

26. Logical 'and' and 'or' operations

Another type of logical operation is the operation of shifting. Basically, shifting involves the displacing of the bits which form a given information pattern one or more places to the left or right. The effect of a shift depends on the use to which the programmer puts it, but clearly one such use is to multiply a number by the radix of the number system it is represented in. For example a decimal number e.g. 123, is multiplied by 10 for each place it is shifted to the left, e.g. $1230 = 123 \times 10$, $12300 = 123 \times 10^2$.

Floating point and fixed point representation
A major problem in scientific computing is representing large numbers to sufficient degrees of accuracy. A string of numerals, e.g. '12345' can be used to represent different numbers, depending on the position of the decimal point, e.g.

$$1 \cdot 2345 \quad \text{or} \quad 12345 \times 10^{-4}$$
$$12 \cdot 345 \quad \text{or} \quad 12345 \times 10^{-3}$$
$$123 \cdot 45 \quad \text{or} \quad 12345 \times 10^{-2}$$
$$1234 \cdot 5 \quad \text{or} \quad 12345 \times 10^{-1}$$
$$12345 \quad \text{or} \quad 12345 \times 10^{0}$$
$$123450 \quad \text{or} \quad 12345 \times 10^{1} \text{ etc.}$$

An alternative way of expressing each of the numbers on the left hand side of the table is by following the string of numerals with the value of the power of 10 shown on the right hand side, e.g.

$$1234 \cdot 5 \quad = \quad 12345, -1$$
$$123450 \quad = \quad 12345, 1$$

Representing numbers in this way allows the computer to deal with extremely large or extremely small values: e.g. '12345, 99' would require a

92

lot of space to write, and, if held as a binary integer, would require a vast number of bits. In this simple form, however, this large number can be held as two relatively small ones. This notation is known as floating point notation, since the position of the decimal point in the string of numerals is implied by the value of the exponent or characteristic attached to it. In fixed point arithmetic, the position of the decimal point is pre-determined, so the size of number which can be represented depends on the number of character or bit positions available. Most processors have a limit on the physical number of bits or characters on which arithmetic operations can be performed: when the values required exceed these physical limits, then floating point notation must be used for number representation. Requirements for this type of number are usually restricted to scientific applications, and special purpose software and hardware are available for handling calculations using floating point notation.

3.1.6   Timesharing and multiprocessing

Peripheral transfers

Data is transmitted between peripheral units and the processor over cables known as *channels*. The connection between the processor and its various input and output channels is called the processor's interface. The significance of standard interface, which enables a processor to accept data along a channel, from any type of peripheral is described more fully in section 3.2.1. Signals from a peripheral unit attempting to enter the processor via the interface cause a momentary 'interrupt' to occur in the processor. The processor can sense the occurrence of an interrupt and can take action to allow a unit of data (e.g. a six-bit character) to be transferred into the processor according to the program instructions. The importance of the interrupt concept lies in the ability of a computer to continue with other activities between the occurrence of interrupts. Earlier computers without this facility had to wait for a data transfer to finish before any other processing could take place. Since the relative speed at which data can be input is considerably slower than the speed at which it can be manipulated, this meant that the processor was idle for a large part of its time.

Multiprogramming

The concept of interrupt in peripheral transfers can be used not only to improve the efficiency of a single program, but also to allow for the sharing of processor time between more than one program. The most common pattern of events for any single program is a succession of 'calculation' stages interspersed with stages of reading in or writing data. The propor-tion of time between these two types of activity will vary considerably between different programs: some will spend most of the time with input and output operations; others will do little in the way of input/output operations but will require a great deal of access to the processor's control unit for processing data. With multiprogramming, the resources of the processor can be shared by several programs. While program A is waiting for a data transfer to be completed, program B can be activated to use the processor's control facilities. When program A's data transfer is completed, it becomes ready to continue processing. Program B can then either be suspended so that program A can immediately continue, or program B can continue until it requests a peripheral transfer. At this stage program B will be suspended and program A can start again. Control over the sharing of time between programs can either be performed by hardware, or more usually by a special program which is always resident in the processor, sometimes known as an 'executive' program.

Priorities
A further feature of multiprogramming is the allotting of priorities between programs. The program with the highest priority awaiting activation is always given the next turn with the control unit. Normally, programs requiring relatively little processor time will be given a relatively high priority as they will not require very much time. Programs requiring a relatively large share of processor time will be given a lower priority to prevent them monopolising available time.

Security
Security between programs is another feature of multiprogramming. It would clearly be disastrous if two programs in a processor corrupted each other by overwriting or interfering with each other's areas. Security can be achieved either by hardware or by software. Hardware lockout, as it is called, is a device which prevents a program addressing any part of store outside the limits set when that program is entered into the processor. Software performs a similar function, using the executive program to check that no store violation occurs. Similarly, control is exercised over the allocation of peripherals to different programs within the processor. Timesharing of resources may also occur within individual programs, by defining parts of an overall program into sub-programs, each of which can be timeshared with other sub-programs within the main program. In this case, responsibility for security will remain within the main program, since automatic security checking normally applies only to main programs and not to individual sub-programs.

3.2 PERIPHERAL UNITS

This section describes the peripheral units of a computer system. Peripheral units are the limbs and sense organs of a system. The processor, as the brain, performs the operations on raw data which converts it into information: peripheral units are needed to feed this raw data into the processor, store intermediate states of this data for further processing, store instructions not immediately required by the processor, and record and present the results of the processor's activities in a form intelligible to the people requiring them.

3.2.1 Interface

General

In early computers, peripheral units and the central processor were generally designed as a complete system. Control of the peripheral units and the transfer of data between them and the central processor was performed by peripheral control units integrated with the central processing unit. Thus, if, for example, a user required a faster arithmetic unit, he would have to exchange not only his processor but some, if not all, of the peripheral units associated with it. Later developments now enable peripheral units to be designed to operate with any central processor, within the limits of the particular range of computers involved. This is achieved by building the peripheral's control unit within the unit itself. Data is transferred to the central processor by means of a link which is known as the *interface*. Regardless of peripheral type, data travels to the central processor in a standard format, enabling the processor to accept input from and generate output to any peripheral. The number of peripherals that can be attached to a given processor depends on the number of interface channels available on the processor. But a user may plug in additional peripherals up to this limit, or change processors but still retain his peripheral units, plugging them into the new processor.

Transfer modes

Information transferred between a peripheral unit and the central processor is generally of two types:

1. Control information, enabling the processor to determine the status of the unit and to initiate and terminate its operation.
2. Data being input to or output from the processor.

The rate at which data is transferred to the central processor varies, but in general two modes of transfer are used, a 'character' mode in which data is transferred one character at a time, and a 'burst' mode, in which several characters are transferred together. The former mode is used with so called 'slow' or 'character' peripherals such as card readers and paper tape punches. The latter mode is used with fast peripherals or backing stores such as magnetic tape and discs.

3.2.2 Card reader

General

A card reader is an input peripheral unit designed to accept data from

95

punched cards. Different sizes of card are in use, but in general cards used with computer systems are 80 column cards or, more rarely, 40, 45 or 90 column cards. The nature of punched cards is described more fully in section 3.7.1. Card readers are supplied with speeds of from 100 up to 1000 or more cards a minute.

Card codes

As described in section 3.7.1, data is recorded on punched cards by means of various coding systems. Various combinations of holes represent numerals, letters and symbols. The card reader detects the pattern of holes in the card and transfers this data to the central processor. The code in the punched card must be converted into the internal code used by the processor, as described in section 3.1.1. This conversion from *external* to *internal* format can be performed in different ways, depending on the type of card reader. These are two main ways of performing conversion: hardware and software. Normally the systems analyst need not be concerned with the details of how this decoding is performed: data will be punched in the standard card code as prescribed by the manufacturer, and the card reader will be designed to accept this code and convert it into internal format. Card code conversion becomes significant when non-standard codes are considered, either in order to read data punched in a non-standard code, or to detect non-standard codes as errors. In the case of hardware decoding, some card readers are supplied with a switch enabling different card codes to be selected manually before data is read.

Automatic rejection of cards punched in a non-standard format is another feature of some hardware conversion card readers. An alternative to hardware conversion of non-standard codes is to read cards in a binary image mode. In this case a 'mirror image' of the pattern of holes in the card is transferred into the central processor as a bit pattern. This pattern can then be accessed by program and either converted into specific internal codes, or cause rejection of the data as invalid.

Operation of card reader

A card reader operates by moving a punched card from a storage container known as a *hopper*, along the *card track* where it is checked and read into a *stacker* (see figure 27). The *card track* contains various positions known as stations. The number and operation of these varies from machine type to machine type, but there will always be a *reading station* at which data is sensed by detecting the presence or absence of holes in the card, and one or more checking stations where data is again read from the card and compared with the data from the reading station as a check operation. A further stacker may be provided for holding rejected cards. The timing of cards along the card track is crucial to the accurate reading of cards and is effected either mechanically or electronically by means of photo-electric cells.

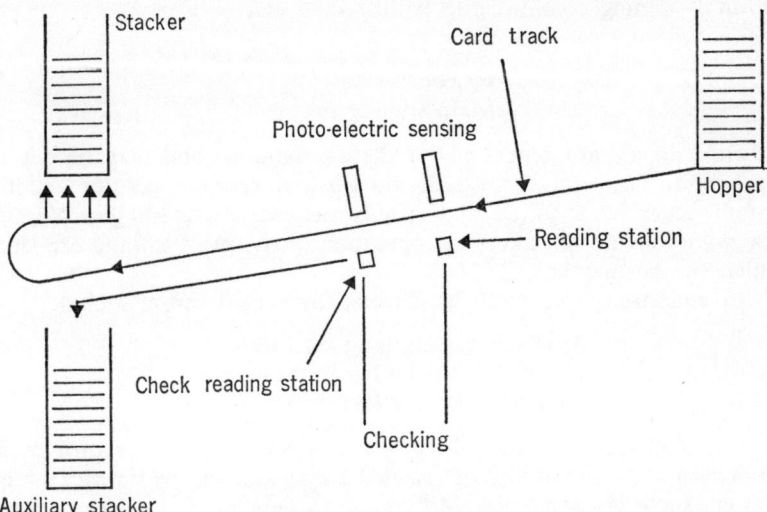

27. Operation of a card reader

Characteristics of card readers
The detailed mechanics of the operation of card readers need not concern the systems analyst. However, certain characteristics are significant in designing systems in which use of a card reader is required. There are two basic types of card reader, the *cyclic* reader and the *on demand* reader. With the cyclic reader, a card can be fed at only one point in the basic feed cycle of the mechanism. This means that the time taken to access data from the card may vary from one card to another. Thus the amount of processing between card read instructions is critical, since if the maximum access time is taken for each read, the total reading speed may be drastically reduced. Thus programs written to read data from cyclic readers must be designed to minimise this problem, possibly by the use of buffering techniques, discussed more fully in section 3.4.7. The importance of minimising idle time must be considered in relation to the general techniques of time sharing used in the particular computer system; if multiprogramming is possible the speed of card reading may depend on the relative priority of the card reading program rather than the type of card reader. With on-demand card readers, a card is fed immediately on being given a card read instruction. This means that there is a constant access time for each card read. This characteristic minimises idle processor time due to delays in the card cycle. However double buffering may still be employed in order to time-share processing and input procedures.

Error checking
The operation of a card reader may be suspended when certain conditions, both normal and exceptional occur.

97

D

Normal running conditions to be expected are:

> Hopper empty
> Reject stacker full
> Main stacker full

Hardware signals are generated for these conditions and may be detected by program. The analyst must decide whether special operator action or program action is required and specify programs accordingly. Normally these conditions do not affect the operation of the program and are simply rectified by the operator.

Error conditions which can be detected by a card reader include:

> Misfeed or jam along card track
> Error detected by hardware checks
> Unit of data lost in transfer.

The analyst must consider the characteristics of particular card readers with regard to the handling of rejected cards and the occurrence of malfunctions such as card jams. Most card readers have a reject pocket or stacker into which cards which have failed hardware checks are directed. The analyst must specify action to be taken when such conditions occur, and be aware of the reasons for card rejection; these can vary from card reader to card reader, but are generally due to card holes being displaced, damaged cards or cards not falling within the tolerances of the reader, or due to invalid card codes being punched. In the case of card wrecks or card jams, it is important to establish whether or not data from the damaged card or cards has been read successfully. This again depends on the type of reader; whether cards are read end on or edge on, and whether the card has passed over the complete card track or has been wrecked before completing its passage past the reading station. Systems using card readers must thus cater for all eventualities and actions to be expected from particular types of reader.

3.2.3 Card punch

General

A card punch is an output peripheral unit designed to punch data into cards. As with card readers, the size of card is generally 80 columns. The nature of punched cards is described more fully in section 3.7.1. Card punches are supplied with speeds of from 30 to 400 cards per minute.

Card codes

The various coding systems by which data can be recorded in cards is described more fully in section 3.7.1. A card punch causes data held in *internal* form in the processor to be punched into cards in the appropriate card code as *external* form. Normally, the card punch will punch data in the standard code used with the computer system, and non-standard punching will not be possible. Some punches have hardware devices enabling different card codes to be selected manually. It is also possible to punch in different codes by performing translations by software from one internal format to the internal format of the punching required.

Operation of card punch
A card punch operates by moving a card from a *hopper* along a *card track,* where data is punched and the accuracy of the punching checked, and into a *stacker* (see figure 28). The card track contains a *punching station* at which holes are punched into the card by means of special knives: these may be a bank of eighty knives for punching rows of holes at a time, or twelve knives for punching columns at a time. There is also a sensing device for comparing the data in the card after it has been punched with the original data, thus checking the operation. A further stacker may be present for holding rejected cards, or such cards may be automatically offset in the main stacker. As in card readers, the timing of cards as they pass along the card track is controlled either electronically or mechanically.

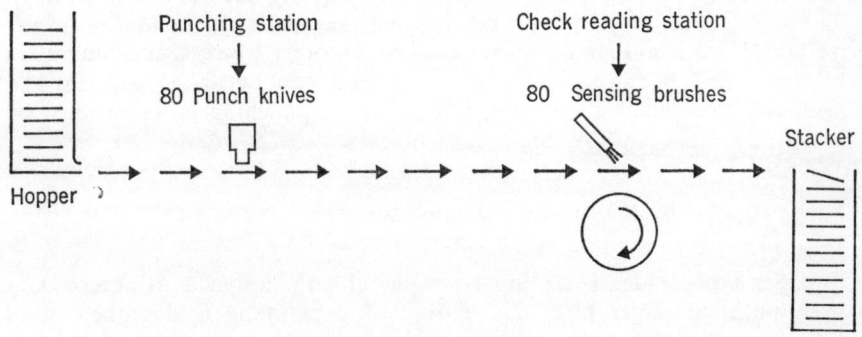

28. Card punch operation

Characteristics of card punches
As with card readers, the characteristics of two basic types of card punch must be considered by the analyst. Card punches are also of the cyclic or on demand type, and programming considerations are similar to those described for card readers. In addition, with certain types of card punch, the card which has been punched remains in the card track and is not checked or fed into a stacker until the next feed instruction is given. It is thus normal practice when specifying programs to cater for this condition by giving an additional feed instruction at the end of run to clear the card track: operators may otherwise forget to remove this last card, and in any case it may otherwise not be checked.

Error checking
The operation of a card punch may be suspended when certain conditions,

both normal and exceptional, occur. Normal running conditions to be expected are:

> Hopper empty
> Reject stacker full
> Main stacker full

Hardware signals generated by these conditions may be detected by program. As with card readers the analyst may specify program action, or rely on rectification by the operator. This latter course is usually adequate.

Error conditions which can be detected by the card punch include:

> Misfeed along card track
> Disagreement between data punched and data subsequently read
>   from the card
> Unit of data lost in transfer.

The analyst must ascertain the action taken by different card punches on detecting errors. These include feeding rejected cards into a reject stacker, or off-setting the error card, and possibly the following card, in the main stacker. Again, with certain punches when this condition occurs the error card is automatically re-punched. In other cases, repunching must be performed by program. When a misfeed jam occurs it will also be necessary to repunch the damaged card, and a punching program must be specified to take care of these conditions.

3.2.4   Paper tape reader

General

A paper tape reader is an input peripheral unit designed to accept data from punched paper tape. The nature of paper tape is described more fully in section 3.7.2. Paper tape readers operate at speeds of from 300 to over 1000 characters per second.

Paper tape codes

As described in section 3.7.2, various types of code are used for recording data in paper tape, distinguished by the number of holes required to represent each character. Paper tape may be 5, 6, 7 or 8 track, and readers are available for all types of paper tape. The paper tape reader detects the pattern of holes in the tape and transfers the data to the processor. Certain punchings are used as control characters rather than data characters: these indicate for example, the beginning and ending of blocks of data, the deletion of characters, i.e. a symbol to be ignored when read, and special control codes used with other devices which can read paper tape, such as a teletype machine. The analyst must establish the type of paper tape and associated code the paper tape reader can accept, and decide whether data will be input using all available paper tape codes, or in a limited sub-set of these codes.

Modes of paper tape reading
With paper tape, data can be punched either continuously, or in blocks separated by gaps of unpunched tape; further, the character set which can be punched in the tape may be larger than that allowed for by the internal format of data in the processor. Because of these factors, there are several modes of reading data in which the reader operates. These modes are usually selected by the input program. The analyst should be aware of the various input modes available, and specify the one most suitable to the particular operation and method of recording data used. Some examples of modes of reading with 8 track paper tape on a machine whose internal form allows a 6-bit character format are given below.

Read a specified number of data characters
This requires the input instruction to specify the number of characters to be input.

Read data characters until a stop character is reached
This requires data to be input until the reader recognises a control character indicating the termination of the read operation.

Read data and control characters
This requires the input of all characters including special control characters indicating 'delete' and 'blank'.

Read all punching combinations
This requires a binary image of the input data to be transmitted to the processor. This could be used, for example, when reading, 5, 6, 7 or 8 track paper tape. The image would be converted by specially devised tables into the appropriate internal character form.

Characteristics of paper tape reading
A paper tape reader consists basically of a tape dispenser which holds the reel of tape to be used, a reading station consisting of a photo-electric device for sensing the holes as it passed through the reader, and a receptacle for holding the outputs. Most paper tape readers present characters on demand, which means in effect that the reader can stop between individual characters. Some readers can only accept blocks of data on demand and can only stop on a gap between blocks. Special wiring panels which can be interchanged by the operator to give different interpretations to special codes and control codes described above are another characteristic of some readers. It should be noted that because reading of paper tape is generally performed by a light sensitive device certain colours of paper may not be acceptable. The analyst must establish what types of paper can be operated on the reader.

Error checking
The codes in which paper tape is punched normally include a parity bit (see section 3.1.3); the reader will check that each code has correct parity, and generate an error signal if parity failure occurs. This signal will either cause the reader to stop automatically, or will generate a signal detectable

101

by software, for program action. An input program must be specified to deal with such signals either by re-reading the offending character or block, or attempting to correct the character in store.

3.2.5 Paper tape punch

General

A paper tape punch is an output peripheral unit designed to punch data into paper tape. As with paper tape readers, a punch may produce 5, 6, 7, or 8 track paper tape. The nature of paper tape is described more fully in section 3.7.2. Paper tape punches operate at speeds of from 100 characters per second.

Paper tape codes

These are discussed more fully in section 3.7.2 and section 3.2.4 on Paper Tape Readers. Paper tape punches as with readers generally operate on one standard code, but some may also punch data in other types of code. In this case special programs are required to arrange the data in an appropriate internal format. A parity bit is automatically added to each code as it is punched (see section 3.1.1).

Modes of punching paper tape

As with paper tape readers, the nature of paper tape means that data may be produced in several different modes, according to the requirements of the user. The modes available will differ according to the type of punch and the output code standard used, but the analyst must be aware of the output modes available, and the subsequent use of the output tape. Some examples of punching modes available with an 8 track standard code are:

Punch a specified number of data characters
This mode requires the output instruction to specify the number of characters to be output. Any shift characters will not be punched, but depending on the type of punch may or may not be included in the count of characters.

Punch a binary pattern
This mode outputs a binary image of the input character in internal format. This mode can also be used to output data in 5, 6, or 7 track paper tape.

Punch blanks
This mode outputs sprocket holes only, and may be used to output inter-block gaps or trailers at the end of a reel.

Characteristics of paper tape punching

A paper tape punch consists basically of a tape dispenser, which holds a spool of blank tape, a punch head, and either an output spool for winding on punched tape or a bin into which punched tape falls loosely, to be rewound manually after punching is completed. As with readers, special panels may be available which can be changed by the operator to enable

102

different codes or types of parity (odd or even) to be generated. It should be noted that because of the restrictions on reading paper tape by photo-electric devices, certain colours of paper tape are not acceptable.

Error checking
After punching a character, the punched code is generally automatically re-read and compared with the original internal code. If the comparison fails, the punch will either stop automatically, or a signal is generated which can be detected by program for appropriate action to be taken.

3.2.6. Printer

General
A printer is an output peripheral unit designed to present data in printed format. Print is usually confined to upper case letters, numerals, and a number of special symbols, although special character sets are provided for special circumstances. Printers vary according to the number of characters which may be printed on a line, and the speed at which printing takes place. Printing may be onto continuous form stationery, either in single or multipart sets of various types, or onto other media such as continuous card stationery. Many different types of printer have been developed, ranging from electromechanical machines such as barrel and chain printers, to electrostatic page printers and electron beam recording devices. However, the majority of printers in use are elecromechanical printers which output a line of print at a time, and are known as line printers. Line printers print from 80 to 160 characters per line, and at speeds of 100 to 2000 lines per minute, with character sets of up to 64 characters. Characters are usually spaced at 10 per inch, and 6 or 8 lines per inch may be printed.

Types of line printer
The characteristic which distinguishes a line printer from other types of printer is that a line of print is output during each cycle of its action. Different types of line printer either assemble the line, character by character, during the print cycle, or cause the complete line to be printed simultaneously.

Generally the different techniques used to set up and print the line need not concern the systems analyst. However, programming techniques may differ for different types of printer. With a *chain printer,* type is carried on a chain or belt moving across the paper. When the appropriate character is opposite the required print position, the print instruction causes the print hammer to strike. Careful programming will optimise the time taken for all characters in the line to be output, by reducing the number of passes of the chain over the face of the paper. A *barrel printer* consists of a rotating cylinder, on whose surface the full character set of type is present for each character position on a line of print. A complete line

103

of print is set up on one rotation of the barrel, print hammers striking the paper as each required character on the barrel is opposite the appropriate character position. Thus program action will not lessen or increase the speed of printing. However, because of the arrangement of characters or the surface of the barrel, by limiting the characters to be printed to a subset of the full character set—e.g. letters and numerals only—a complete line can be achieved in less than a full revolution of the barrel, thus effectively speeding up printing. To achieve this speeding up a special mode of instruction might be specified: not all printers have this facility.

Operation of a line printer

Figure 29 shows the general arrangement of the main components of a barrel line printer. These are the *print barrel* which contains the complete character set at each print position; a carbon ribbon, generally of silk, which covers the complete width of the print barrel, and is automatically wound and unwound from two spools; the *paper,* which is moved from an *input stacker* past the print hammers to an *output stacker* synchronised with the line-by-line printing cycle; and the *print hammers* which are activated as each required character moves into position at the required print position on the line. The hammers strike the paper surface, pressing it against the carbon, forming an impression on the paper of the type on the barrel. Paper movement is controlled in two ways. The program instruction causing a line to be printed may specify paper movement of a limited

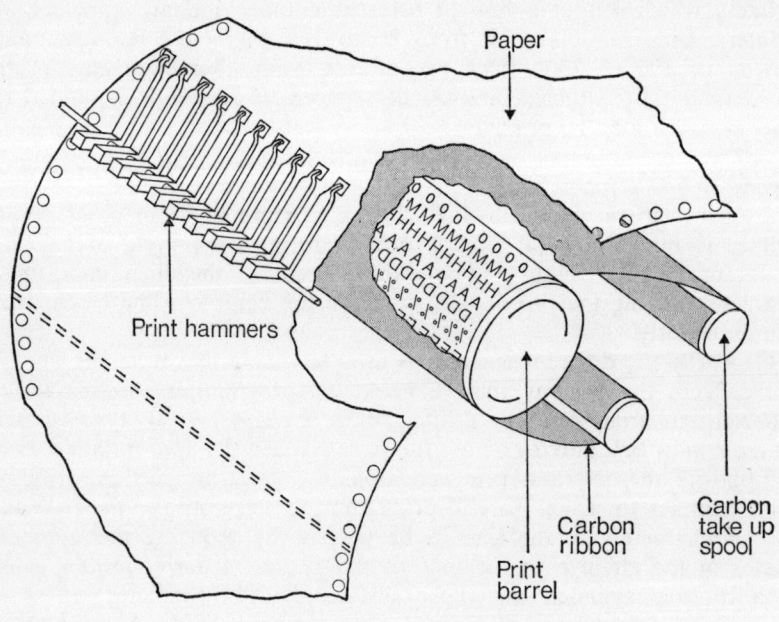

29. A barrel printer

number of lines to precede or follow the printing of the line. Since the number of lines which can be advanced by this technique is usually limited to two or three, and in any case is less than the total capacity of a normal page of print, further paper movement would require the printing of a dummy or blank line, requiring a full additional print cycle.

Printer control loop

A second method of controlling paper movement is by means of a device known as the printer control loop. A program instruction can cause paper movement or *paper throw* to commence. Synchronised with this movement is the movement of a loop of plastic or paper tape punched with a number of holes in up to 8 tracks or channels. A photo-electric cell detects the presence of holes, and will terminate the paper movement when a hole is detected in the appropriate channel. Normally one channel is by convention assumed to represent this first print line of a document, known as 'top of form'. A paper throw instruction specifying this channel will cause the paper to be positioned at the top of a page. Other channels may be specified by the programmer or analyst for specific purposes for given documents, for example for controlling the start lines of various variable length sections in an invoice, such as customer address, list of items and grand total. Use of a paper tape loop can speed the operation of a print program where printing of different types of data has to start at fixed points on a page. The analyst must specify the requirement for a paper tape loop where this would result in such time saving. The detailed design of the loop, and the allocation of control channels will normally be left to the programmer, who will supply details with program documentation.

Error checking

Most printers generate a signal detectable by program when the paper in the input stacker falls below a certain level. Program action can then be taken to suspend printing at a convenient point to allow the operator to load a further supply. Other conditions, such as malfunctions of the ribbon feed, or of jamming of the paper feed will cause the device to stop, without necessarily generating a program detectable signal.

3.2.7.   Typewriter

General

Typewriting devices may be used as on-line peripheral units, acting for both input and output. A typewriter peripheral is in appearance similar to a conventional typewriter, with a standard alphanumeric keyboard, together with additional controls described below. Input is either by manual operation of the keyboard and therefore limited by the speed of the operator, or in some cases punched paper tape may be used to 'drive' the typewriter at somewhat faster speeds of up to 10 characters per second. Output to the typewriter can be performed at up to 10 characters per second.

105

Types of typewriter

Typewriters are used with a computer system for two distinct functions, although the basic device is very similar for both functions. These are as a console or control typewriter, and as an interrogating typewriter. The former function is used in conjunction with an executive or supervisor program or more general operating system, as described more fully in section 3.6. The typewriter acts as the input medium for control instructions from the operator, loading and activating programs, allocating periphereal units, initiating control functions such as store dumps and file initialisation. Output similarly will consist of information to the operator about the state of peripherals and programs in the system, and may also consist of messages from operating programs requiring operator intervention. When used as an interrogating typewriter the device is under the control of an interrogation program. This will generally be associated either with information retrieval from large, probably direct access, data files, or with some form of conversational programming system (see section 3.6.6.). In both uses, the user accesses the program in question and types messages: the program operates on the message and replies by outputting the required reply, either the data required, the result of a calculation, or the acceptance of a program language statement.

Method of operation

Typing is generally limited to upper case letters, both for input and output. In addition to the normal typewriter switches, additional keys or switches are present. These provide communication with the central processor, informing it that the user wishes to access the processor. Depending on the type of typewriter, and the program operating it, the signal generated will either give immediate priority to the typewriter messages or cause the circuit to wait until a suitable moment before allowing the user to input his message. When the processor is ready to accept data this is signalled either by a light being switched on or a message typed out. Messages are typed completely by the user, a 'hard copy' appearing on the paper. When the user is satisfied this message is correct he then activates another switch or key which causes the complete message to be transferred to the processor. Alternatively each character typed by the user is transferred directly to the processor. Other special controls enable the user to delete or erase an incorrect message and repeat it correctly. In addition console typewriters have special controls indicating the state of programs in the processor, showing when such programs are active.

3.2.8.   Visual display units

General

A visual display unit or graphic display unit is a peripheral device which consists basically of a screen on which a picture formed by program can be displayed. Visual display units also act as input devices, in that control over the output picture may be exercised either through a keyboard or by means of a light pen. The type of picture displayed may be confined to a set of alphanumeric and special characters, or may consist of any pattern of lines and curves forming a diagram, design or graph.

Types of visual display unit
Visual display units may be grouped into two main categories: those used as interrogating devices, and those used for design work. The division of the hardware between these two categories of operation is not precise, but in general interrogating visual display units use a typewriter keyboard for input communication, while visual display units for design make use of a light pen as well. The size of a visual display unit is measured either by the number of characters it can display—e.g. 13 lines of 40 or 80 characters per line, or by the matrix of 'dot positions' available, analogous to the scale of graph paper—e.g. a matrix of 1024×1024 dots. 'Pictures' are formed by lines joining dots on the matrix. The pictures or text is formed by a cathode ray display on a phosphor coated surface.

Method of operation
The surface of a visual display unit can be considered as a matrix of dots. Each dot can be illuminated by means of a beam of electrons. The choice of which dots to be illuminated, and hence the picture or pattern of characters to be displayed, is made in response to instructions from the output program. In the case of an interrogation visual display unit, with output confined to letters and symbols, each character of an output message identified by its internal character form is translated into the 'dot pattern' for the character on the visual display unit. From the point of view of programming, the visual display unit operates in much the same way as any other character peripheral. Input is accepted by typing a message on a typewriter keyboard similar to that used with interrogating typewriters described in section 3.2.7. In some cases the keyboard is restricted to numerals only. The program controlling the visual display unit recognises input messages, accesses and processes any required data and responds by displaying the result on the screen in the same way that an answer would be typed by an interrogating typewriter. The input message is not automatically displayed on the visual display unit screen but of course the controlling program may output the message if required. Visual display units used for design purposes operate somewhat differently. In this case individual dot positions are available for illumination by the user program, each dot being identified by its (x, y) co-ordinates. In this mode the picture is built up on the unit in a series of lines formed by illuminating consecutive dots. The light pen used in conjunction with this mode of operation consists of a light sensitive pencil-shaped device, connected to the visual display unit. When pointed at a particular 'dot' position on the surface of the unit, and that dot is illuminated, a signal is sent back to the program, which can thus determine the (x, y) co-ordinates of the dot over which the

107

the light-pen is poised. In addition, a keyboard or switches are provided so that the user can control the action the program is to take: the 'dots' under the pen can be caused to remain illuminated, be extinguished, or cause the program to perform any required operation. In this way the pen can be used to 'sketch' lines on the surface of the tube, to modify lines on a picture already displayed, or to act in an interrogation mode, e.g. by pointing at a question displayed in character form on the screen. Once a picture has been built upon the screen, it can be stored by the program recording the co-ordinates of the illuminated dots. Manipulation of a picture, such as rotation, altering scales, creating perspective, are all performed by mathematical operations on a set of co-ordinates. Hard copy, i.e. a permanent record of the picture, can be made by transferring these co-ordinates onto another device, such as the digital incremental or graph plotter described in section 3.2.9.

3.2.9 Graph plotters

General

Graph plotters, also known as digital incremental plotters, are output peripherals designed to produce graphs and diagrams by means of combining the movements of a moving pen and paper. The output will consist of any pattern which can be made by a combination of straight lines joining imaginary dots, and can thus consist of letters, characters or numerals as well as curves and lines. The size and speed of a graph plotter is measured in terms of the number and size of increment steps available, the width of paper over which plotting takes place, and the speed at which the paper and pen move. Typical sizes are: paper size 11 in. × 120 in. speed 300 steps per second, step size 0.01 in.; paper size 48 in. × 72 in., speed 450 steps per second, step size 0.005 in.

Types of graph plotter

There are two main types of graph plotter, differing in the method by which the pen traces out its path. These two types are known as the *drum* type and the *flat bed* type. With the drum type, the pen moves back and forth over a moving roll of paper, which can also move back and forth under the pen. In the flat bed type, the paper remains stationery while the pen moves across the surface of the paper in two directions.

With the drum type, the length of output graph is virtually unlimited, while the paper for the flat bed type is fixed in size. Special continuous stationery is required for the drum plotter: the flat bed type will take any suitable sized paper.

Method of operation

Graph plotters produce output diagrams by moving a pen across paper in a series of small movements or steps. This movement is formed as a result of the simultaneous motion of the pen and paper (in the drum type) or of the pen alone in two directions (in the flat bed type). In effect, the direction in which a single 'step' line can be drawn is along any of the lines shown in figure 30.

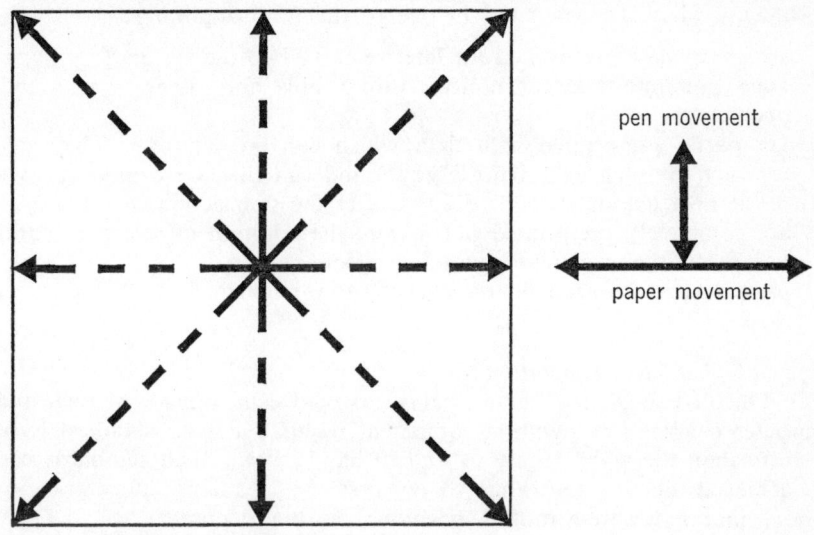

30. Step directions – graph plotter

Any curve or line which cannot be followed exactly by a line in one of these directions is made up of a series of small steps approximating to the required line. Instructions to the plotter enable the pen to be raised or lowered (i.e. lines can be started or ended at any required position on the paper), and allow the pen to be moved continuously along a path made up of a series of (X, Y) co-ordinates. The formation of curves and lines is simplified by extensive software aids, which also provide standard routines for generating alpha-numeric characters to be used as text and annotation. Graph plotters are relatively slow devices, and are frequently operated off-line from magnetic tape or paper tape onto which the series of instructions required to generate the output have been transcribed by a separate program. Some plotters operate on a fixed step size of from 0.01 in. to 0.005 in.; in others, the step size may be varied by program, usually in increments of 0.005 in. The smaller the step size the more accurate is the approximation, but the slower the output.

3.2.10 Character recognition equipment

General

Card readers and paper tape readers must rely on accurate transcription from original documents into punched cards or paper tape, and this transcription process can involve substantial costs. This manual transcription

109

of data into machine language can, however, be eliminated by the use of machines capable of scanning the source documents, transcribing the data into machine language and making it available for processing. Such machines utilise the techniques of character recognition, and will handle documents which are prepared in one of the following ways:

1. are completely pre-printed for later re-entry into the system (e.g. insurance premium renewal notices, utility bills and other 'turnaround' documents)
2. are partly pre-printed with data which can be determined before a transaction (such as a bank cheque) and can then be completed as a result of a manual keyboard entry after the transaction
3. are completely pre-printed but require the addition of selective marks to indicate the particular transaction before re-entry
4. are prepared on a keyboard machine as for punched cards and paper tape.

Types of character recognition
It is difficult enough for human beings to read each other's writing, and computer character recognition equipment requires a more standard form of entry than the wide variety of human handwriting. Such standards can be achieved in the following ways: first, by reducing all entries to formal marks in pre-arranged positions on the document so that the machine can scan for the presence of a mark or no mark and give the presence or absence a particular significance known as *mark reading*. Second by the use of highly stylised character shapes printed in ink which contains magnetisable particles and which are recognisable from their verticle projection. Such systems, known as *Magnetic Ink Character Recognition* or MICR, use character styles designed primarily for machine recognition and have very slight concessions to any requirement to be readable by human beings. Two styles dominate (E 13B and CMC 7 as illustrated in figure 31) and both require exacting standards of printing. It is probable that MICR use will remain confined to banking and similar applications, and the wider range of commercial applications will call for a less precise requirement in printing standards.

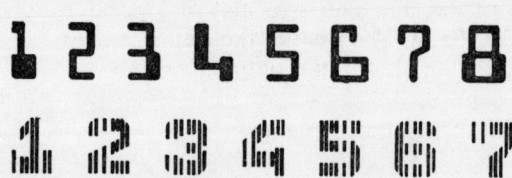

31. Magnetic ink character recognition styles. Top: E13 B. Bottom: CMC7

Such a requirement is met by the third method, *Optical character recognition* systems (OCR) which use character styles more like conventional ones printed in conventional ink, usually black. Figure 32 shows one of these styles, the I.S.O. "B" font. Optical reflectance is measured and data about each character is built up until it is recognised and accepted.

```
This is I.S.O. "B" font which looks like a normal type face.
```

32. An optical character recognition style

Method of operation
Five main activities are involved in the character recognition process. First, the character or mark presented to the machine must be converted into suitable electrical signals for processing, and this will be done by a magnetic or photo-electric device of some sort. Next, it is probable that the signal will then need to be modified to produce a signal more suitable for precise analysis, and such modification may standardise the amplitude of the character signal or improve the contrast and quality of the character. Thirdly, the standardised signal will be processed to extract a number of characteristic properties which enable it to be identified. On the basis of these characteristics a decision will then have to be reached as to the identity of the character or, if it is not possible to identify it, to reject it as being unreadable.

Finally the result of this decision must be communicated to a suitable output unit to provide the output in an appropriate form.

The recognition process takes place at very high speed, characters on each document being recognised, and transferred to the output unit one after another. The limitations in the speed of operation for such equipment are usually related to the speeds at which documents can be mechanically handled. This operation usually entails the automatic extraction of single documents one after another from a batch in an input hopper, and the transportation of each document past a reading head to an output stacker. If output is required to some intermediate medium (e.g. paper tape) then the operation of the output device may itself constrain the overall speed of operation.

Usually the data read from documents is transferred at electronic speeds into a central processor, where double buffering techniques (see 3.4.7 and 4.5.3) are used to handle input data so that the document reader can be kept operating at full speed.

Document feeding speeds vary from between 1,000 to about 2,000 documents per minute.

Some document readers have two output stackers, one to hold valid documents, and another to hold rejected documents that contain one or more marks or characters which cannot be recognised by the device.

Uses of MICR in data collection

MICR techniques are widely employed in the field of banking and uses outside this area are not widespread. The main limitation is probably the high standard of accuracy required in printing of the special characters employed. This factor tends to suit applications where the majority of data is put onto the document when the original document is first printed. For example, bank cheques contain

> Cheque No.
> Customer's Account No.
> Customer's Branch No.

Encoders can be used to print transaction details onto the documents after their initial issue, for example bank cheques are encoded with the value of the cheque after the cheque has been presented to the clearing office. This particular system has had world wide acceptance and the investment in such techniques has been worthwhile for this one application alone. However, the capital cost of encoders as compared with OCR printing devices, somewhat limits the use of MICR techniques in areas where printing is needed at many distant transaction points.

One particular development has been the Document Sorter/Reader, a device which reads cheques and sorts them into a dozen or more pockets at high speed, according to characters appearing in a specified field. Thus cheques can be sorted into batches relating to particular branches and accounts, in a manner similar to using a punched card sorter (see 3.8.1).

Uses of OCR in data collection

The wider tolerance permitted in the creation of characters for OCR reading, has meant that this technique can be widely applied in business and commercial systems. The ability to produce characters in OCR fonts on line printers has particularly hastened the acceptance of turn-around systems in which documents created by the computer, are circulated in some external routine where characters or marks are applied to them, before re-entering the system as transaction details.

The relative simplicity of OCR encoders has also led to the adoption of OCR techniques in retail systems, recording transactions at point of sale. Tally readers have been developed to read transaction details from tally rolls printed by cash registers.

Uses of mark reading

Mark reading is often used in conjunction with OCR techniques to provide transaction data on turn-around documents, for example in meter-reading where fuel consumption details may be recorded as marks on a previously prepared call sheet: marks are used as a method of coding numeric values on the call sheet. Mark reading is also suitable for census documents in which the presence or absence of a mark provides answers of the YES/NO type.

112

3.2.11   Communications equipment

General

The basic characteristic of data communication equipment is its ability to transfer data speedily to and from a remote computer and so provide up to date information which will, for example, enable management to receive up to date reports on a current situation from remote locations, and in turn to communicate instructions to those locations.

In early applications the data was collected at the computer installation via telegraph circuits using paper tape which could then be used as a direct input to the computer at the receiving terminal. Telegraph circuits are capable of handling data at speeds up to 6 characters per second, which is relatively slow compared to the high speeds of data manipulation in the internal memory of the computer.

Further developments in data transmission systems have followed. Telephone circuits turned out to be convenient media for the transmission of data, and transceivers (i.e. transmitter/receivers) providing access to these circuits via punched cards or paper tape are now common in large computing installations. Telephone circuits are used to carry data at speeds of up to 500 characters per second, and often permit several independent signals to be transmitted over the same line simultaneously.

Modulation

There is no problem in presenting or collecting data when the circuit connecting the transmitter and receiver is very short. Unfortunately, unless the circuit is short and unless the speed of signalling is slow, the pulses will be so attenuated and distorted as to be unrecognisable at the receiving end.

The problem is overcome by transferring the signal from a series of discrete pulses into a continuous high frequency signal, less liable to distortion, known as a *carrier wave*. By changing the pattern of this wave (a process known as *modulation*) information may be transmitted over long distances. A modulator is a device used in conjunction with a transmitter, to superimpose the data on the carrier wave. A demodulator is a device used at the receiving end to decode the data. A modem is a combination of a modulator and demodulator.

Line systems

Data is transmitted by communications equipment along lines such as telephone lines. Transmission speeds are measured in Bauds which is equivalent to bits per second. Thus if a character has 8 bits, a speed of 800 Bauds is 100 characters per second. (The Baud was named after the French telecommunications engineer Emile Baudot).

The various types of line system, shown figure 33, are as follows:

1. *Simplex:* A line system along which signals can only be sent in one direction.

113

2. *Half duplex:* A line system along which signals can be sent in either direction, but in only one direction at a time.
3. *Full duplex:* A line system along which signals can be sent in both directions simultaneously.

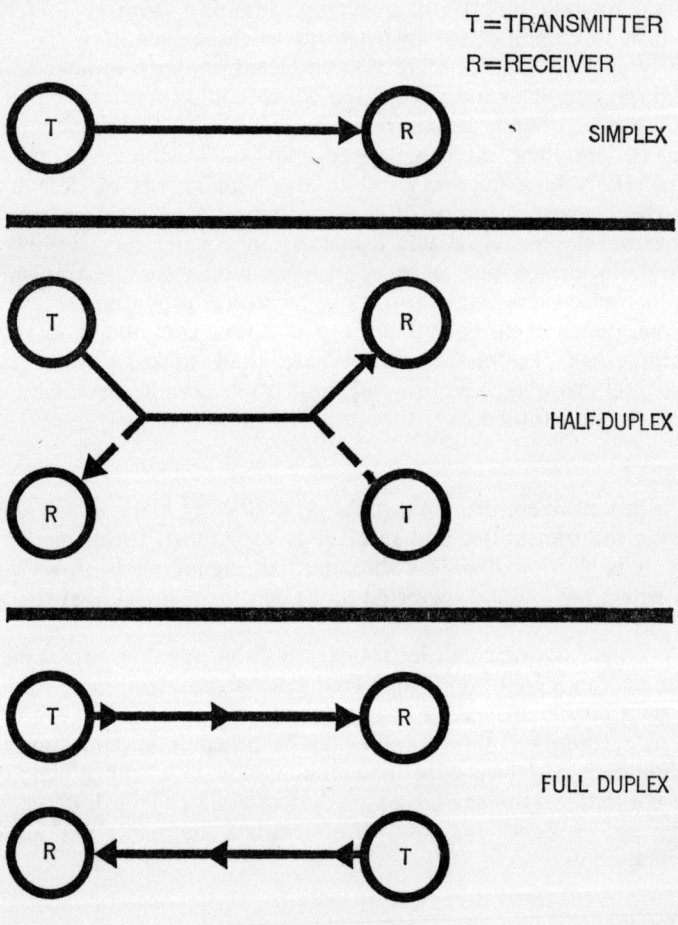

33. Line systems

Message switching

This involves the use of a computer as a 'Store and forward' centre in the acceptance, routing and transmission of information messages between remote locations. The data flow pattern in this case involves two-way message traffic between a number of terminals and a central switching centre. The same circuit is shared by more than one channel, and a device

114

known as the communications multiplexer allows each channel in turn to transmit part of its message along the shared circuit for a short time.

On-line and off-line working
In an off-line mode the data transmission system is not connected directly to the processing system, and the terminal input/output media are generally punched cards, paper tape or magnetic tapes. The on-line mode connects remote terminals via the data transmission link directly to the computer.

Error detection units
These are units used for the automatic detection of errors caused by interference on circuits. Generally errors are more likely on a public network circuit than a private system. There are several methods of error detection in use. They include: —

1. Duplicate transmission of data or immediate return transmission. This is a fairly reliable method but the cost in equipment and time is often considered excessive.
2. Single character parity. This is a commonly used method.
3. Two dimensional parity, which has character parity and block parity. This enables automatic error correction but it is more usual to request retransmission in common with the other methods.

GPO data transmission facilites
The GPO have the monopoly in the United Kingdom of the provision of public communication services. Datel Services are a combination of a particular type of line and where necessary a modem unit to provide a customer with a data transmission facility for a stated speed range.

3.3 STORAGE

A computer is a tool which process a raw material, basic data, into a product, information—a process which is regularly performed by a human being though sometimes more slowly. When a human being is performing this type of task, for example an office manager dealing with a letter, he may well want to refer to other information files, and to work out some calculations on a scratch pad. Both the storage of information in files, and the use of a scratch pad, are operations also performed by computers, requiring *storage units*. This section describes the two basic types of storage unit used with computer systems, the scratch pad type of storage, known variously as high speed storage, main memory or immediate access storage and composed usually of magnetic cores, so also known as core storage;

115

and the file storage units known collectively as backing store. The main backing storage devices described are magnetic tapes, drums, discs, and magnetic card files.

3.3.1   Core storage

General

Core storage is the most common form of high speed or immediate access storage used as the main storage medium in a computer system. It is in core store that programs are loaded when being used, and into core store that the data which is currently being processed is progressively read from output peripherals or backing store. As the processing is carried out, the results are written out from core store to output peripherals or again to backing store. Core storage differs from other storage because the time required to access any item of data held anywhere within the store is the same, and also extremely small, usually a few microseconds. In all other storage devices the access time is considerably greater, since some mechanical action is required, whereas the selection process in core storage is completely electronic.

Description of magnetic cores

Core storage is so called because it is made up of thousands of small magnetic cores or rings, made of ferrite. These tiny rings are threaded on a complex pattern of minute filaments. These filaments or wires make up sets, each of which represents a particular unit of data. As signals pass along a wire and through the magnetic core, the magnetic polarity of the core can be made positive or negative. Similarly, the current state of polarity of a given core can be detected by means of another set of wires and hence transferred to another core or cores.

The magnetic states of positive and negative can be used to represent the two values '0' and '1'. Operations on patterns of '0's and '1's form the basis of all computer operations, as described earlier in section 3.1.5. Figure 34 shows a core with three wires passing through it, A, B, and X. Any signal passing along 'A' will cause the core to adopt one polarity, say positive. The core will remain positive until a signal passes along 'B', when it will become negative. If a signal passes simultaneously through 'A' and 'B', the state of the core will be unchanged. In this way the state of the core, and hence the value it represents, can be changed in accordance with the various operations performed by the computer. The wire 'X' allows the state of the core to be detected when required, and transferred to other cores.

Operation of core storage

For a systems analyst the details of how a core store actually works are not important, and the following simplification is quite adequate. Core storage can be imagined as a huge sheet of squared paper. Each box or

116

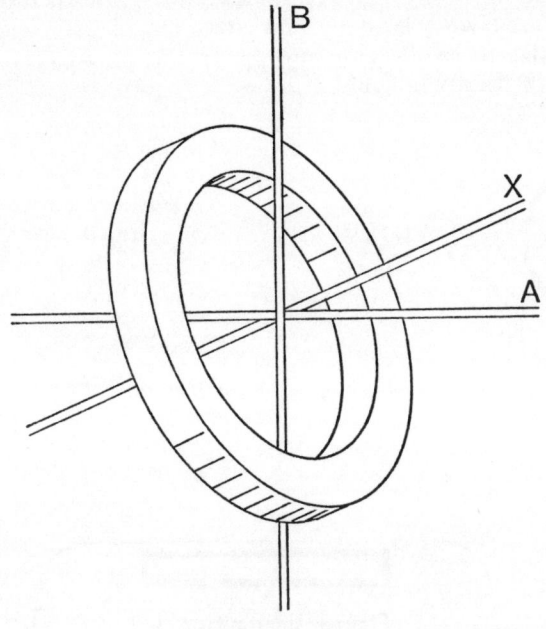

34. Magnetic core

square is numbered serially, starting at '0', rather like a row of pigeon holes. Each box can contain either the digit '0' or the digit '1', and the value of this digit can be altered or transferred by electronic operations, and the time taken to reach any given element is the same as that required to reach any other one.

3.3.2 Magnetic tape

General

Magnetic tape is a continuous narrow strip (usually $\frac{1}{2}$ in. or 1 in. wide) of plastic material, the surface of which is coated with magnetic oxide. Magnetic oxide is capable of being given a magnetic charge, which it retains almost indefinitely and since all data handled by computers is represented by means of a pattern of the digits '0' and '1', this property can be used to retain data required for computer processing.

Operation of magnetic tape unit

Magnetic tape is stored in lengths, usually of 2,400 feet, wound onto reels of about 10 in. diameter. In order to record data onto the reel, or to read data from a reel, a device known as a magnetic tape unit or a *tape deck* is used. This consists basically of three parts, the input reel, a read/write head, and a take up reel. Tape is wound from the input reel onto the take

117

up reel at extremely high speeds, for example 75 in. per second, passing through the read/write heads, which either record data onto the tape, or transfer the data from the tape into the central processor. This process is shown diagramatically in figure 35.

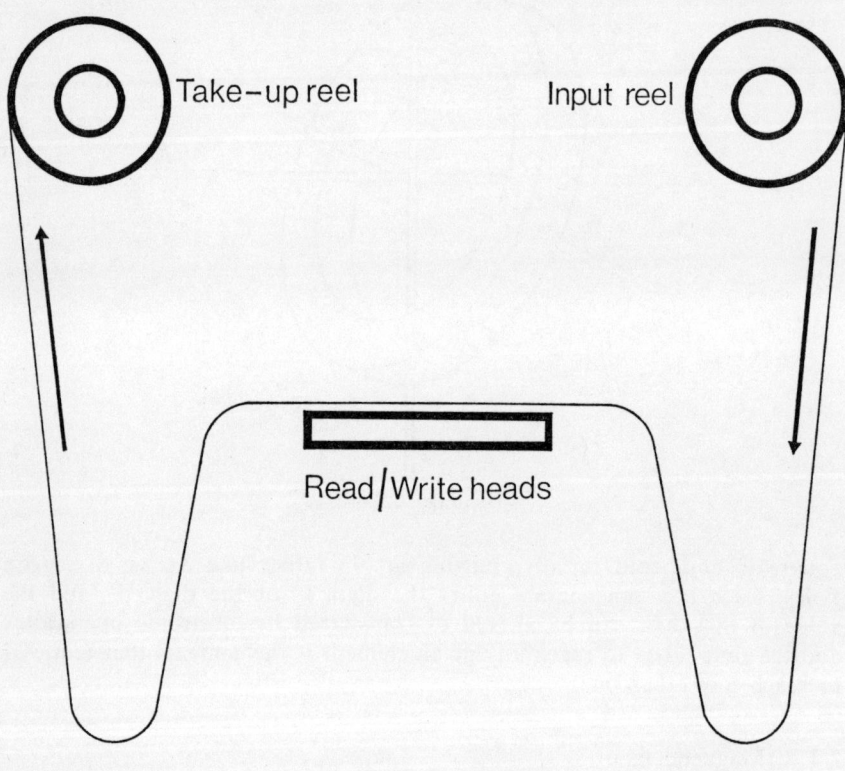

35. Magnetic tape deck

Sophisticated and ingenious devices are incorporated in a tape deck to enable the tapes to be moved at high speed without distorting the surfaces or damaging the oxide, but in principle the operation of a tape deck is very simple, and similar to a tape recorder used for recording music and speech.

Characteristics of magnetic tape
Information on magnetic tape is recorded in a series of tracks running lengthways along the tape, as shown in figure 36.

An individual character on the tape is made up of the bit pattern obtained by taking a 'cross-section' through all the tracks at any position along the tape. A typical recording density for magnetic tape might be 556

118

such character positions per inch. At a speed of, for example, 75 in. per second, 41,700 characters per second may be transferred to or from tape.

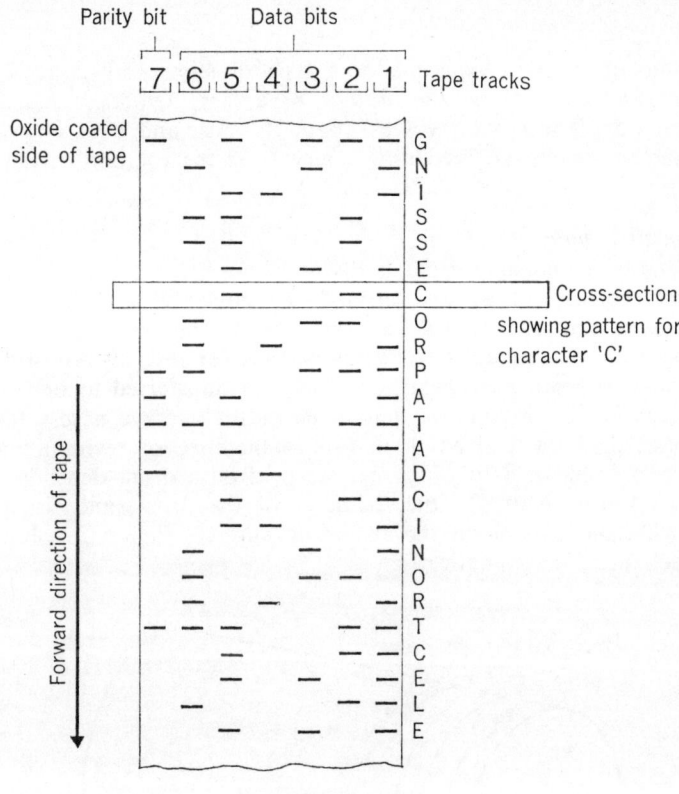

36. Data recording

Error checking

Data on a magnetic tape is usually grouped into units known as blocks. A block is the amount of data which is transferred to or from tape as a result of a single 'tape read' or 'tape write' instruction from the processor. Within a given block transfer of data, two types of automatic check are made, parity checking and block checking. For a description of parity checking see section 3.1.1. Block checking is done by 'adding up' the number of 1 bits along each channel in the block, and arriving at a block total. This is compared with a total obtained in a similar way when data was written to the tape, and recorded on the tape with the block data.

If any transfer failure occurs, the hardware will usually attempt to repeat the read or write operation a fixed number of times. If failure continues, the operation stops, and a hardware signal is set which can be detected by program. Further action will then depend on the requirements of the program.

3.3.3 Magnetic drums

General

A magnetic drum is an example of a type of store known as a cyclic store. Data is recorded on the surface of a large cylinder coated with magnetic oxide. The store revolves about its axis, and data is available for reading or writing at fixed time intervals as the rotation of the drum takes it past a read/write mechanism.

Operation of a magnetic drum

The surface of a magnetic drum is divided into strips, known as 'tracks'. Each track consists of a band around the circumference of the drum. Opposite each track is placed a read/write head. The drum rotates at high speed about its axis, at up to 7000 revolutions per second. As each track passes under its read/write head, data may be transferred to or from the track via its head. This means that in effect the average access time for any item of data on a given track is half the time of revolution of the drum. The amount of data which can be held on a drum depends on the number of tracks available, the diameter of the drum, and the density with which data is held on the recording surface. Figure 37 shows the drum mechanism diagramatically.

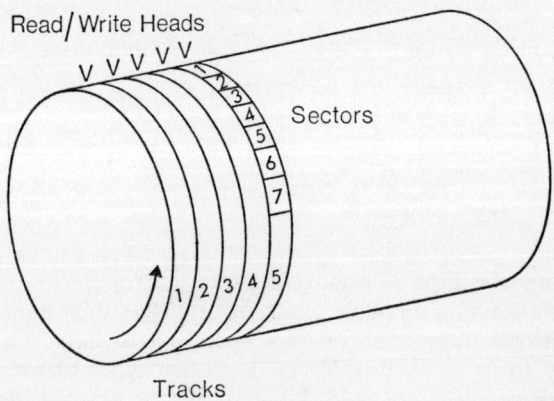

37. Drum storage

Organisation of data on a drum

Data on each track of a drum is recorded as a series of characters, each character being composed of a fixed number of bits. For addressing purposes, the positions round a track are serially numbered, from 0 up to the maximum which can be accommodated, say 1023. Thus any individual location on the drum can be identified by giving its serial or 'sector' number and its track number. Certain track and sector numbers are reserved for control information required by hardware. Data on a drum can be considered to be occupying one continuous storage area, and transfers can be of any length. However data on a drum can also be organised as serial files or direct access files depending on the file storage software available with a system.

Error checking
Parity checking on each individual character is provided as described in 3.1.1. Where data is recorded in blocks when a serial file method of organisation is adopted, software checks on block parity may be provided by some systems: such checks are not normally part of the hardware system.

3.3.4 Magnetic disc storage

General
A magnetic disc storage device is a method for achieving fast direct access to any item stored in the device, without having to wait for unwanted items to move past the read/write mechanism. Magnetic disc devices are the most commonly used storage media for direct access files.

Operation of a magnetic disc
A magnetic disc store consists of a number of flat discs, each resembling a gramophone record. Each disc is coated on both sides with a magnetic oxide recording surface, and associated with each surface is a read/write head which can move over the whole surface of the disc, in the same way as a pick-up arm moves over the surface of a gramophone record. Data on the surface of the disc is organised in bands, and the read/write reads can move into position over any required band. Figure 38 shows this arrangement diagrammatically.

The set of read/write heads form a comb-like arrangement. All the heads are rigidly fixed together, so that all heads are poised over the same band number at any one time. The heads remain off the surface of the discs when not in operation, and move above the surface when the device is operating. Discs rotate at speeds of up to 2,400 revolutions per second.

In some units known as *fixed disc stores*, the discs are permanent and cannot be removed from the unit. *Exchangeable disc stores* on the other hand allow the user to remove sets of discs, known as *cartridges* and replace other cartridges in the unit. The capacity of a disc store depends on the size and number of disc surfaces, and the packing density on the surface. Fixed disc stores usually have greater capacity than individual exchangeable disc cartridges. A typical fixed disc may hold up to 400,000,000 characters of information, a cartridge up to 8,000,000 characters. Access time depends on the speed at which the disc rotates, and also on the software philosophy for data organisation, which plays an important part

121

in the way data is retrieved. Some examples are: fixed disc store: 150 ms; exchangeable disc store: 100 ms.

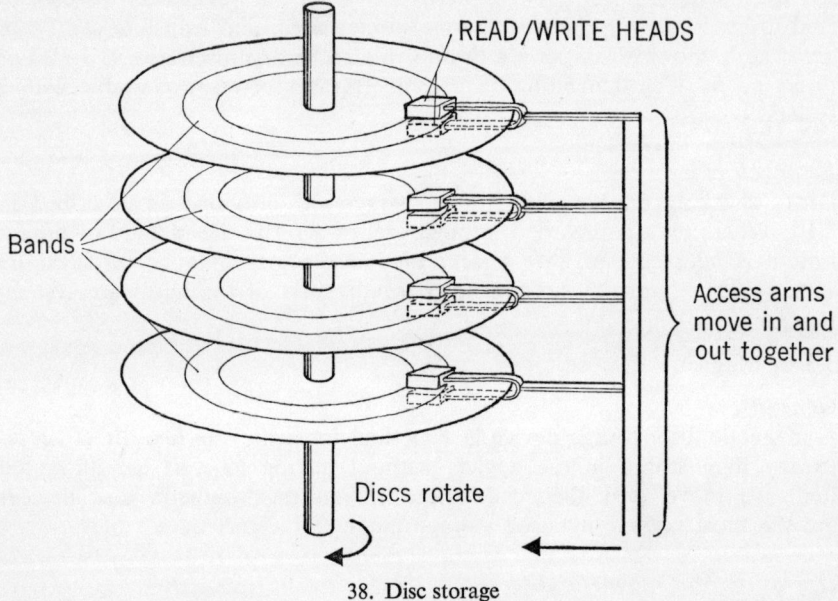

38. Disc storage

Data organisation

Data is recorded on the surface of a disc in a series of concentric bands. Each band is divided into equal blocks, as shown in figure 39.

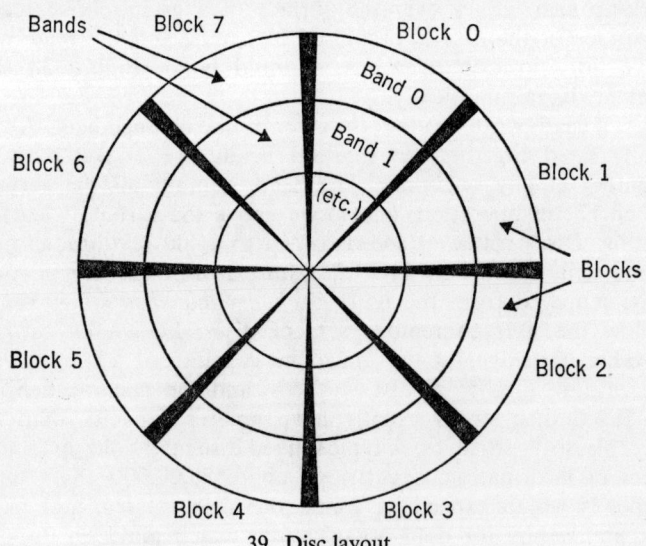

39. Disc layout

Each block has a fixed capacity in characters, and a short area of unused band separates each block. Since the number of characters in each block or each band is the same, the packing density increases as the diameter of the band decreases towards the centre of the disc. In some systems the recording density may also increase, the outer bands either containing more or larger blocks of data.

Since the object of disc storage is to function as a direct access device, data is organised on the disc unit in such a way as to minimise access time. Any mechanical movement increases access time, and a major source of delay is incurred when the read/write heads have to be repositioned. All the heads move together, so at any one time all the heads are poised over the same band in each disc. Once a band position is reached, data can be transferred continuously to and from the same band on any or all the discs, each head being activated in turn. Thus for programming purposes the 'cylinder' formed by all the bands of the same decimeter can be considered as a continuous area of store, known also as a 'seek area'. This is shown diagrammatically in figure 40 and described in detail in section 4.6.

Error checking
Each block on the disc has related to it a check sum, in the same way that block checks are used on magnetic tape. Any errors detected on data transfer will normally cause attempts at repeating the operation. If this is unsuccessful, a hardware signal detectable by software is generated. In most systems using direct access storage techniques attempts are then made automatically to store the information in another part of the device. Parity checking as described in section 3.1.1 also takes place.

3.3.5 Magnetic card files
General
The magnetic card file (MCF) is a device for providing direct access storage. Data is recorded on flexible plastic cards, about 16 in. by 4½ in. assembled into magazines of up to 256 cards. Magazines can be easily loaded and removed from the device, giving unlimited storage capacity. A device known as the retrieval unit transfers data to and from cards in the unit.

Description of MCF card
The basic unit on which data is recorded on a MCF is a thin plastic card coated with magnetic oxide. Data is recorded on e.g. 64 parallel bands or tracks running the length of a card, and can hold up to 160,000 6-bit characters. As shown in figure 41 notches on the top and bottom edge of the card define the card address, and are used to retrieve appropriate cards from the magazine by mechanical selection methods.

Description of a typical MCF magazine
A magazine is a container which holds up to 256 cards, divided into two decks of 128 cards each. Cards may not be interchanged between decks.

Cards can be inserted or withdrawn from a deck of a magazine by means of slots at the start of the feed and return tracks in the retrieval unit.

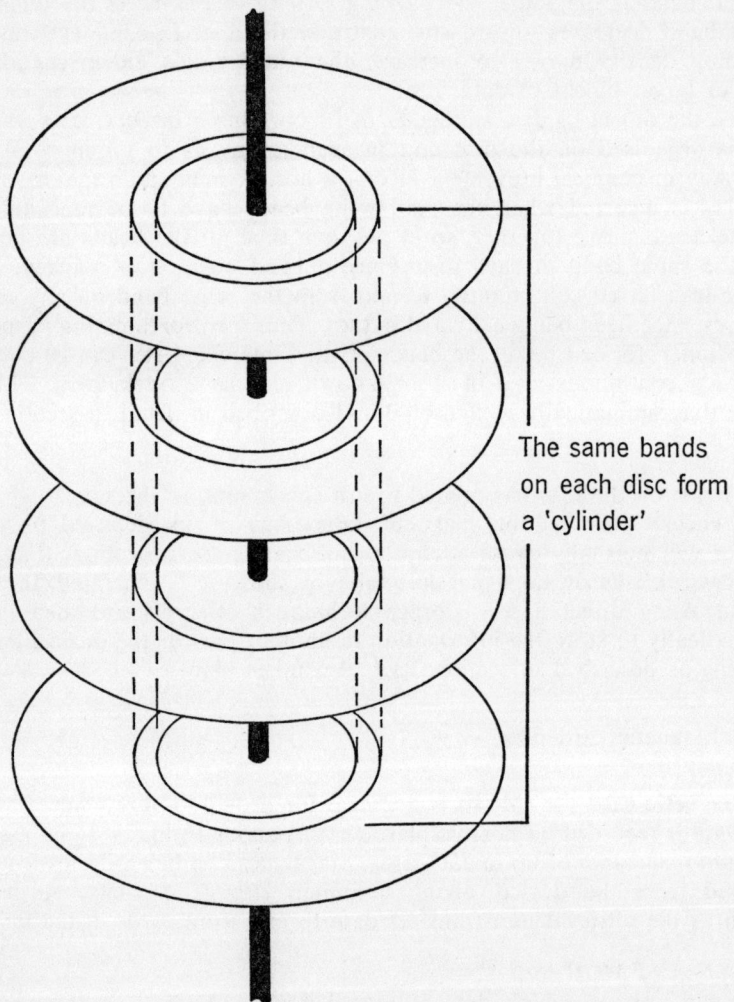

The same bands
on each disc form
a 'cylinder'

40. The seek area or cylinder concept

Description of a typical MCF retrieval unit
Up to 16 magazines may be held in a single retrieval unit, and may be easily removed and replaced by other magazines. Figure 42 shows the main parts of the unit, which are

        (a)   Card feed and return tracks
        (b)   The gate
        (c)   The drum

(d) Read/write break
(e) Photocell.

The cycle of events to complete a transfer operation is as follows. The notches which record the address of the card containing the required data activate a mechanical system of levers, which selects the appropriate card from its magazine. The card passes along the feed track and through the gate onto the drum, which rotates at a constant speed. The card is wrapped round the drum, encircling about 60 per cent of its surface. As the drum revolves it carries the card under the read/write heads, enabling data transfer to take place. The card remains on the drum until all transfers are completed, when a program signal releases it. The card is then deflected by the gate into the return track and is replaced mechanically back into its magazine. On some devices read/write heads are not positioned over all the bands on the card, but over a given number, e.g. four. The whole card can be covered by moving the heads into different positions, e.g. 16 positions for a four band head to cover 64 bands. In some devices more than one card may be held on the drum at one time. The function of the photocell is to check the card address by scanning the pattern of notches on the card.

Identification notches   Identification notches

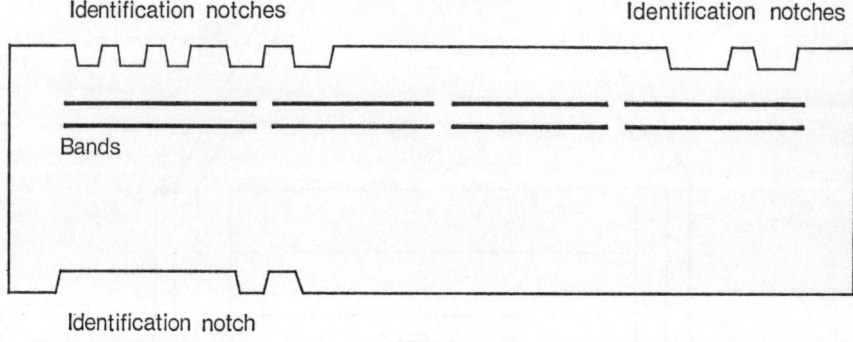

Bands

Identification notch

41. Magnetic card

Data organisation
The way a file is organised in a MCF unit will depend largely on the software philosophy adopted by the manufacturers or developed by the user. Mechanised movement involving frequent card selection to access successive data items will slow access times, so it is usual to treat each card as a 'seek area' in direct access terminology, analogous to the 'cylinder' concept in disc storage devices (see figure 40).

Error checking
Each character recorded on the card has a parity bit, and parity checking is performed as described earlier in section 3.1.1. The photocell also provides a check that the correct card has been selected by comparing the card address pattern with the original selection address.

125

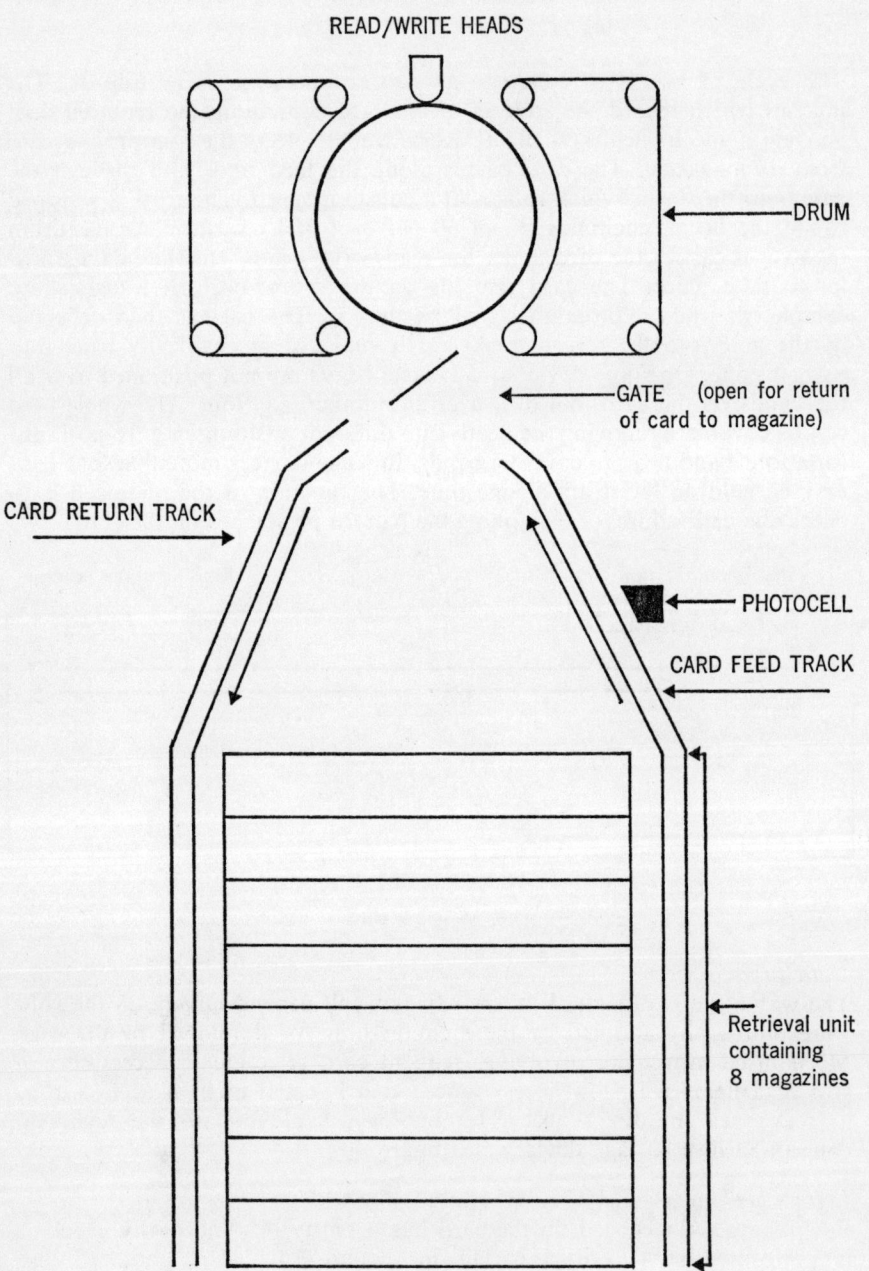

42. Magnetic card file unit

PART 3B   SOFTWARE

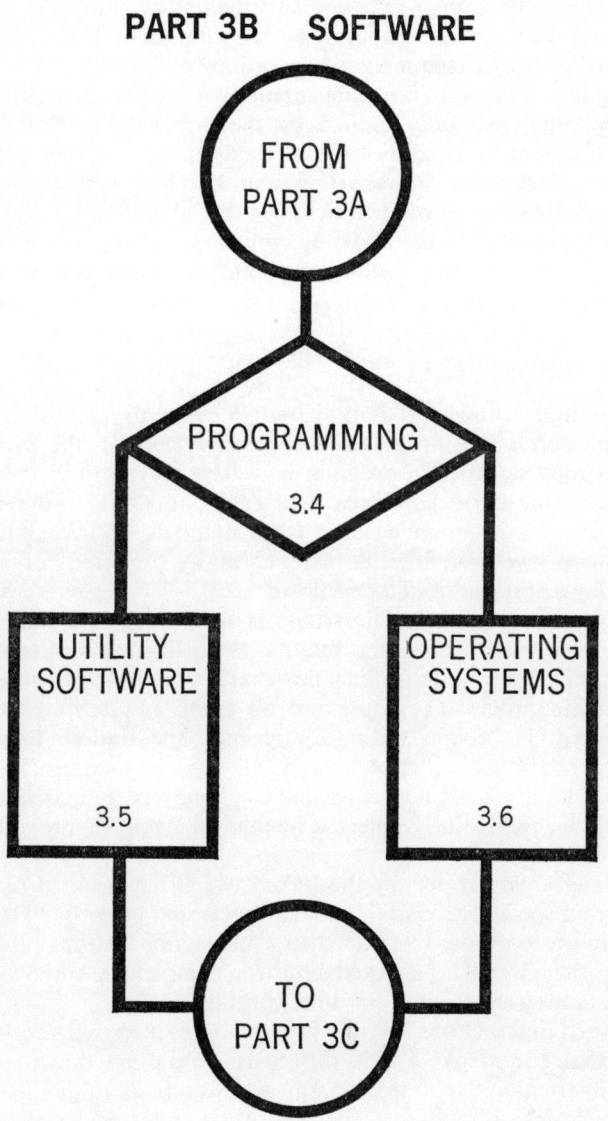

3.4 PROGRAMMING

This section describes the concept of programming, the means by which computers can achieve their extraordinary flexibility. Basically, a program is a set of instructions to be obeyed by the computer. Each instruction is obeyed in turn, and the sequence of operations determines the overall task of the computer. Since programs are infinitely variable, the number of such tasks is unending, and only limited by the ingenuity of those using the computer. This section describes the methods used in planning the logical structure of a program (flowcharting and decision tables); the task of translating this into a format which can be accepted by the computer (coding and languages); methods by which programs are checked and corrected (diagnostics and remote testing); and finally a discussion of programming techniques.

3.4.1 Flowcharting

In programming, a flowchart is prepared in order to show diagrammatically the logical relationship between successive steps in the program.

The steps represented in a program flowchart may vary in level of detail from a one-to-one correspondence with each individual program instruction to sections of program needing large numbers of instructions.

At least two levels of flowchart are normally prepared for a program, an *outline* flowchart and a *detailed* flowchart.

The purpose of an outline flowchart is to identify the main functions to be performed by the program, relating them to the requirements of the program specification. The outline flowchart will relate input and output functions, main processing loops, and all entry and terminal conditions. If the program is broken up into segments, the outline flowchart will indicate the relationship between segments and any control routines. In some cases a separate segmentation flowchart may be prepared as a higher level of outline, particularly where a lengthy or complex program is being written.

The outline flowchart acts as the link between the detailed requirements of the program specification and the resulting coded program. The flowchart should therefore be cross-referenced to each section of the specification in such a way that it can be checked both for completeness and consistency.

A detail flowchart is used as the working document from which the detailed coded instructions of the program are prepared. It is therefore necessary that the steps of this flowchart should be directly related to individual instructions or simple groups of instructions in the final program. Any programming techniques to be used, such as loops, modification, subroutines, switches, must be indicated symbolically on the flowchart so as to provide clear, unambiguous and logically consistent directions for coding. By the use of cross-references from the detail flowchart both to the outline and to the resulting coded program, the program is made easier to understand, and a direct link from the program specification to the part of coding and implementing the specification can be maintained.

It must be emphasised however that a flowchart must be intelligible. Individual steps, while they may be coded directly as a single instruction, must be annotated, not with the operation code of the instruction, or a direct interpretation for the code, but with a note of the logical purpose of the step. A step shown as

ADN O COUNT

or   *Add to count*

is meaningless:

Increment count of records read by 1

conveys some useful information.

An example of an outline flowchart is given in figure 44, and of a detail flowchart in figure 45.

Flowcharts form an essential part of the documentation of a program. By suitable crossreferencing or indexing techniques, the listing of the coded instructions of a program may be linked through detail and outline flowcharts to the program specification and ultimately to the overall system definition. In this way the effects of any alteration of a system on individual programs or parts of programs can be quickly checked.

Because flowcharting is used for representing the logic of a program symbolically as a series of successive steps, it is used in the preparation of programs which are coded as a series of instructions performed in sequence. Alternative techniques for planning programs may be adopted if the programming language being used imposes a special sequence of operations, as for example in the use of generators for particular types of program (see 3.4.3). Another technique used for linking the requirements of a program specification with the resulting program is the use of decision tables (3.4.2).

Symbols are used for the representation of the different types of step in a flowchart. Several generally accepted sets of symbols are in use: one such set is shown in figure 46.

3.4.2 Decision tables

General

In section 3.4.1 on flowcharting, one method for representing the logical structure of a complex procedure was described. Another method for achieving the same objective is by means of decision tables. Using a table for recording information is familiar in every day life: timetables, price lists, interest rates, tax tables are all well known. A decision table is an extension of this principle, enabling alternative conditions to be related to required actions in a way which is clear and unambiguous.

E

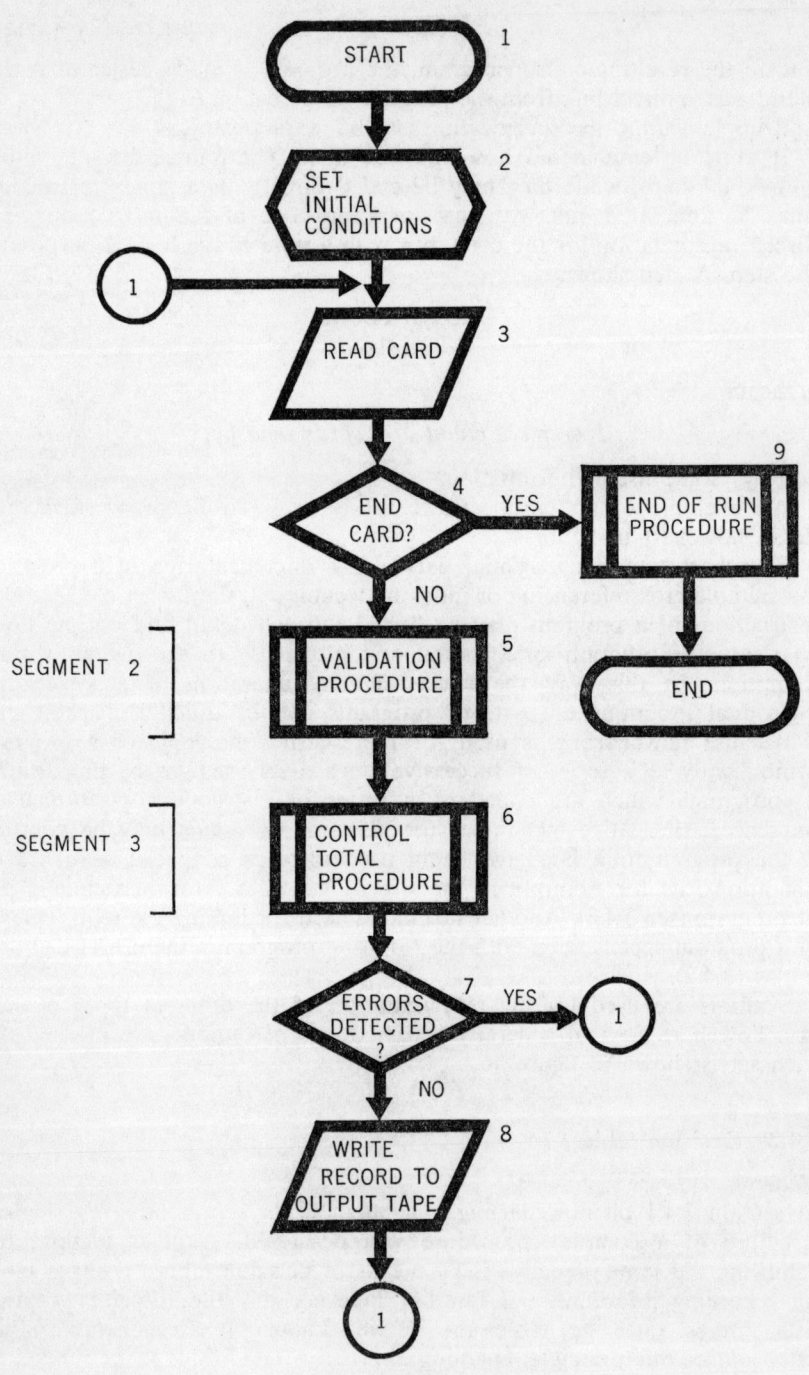

44. Outline flowchart

130

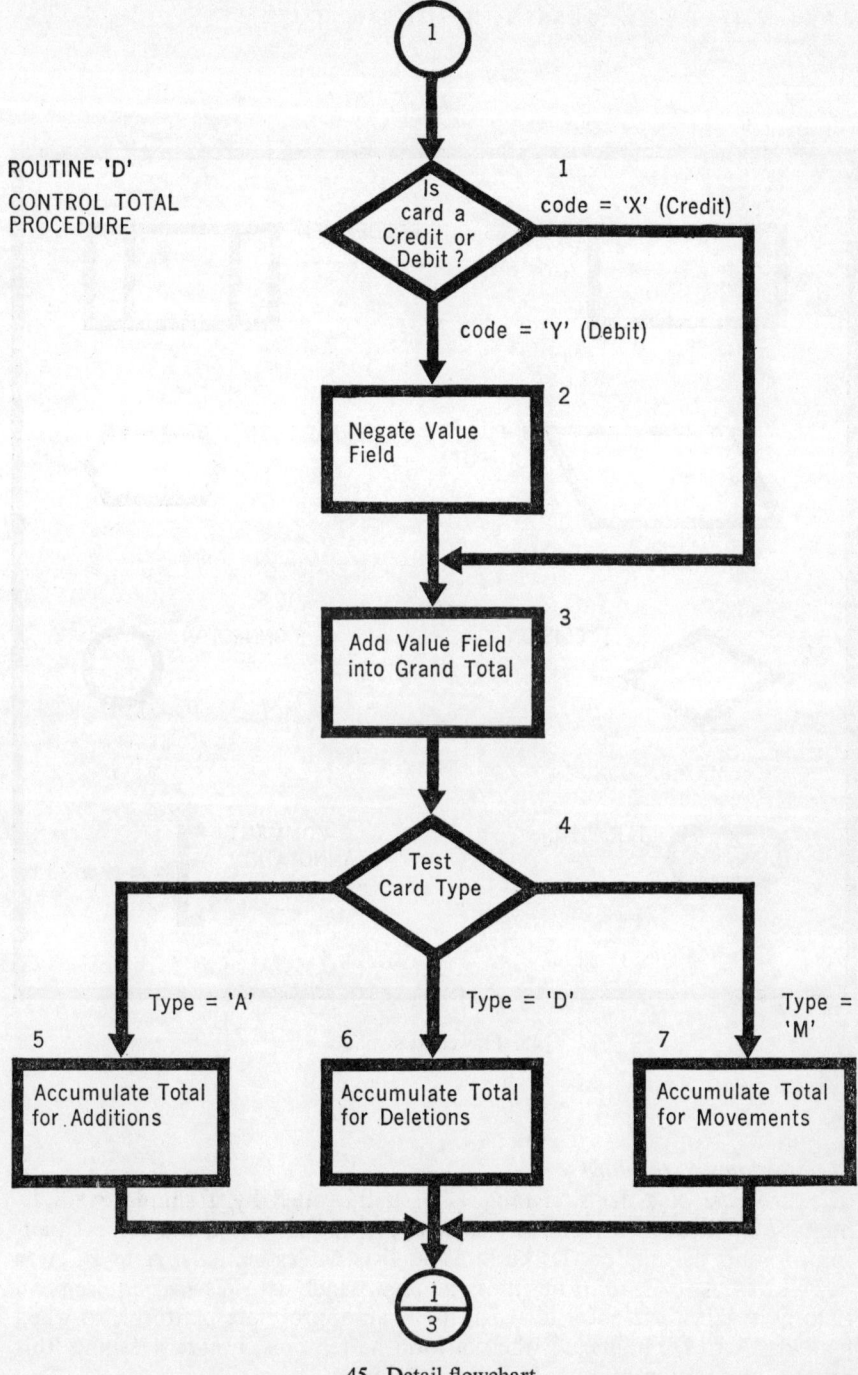

ROUTINE 'D'
CONTROL TOTAL
PROCEDURE

45. Detail flowchart

131

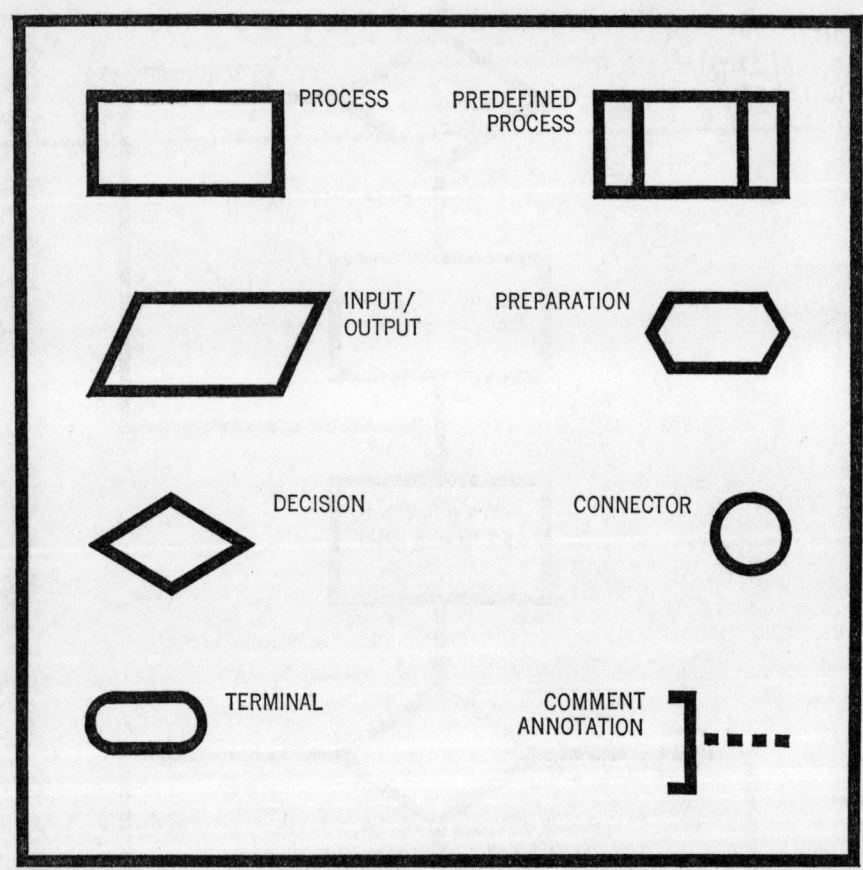

46. Flowchart symbols

Decision table structure
The structure of a decision table is best illustrated by a simple example. Suppose a traveller at a railway station, A, wishes to find the correct platform for his destination, D: he looks at lists of destinations, A to B, A to C, A to D, A to E, until he finds the item which satisfies his requirement, A to D. Against the entry for each list is an appropriate platform, so when he finds 'A to D', he locates the platform he requires. Figure 47 shows this process of decision making in a tabular format.

132

A to B	Yes	No	No	No
A to C	No	Yes	No	No
A to D	No	No	Yes	No
A to E	No	No	No	Yes
platform 1	X	–	–	–
platform 2	–	X	–	–
platform 3	–	–	X	–
platform 4	–	–	–	X

47. Decision-making at a railway station

From this table it can be seen that platform 3 is the required platform for the journey from A to D: each condition is shown, together with the appropriate *action* necessary. A decision table is always set up as an arrangement of four blocks, as in figure 48.

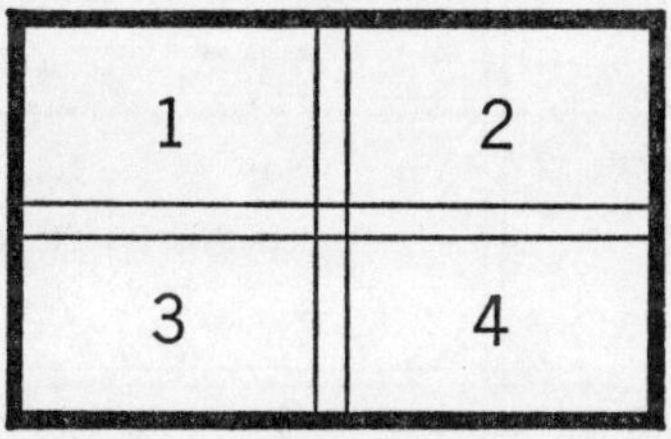

| 1 | 2 |
| 3 | 4 |

48. Decision table blocks

Block 1 is known as the *condition stub,* and contains a list of answers to the question, 'What are the conditions to be examined?'
Block 2 is known as the *condition entry,* and for each entry in the condition stub list contains a list of answers to the question, 'What values does the condition have?'
Block 3 is known as the *action stub,* and contains a list of answers to the question, 'What action must be taken when a condition is fulfilled?'
Block 4 is known as the *action entry,* and for each entry in the action stub contains an answer to the question, 'What value does the action have for the equivalent condition entry?'

133

All decision tables are based on this structure, and the structure of extremely complex problems may be shown in this way. The combination of condition entry and its equivalent action entry form what is known as a decision rule, and this can be shown by a further example. One field where fairly complex combinations of conditions must be examined before one of a large number of available actions takes place is the field of motor insurance. In a highly simplified example, we might outline four basic rules.

Rule 1: Any previous convictions: no insurance policy allowed.
Rule 2: No convictions, but aged under 30: policy allowed with £10 excess.
Rule 3: No convictions, aged 30 or over, any driver: policy allowed, with £10 excess.
Rule 4: No convictions, aged 30 or over, one driver only: policy allowed with no excess.

These rules can be exhibited simply in the decision table shown in figure 49. In this table the *condition stub* lists the various conditions relevant for assessing the applicant's risk factor. The *condition entry* gives values to these factors: there can either be simply Yes and No, or an actual value, such as 'less than 30'. A dash (—) shows that for a particular rule, the values of some conditions are not significant: in the example, if an applicant has been convicted, his age is irrelevant. The *action stub* lists the available policies, or the decision not to give a policy at all. In the action entry the type of policy allowed for each set of condition entries is indicated by an 'X', so that the four decision rules are described.

	Rule 1	Rule 2	Rule 3	Rule 4
Any previous convictions?	Yes	No	No	No
Age?	—	less than 30	30 or over	30 or over
Any driver	—	—	Yes	No
No policy allowed	X	—	—	—
Policy with £10 excess	—	X	X	—
Policy with no excess	—	—	—	X

49. A four rule decision table

Limited entry tables are those in which all questions can be answered Yes or No. A table in which the condition entries contain values rather than Yes or No answers is known as an *extended entry* table. Where values and Yes and No answers are combined, the table is known as a *mixed entry* table.

A complex set of operations can be shown by a number of linked tables, by adding another action entry indicating that further conditions are applied in another table, as shown in figure 50.

Table No. 2	Rule 1	Rule 2	Rule 3
Security available	Y	N	N
Credit allowed	–	< £100	< £100
Special clearance	–	Y	N
Pass order	X	X	–
Return to customer	–	–	X
Go to Table No.	4	3	5

50. A mixed-entry decision table

Programming techniques

Decision tables provide a powerful tool for the representation of complex procedures. A table prepared by an analyst and supplied to a programmer is a clear method for expressing requirements. Programmers can then prepare conventional flowcharts from the table before coding the program. However some languages (see 3.4.4) have been developed which enable decision tables to be input directly as language statements. Routines known as 'pre-processors' then convert the tables into a series of source language statements which can be accepted by the compiler being used. Decision tables are sometimes a temptation to analysts and programmers to identify excessive detail: but the use of decision tables to assist in program and system specifications has several advantages over conventional flowcharting. Among these are: the discipline imposed on the analyst to prepare a complete and accurate problem description; efficiency in communication between analyst and programmer; standardisation of program and system documentation; automatic program generation from decision tables; and, not least, the ease with which tables may be typed, written and altered when compared with the effort needed to alter and re-draw flowcharts.

3.4.3 Coding

Instructions

The detailed logic of a program is defined at the flowcharting stage or by means of decision tables. Coding is the task of converting the logic into a set of instructions which can be obeyed by the computer. The set of instructions available to the programmer for this purpose may range from simple machine code, which is directly interpreted by the computer, to high level languages requiring conversion into machine code before the program may be run. (See section 3.4.4 on languages.) The choice of language will determine the organisation of the program, both at the flowcharting stage and hence at the coding stage.

Programming in a basic machine code will require a greater degree of detail in the flowchart, since the programmer has to code more instructions: the use of a high level language enables the programmer to take short cuts by using single language instructions resulting in many machine code instructions. The rules of program writing in different languages require the programmer to set out the coding in certain standard formats, and to organise the arrangement of data areas and instruction areas according to the requirements of the language chosen. However, certain basic principles of coding will apply generally, regardless of the particular language chosen.

Basic principles

The overriding principle in coding is *simplicity*. Programs have to be used and frequently this means that a program will have to be used by someone other than the original programmer. Since there is no single programming solution to any particular problem, any program is to some extent the original creation of an individual, and thus displays the idiosyncrasies of that individual. The problem of explaining how a program works is simplified to some extent by a good flowchart or decision table. The process can be furthered by making sure that the sequence of instructions chosen to perform the steps presented on a flowchart is not only correct, but the simplest sequence possible.

Simplicity is achieved by careful program organisation. This means that the program must be divided into logically related parts, such as initialisation, input, processing, output and exception conditions. This organisation will be reflected in the flowchart logic, but must be adhered to at the coding stage. Since there is no one way of coding any sequence of instructions to perform a given operation, the programmer must as far as possible avoid any complicated and obscure methods for performing an operation, even if such sequences are 'clever' in minimising storage or in speeding up the program operation. The saving of a few microseconds may not be significant as a proportion of the overall program time!

The programmer is tempted to deviate from simplicity by two conflicting requirements: speed and space. A program should as far as possible operate as *fast* as possible and take up as little expensive main *storage* as possible. Unfortunately these two requirements are often mutually incompatible. Space may be saved in two ways: by avoiding repetitious coding sequences by using subroutines and loops to perform sequences of instructions occurring repeatedly throughout the program, and by economising on areas allocated for intermediate storage of data by sharing input and output areas. However, the use of subroutines requiring setting up of parameters, or of loops requiring counts and instruction modification may mean an increase in time. Conversely, speeding up a program by means of double buffering of input and output and the avoidance of time consuming modification may mean an increase in store used. In situations where either of these factors is critical, the systems analyst must specify to the programmer whether speed or storage must be sacrificed. If the position is not critical, the programmer should choose the simplest methods, or the methods which will be most easily understood by other programmers.

Segmentation and overlays

In order to enable a large program to be run with the minimum of storage used at run time, programs may be divided into sections known as segments or overlays. Each such section may be considered as an independent program, which performs part of a long routine: when its job is over, it calls in the next section, which replaces the previous section in storage. Certain parts of the program or data areas remain common to all segments so that communication can take place between different parts of the program.

The technique of considering the program in sections can also be used even if actual overlaying is not required: each part of a program can be written and tested separately as an independent routine, the completed parts being brought together when all are tested.

The decision on how to segment a program and whether overlaying will be necessary will depend on the amount of storage available to the completed program, and the amount of coding necessary to complete the program. This should be estimated before programming starts, so that segmentation can be planned at the outset. If a program is written as one segment and finally proves too large, it will be difficult to split into segments. If it is segmented before coding starts, the segments may be turned into overlays if this proves necessary, or can be combined into a single overlay.

Communication

A good program is not merely one that works correctly: it must be clearly annotated and documented so that it can be readily understood by someone other than the original programmer. Good flowcharts or decision tables will provide a clear link between the program and the specification. The coded program must provide a link to any flowchart and must contain explanatory notes on the coding used. Most programming languages allow the programmer to insert comment or narrative in the coding. This narrative has no effect on the resulting machine code, but allows the programmer to insert notes and references which appear on any listing of the program. Communication between program and flowchart can be achieved by reference numbers appearing on the flowchart symbols. Each section of coding may also be accompanied by a short description of the function of the coding, in the same way that the flowchart boxes have narrative included. As emphasised in section 3.4.1, this narrative should not merely repeat the coded instruction. For example

137

E*

ADD X1 TO X2

is unhelpful, but

ADD BONUS TO NETT PAY

is informative.

Documentation

The coded program is itself an important document, and (if properly cross-referenced to the flowchart and specification) will itself be a clear aid to the understanding of the program. Additional documentation which may be prepared at the coding stage includes full lists of all program switches, entry points, halts and messages produced by the program. Full operating instructions are also an essential part of the program's documentation.

3.4.4 Languages

As explained in 3.4.3, programs consist of instructions which when operated on by the computer result in the required function being performed. These instructions when present in the central processor consist of patterns of digits which can be represented as a combination of numbers or symbols. At this level the instructions are known as basic or machine code. Programs may be written directly in this code, but the process is tedious and difficult. For this reason, most programming is done using a *language* which can itself be converted into machine code by means of another program variously known as a compiler, assembler or generator.

Low level languages

Low level languages are the simplest form of language, consisting of instructions each of which has an exact equivalent in the machine code. The simplification of programming is achieved by making the instructions *mnemonic* (e.g. using the letters ADD for the addition instructions). Areas of storage are given labels, which are referred to within the program, both for addressing data storage areas and program instructions. The actual numerical addresses of the storage areas are inserted when the program is converted to machine code by the compiler itself.

Low level languages may also incorporate certain software facilities such as the ability to call into the program standard subroutines and housekeeping packages for input and output operations. These are described more fully in section 3.5.1.

Because low level languages have a one-to-one relationship with machine code, they can usually only be compiled into the machine code for a particular computer or range of computers.

High level languages

Unlike low level languages, high level languages enable the programmer to use single instructions which on compilation may result in large numbers

138

of machine code instructions. This has the effect of simplifying programming, since fewer instructions have to be coded in order to produce the required result.

Most high level languages have been devised so as to provide a common standard which can be used on more than one type of machine. This is done by providing separate compiler programs for each machine. The coded or *source* program can be converted into the required machine code or *object* program by using the appropriate compiler. High level languages are generally simple to use and easier to amend and correct than low level or machine code languages, since the programmer can write instructions which have a logical interpretation in the context of the problem being programmed. Most high level languages are thus problem-oriented, that is, designed to solve programs of a broad general type.

Commercial languages

Commercial languages are high level languages designed primarily for commercial data processing. The most widely used commercial language is COBOL (Common Business Oriented Language). COBOL was originally developed under the sponsorship of the US Department of Defence, to overcome the problem of compatibility of programs on large numbers of different types of machine. A COBOL source program is divided into four *Divisions*, the *Identification* Division, in which the programmer describes the program, the information appearing on the subsequent listing; the *Environment* Division which deals with the specification of the computer on which the object program is to be run including such information as the size of main storage, the number and types of backing storage and other peripheral devices to be used; the *Data* Division in which the programmer allocates alphanumeric names to the units of data on which operations are to be performed, and defines files, records, working storage and constants to be used by the program; and finally the *Procedure* Division which gives the steps required to solve the problem. These will be expressed in statements resembling a highly formalised form of normal English, taking the form of simple imperative statements, and conditional statements preceded by IF.

Scientific languages

In the same way that commercial languages provide the possibility for writing programs in a language resembling English, scientific languages use a form of mathematical notation, which is converted by the compiler into the required machine code. Scientific languages are thus used for programs written to solve mathematical, scientific or engineering problems. The most commonly used scientific languages which, like COBOL, have an international acceptance, are FORTRAN and ALGOL; the former was originally specified in the USA, while the latter was originally devised in Europe. Both languages enable the user to identify numbers as *integers* or *real* numbers (see section 3.1.1), and allow the use of arithmetic and logical operators.

General languages

Commercial languages are generally required to process large amounts of data using relatively simple arithmetic operations: scientific languages have limited input and output facilities but are capable of extensive and sophisticated computation. The need for a general purpose language arises from the tendency of commercial processing to require more sophisticated computational techniques (e.g. in forecasting problems) and in scientific processing to handle larger volumes of data (e.g. in statistical analysis routines). The lessening difference in requirement between scientific and commercial problems has led to the development of general purpose languages combining the data handling power of commercial languages with the processing power of scientific languages. One such language is PL/1, developed originally by the IBM users' association, SHARE. The main characteristics of this language are:

1. The language is *modular,* i.e. small subsets of the language can provide adequate facilities for writing complete programs, the user learning more extensive facilities as he requires them.
2. The language has sophisticated error correction routines enabling minor slips to be corrected by program, more major errors being signalled on compilation.
3. Program statements and date can be written in free form, eliminating the need for special coding forms.

Generators

Generator programs are highly specialised languages restricted, even more than commercial or scientific languages, to producing specific types of program. For example, a sort generator will only produce a sorting program; a report generator will only produce a program for printing out a report. For this reason generators are not strictly speaking languages, although the more sophisticated report generators tend to enable the user to accomplish most commercial report programs. Generators are described in more detail in 3.5.2.

Compilers and assemblers

Programs written in a language other than machine code must first be converted to machine code before they can be run. This conversion is done by means of a program known variously as a *compiler* or *assembler*. Originally the term *assembler* was given to the program for converting a low level language, while *compiler* was the term for converting high level languages. However the distinction is now not precise, since many low level languages have some of the facilities of high level languages, particularly the use of macro instructions calling in more than one machine code instruction for some language instructions: in such cases the term *compiler* is used generally, whether the source language is low or high level. Compilers operate by reading in each source program statement and converting the operation mnemonic into the machine code equivalent. The compiler allots absolute addresses to all data items and program instructions, and provides error detection by rejecting any incorrectly formed statements, or any labels which would cause logical errors. Normally the compiler produces a listing of the source program, together with the resulting object program, and a list of storage areas used and errors detected. Compilers and assemblers are produced by manufacturers for their own languages, and for the major international high-level languages. Because of hardware limitations, most high-level language compilers do not provide the full set of possible facilities, but some more or less extensive subset, depending on the capabilities of the particular machine. Thus while in theory any high level program can be converted into any machine code for which there is an available compiler, in practice, when being compiled for a different machine the source program may have to be modified to comply with the particular restrictions of the compiler being used.

Simulators

In order to overcome the problem of running programs written in one machine's code on another machine, a technique known as simulation is sometimes used. A program known as a *simulator* converts statements in one machine code into the equivalent instructions in the other code. The efficiency and comprehensiveness of such a process depends largely on the amount of original similarity between codes. Generally this process is only adopted as an expedient to enable a user to continue using old programs on a new machine whilst re-writing them in the new language. In some cases a hardware unit rather than a program is used to perform the conversion. Such a unit is known as an *emulator*.

3.4.5 Diagnostics

Error types

Programming errors may be divided into two broad categories: *logical* errors and *syntax* errors. The distinction may be shown in a simple example. If we assume we are writing in a programming language which has three operations

> EQL signifying 'equals'
> DVD signifying 'divided by'
> MPY signifying 'multiplied by'

The expression

> SPEED EQL DISTANCE MPY TIME

would be *logically* incorrect. A value for SPEED would be computed, but it would be wrong.

The expression

> SPEED EQL DISTANCE DID TIME

141

would have a *syntax* error. The operation DID is not correctly written and is not a legitimate relation in the language. No result, right or wrong, could be computed at all.

Checking

Checking of a program should be thorough, complete and take place at every stage of programming. Thus the first check will be made at the flowcharting stage. This first check will be two fold: to ensure that all the requirements of the program specification have been dealt with, and to make certain that the flowchart correctly exhibits the logic of the program.

If the flowchart has been carefully referenced to the program specification, checking the first of these requirements should be straightforward. The programmer must make sure he has catered for the input and output conditions, all exception conditions and all processing requirements. If any action for any exception conditions have not been made explicit in the specification the programmer should nevertheless make provision for these conditions, noting the action he has taken in the program documentation. A good program should have no loose ends.

Checking the logic of a flowchart can be carried out by means of a technique known as 'dry running'. An example of each condition the program is likely to encounter is written down, and the condition is operated on by writing down the effect of each step in the flowchart on the condition. In this way obvious errors of logic should become apparent.

Checking of the coded program should always take place before the program is punched onto cards or paper tape and submitted for compilation. If the program has been cross-referenced to the flowchart, this will facilitate checking of program logic, since the flowchart should be correct in that respect. All operation codes should be checked for syntax errors, and a check made on all labels used, and all switches to ensure both that they are set or unset correctly and that they are set initially to the correct condition.

Conventions are adopted to distinguish between commonly confused letters and numerals, e.g. I and l, 0 and O, which may be written I and l, 0 and Ө. Having carefully checked the coding for errors of syntax, a further dry run can take place on the coding, to make sure that the written instructions correctly interpret the requirements of the flowchart. The program is then ready for compilation. Ideally, checking of each stage of a program should be carried out by someone other than the original programmer. This ensures that the program can be understood by someone else—a sure sign that the program is likely to be correct.

Compiler messages

Since the process of compiling is a mechanical one, compilers cannot detect errors of logic: however compilation will enable the programmer to discover certain types of syntax error. The most common of these are format errors, where an incorrect operation code has been used, or if the language requires the instructions to be written in fixed locations on a form, the error may be due to incorrect positioning of the instruction on the coding form. The compiler will be unable to interpret such incorrectly formed instructions, so they will be rejected as errors. The compiler will also detect other violations of the 'rules' for a particular language, such as the omission of 'directives' or messages indicating to the compiler action it must take.

Other errors detected by compilers include incorrect labelling of store or program areas: a program instruction may refer to a labelled area which does not appear anywhere in the program; the compiler is then unable to allocate an absolute address to the label, and rejects it as an error. Errors detected at compilation are usually indicated on the listing of the compiled program. The generated object program will be incorrect, and the program cannot be test run until a correct compilation has been made.

Program testing

Having carefully checked the flowchart logic and coding, and obtained an error-free compilation, the programmer might fondly imagine he has a correct program which will work the first time it is run. He will certainly be disillusioned and for this reason the program must be operated under test conditions to correct as yet undetected logical and syntax errors. These can result because of misinterpretation of the requirements of the program and because of syntax errors which have been accepted by the compiler since they do not violate any of the language rules. Testing of a program is usually done against *test data* prepared by the systems analyst who specifies the program.

Program test data will consist of samples of all input to the program together with schedules of expected results, and should indicate samples both of volume data and of all exception conditions. The programmer may also prepare some test data for his own purposes, e.g. to simulate certain hardware conditions. If the program has been written in segments which are to be tested separately, a test segment may have to be included with the segment under test to supply it with any information normally presented to it by other segments: this will also be under the programmer's control rather than forming part of the test data supplied by the analyst. The program will be considered correct when it has correctly processed the test data under all conditions and produced the expected results. It may well be that the analyst has made an error in the test data. The programmer should always check this if in doubt, as systems analysts are not infallible and do from time to time make a mistake!

All errors detected during testing must obviously be corrected; what is important however is that all corrections made must be fully documented. The final set of program documentation must be complete and consistent, so an error in logic must lead to a revised flowchart, and an error in coding to a corrected source program. It is possible in some systems to correct coding errors directly in machine code either by altering the object program before it is run, or by making alterations at run time. One such technique is known as *patching,* where an error is corrected by replacing the error with a branch instruction to a section of coding added to the end of the program to correct the error. Such techniques are useful as temporary expedients, or for correcting errors rapidly, but are dangerous in that if the source program and documentation are not corrected as well, the programmer may lose track of the corrections he has made and forget that the correct results obtained from a patched object program will not be obtained from the recompiled source program. This is particularly important where programs are subject to audit checks; the master copy held for auditing must be obtained from the source program, and not be a patched object program. The programmer may be assisted in program testing by certain diagnostic aids in the form of software usually provided by the manufacturer. These are described below.

Monitor prints

Monitoring is a technique by which the programmer can obtain a printout of various parts of the program while the program is in operation. This enables him to check on how far the program has reached when error conditions occur. The information is printed each time a monitor point in the program is reached, and consists of certain key areas of program and data storage selected by the programmer. As the program testing proceeds monitor points in parts of the program known to be correct can be eliminated. Choice of monitor points is normally up to the programmer, and will be used, for example, to print intermediate results in a lengthy calculation in order to detect at what point an error occurs. Monitor prints can however be requested as part of the 'expected results' of test data, particularly where the systems analyst wishes to make sure that calculations and operations on data which are to be used as input to other programs and not output directly by the program doing the calculations are correct.

Trace routines

A trace routine provides the programmer with a continuous record of the progress of his program, by printing out each instruction as it is obeyed, together with information about the results of obeying the instruction. Trace routines thus differ from monitor routines, which only give information when certain predetermined points in the program are reached. The trace routine may be limited to certain parts of the program, or to listing the results of certain types of instruction only, e.g. branch instructions. A trace routine will normally only be applied to a program as a last resort for finding an error, since it is very time-consuming. It may also be used if inadequate documentation and flowcharts make the detection of errors in a program written by someone else impossible.

144

Store prints
Another aid to the detection and correction of program errors is the production of store prints. The contents both of main storage containing program and data areas, and of the contents of subsidiary storage devices may be listed on a printer. This enables the programmer to check the contents of storage at the point where an error occurs, as well as at predetermined monitor points. Store prints also form an important point of documentation, particularly prints of data files. For example, in updating programs, prints of amendment and master files will be required as part of the expected results from test data.

3.4.6 Remote testing

Open and closed shop
As well as processing live jobs, a computer installation must provide facilities for the testing of new programs. For the purposes of program testing, the installation may be run either on open shop or closed shop principles. In an open shop, a programmer has complete access to the computer, and may operate and test programs himself, or be present while computer operators run the program under his direction. In a closed shop, programmers are not present while the program is being run, all instructions being supplied in written form.

An open shop system has advantage of speed in turnround; a programmer does not have to wait while his job is being run, and can make decisions on the spot when unexpected circumstances arise, thus speeding the process of detection and diagnosing errors. Disadvantages of this approach are its wastefulness of time and the danger of lack of documentation. This approach wastes time because the computer necessarily remains idle while the programmer considers the problems arising and decides on what to do next. The programmer is also tempted to overcome errors by means of immediate corrections to the machine code program, either by altering the program in store at run time or by amending the object version and reloading. The chances are that he will not record in detail the action he has taken, particularly as he will be working under some pressure. When he returns to the tranquillity of his desk, he will find he is unable to remember the action he has had to take, or the reason he has had to take it, and much of the advantage of speedy correction will be lost.

In the closed shop system, the programmer is required to prepare detailed operating instructions for the testing of his program. This means he must consider carefully in advance the expected performance of the program, and describe precisely his requirements in terms of de-bugging aids such as monitor prints, trace routines and storage prints. The preparation of these instructions is itself a useful discipline, since it makes the programmer think out the likely errors in his program.

Since the operator will not be correcting the errors when they occur the programmer will be making his corrections under less pressure than in the open shop system, which means that he is more likely to document the corrections adequately. Inaccurate or incomplete operating instructions, or careless errors made in submitting corrections will mean considerable delays, since in a closed shop a program will be terminated as soon as an error occurs or the operator is unable to continue; there are no second thoughts allowed. However, because these delays are annoying and frustrating to the programmer, he is likely to try to avoid their occurrence by taking greater care.

In most commercial circumstances, the closed shop method of organising the testing of programs is to be preferred.

On-line testing

Developments in direct access hardware and multiprocessing systems have led to the development of 'conversational' programming, where the programmer creates a program by direct communication with the processor. Errors are signalled back to the programmer by means of a remote terminal as the programmer keys in the instructions, and the programmer is able to correct his program as he goes along.

Processor time is not wasted while the programmer is thinking since the system enables processor time to be shared by other users. Similarly, once a program is completed, it can be tested and corrected on line. The programmer can call for storage prints or monitor prints by interrupting the program as it is running and making his request through a remote terminal. Corrections can be made directly in the same way. This system is similar to an open shop system, but does not waste time in quite the same way. If communication with the processor is made through a terminal producing hard copy, e.g. a typewriter, the programmer will have a printed record of all his actions, and will be able to document his program when he has made his corrections.

This method thus combines many of the advantages of both systems. However, it can only be operated with a suitable hardware system, which is made available to programmers as well as to the operation of live jobs. In some cases a visual display rather than a typewriter terminal may be used. In this case no printed record of a programmer's action is made, and the method may be open to some of the criticism already made of open shop methods.

3.4.7 Programming techniques

General

Program writing can be compared with playing chess: the rules are relatively simple and can be learned quickly, but skill in manipulating the rules to achieve the desired end only comes with experience and constant practice. The purpose of this section is to outline some of the general techniques used by programmers in writing programs. The method by which these techniques are applied in practice to produce working programs form part of the acquired skill of individual programmers.

Modification and loops

A program instruction generally consists of an operation code and one or more addresses or operands. In core storage, the instruction is represented in machine code as a particular bit pattern or arrangement of characters. As such it is no different from any other information pattern, and is thus capable of being itself the subject of all the operations which can be performed on data elements in store. A simple example of modification, as the technique of performing operations on program instructions is known, is the creation of program loops. As a simple illustration of this technique let us suppose that it is required to add a series of numbers held in address locations $N, N + 1, N + 2, \ldots, N + 100$, into a location at address M, or 'Total'. One way of programming this would be to write a series of instructions as illustrated in the flowchart in figure 51. However,

51. Flowchart using too many procedure steps (compare with fig. 52)

if there are a large number of locations this method is both tedious and wasteful of storage, since each instruction takes up a part of the program's store area. By constructing a simple loop, in which the address part of the 'add' instructions is modified by increasing it by one each time the addition is performed, the same result can be achieved in less storage space and with rather less writers' cramp! This is shown in the flowchart in figure 52. The use of a loop simplifies program writing, but the flowchart in figure 52 shows that when using a loop the programmer must make sure the loop has an exit, otherwise the program will never finish going round the loop. In the example, the exit is given by the test to ensure that all numbers have been accumulated. This can be performed in many ways, one of which is simply by examining the value of the modified address in the 'add' instruction. If this is equal to the last address of the area containing the number to be accumulated, then the loop is completed. Another requirement of using modification in this way is that if the loop is to be used again, the programmer must ensure that any modified values are reset to their original condition. This is known as initialising a loop, or de-modification. In some machine codes modification is simplified by the use of special modifier locations, and the instructions format enables the programmer to specify a location containing a value to be added or subtracted from the instruction.

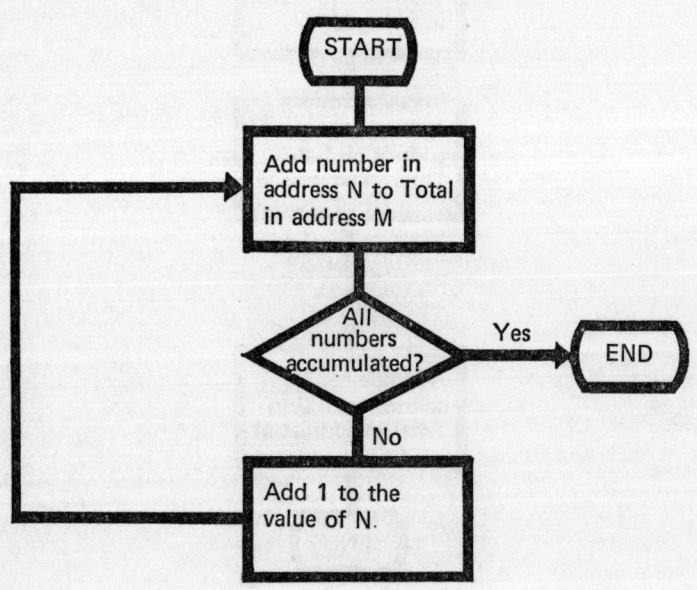

52. Use of looping technique

Subroutines

In most programs there will be some operations which will need to be repeated many times in the course of the program. For example, in a program which is reading data from punched cards, it may be necessary to convert several numeric fields in a card into binary values. (Section 3.1.1 on Information Patterns describes this conversion from external to internal formats.) Conversion of a field into binary is an operation requiring several program steps. These steps are identical in each case, and one way of programming this requirement is to repeat the group of instructions for each field to be converted. An alternative approach is to treat the conversion instructions as a subroutine. Whenever conversion is required, the program 'jumps' to the subroutine, and returns to the main program when the subroutine steps have been completed. Special instructions are available which permit entry and exit to a subroutine, storing the return address to the main program as a link enabling the correct main program path to be followed after the subroutine is completed. This process is illustrated diagrammatically in figure 53.

There are several types of subroutine. A simple subroutine consists merely of a set of instructions with entry and exit points. A subroutine can be made more flexible by allowing it to accept parameters. This means that the actual operation of the subroutine can be varied to suit specific applications of the general operation it is designed to perform. In the example given above of a conversion subroutine, the 'simple' version would expect the data to be converted to be always the same size and located at the same address, and the main program would have to ensure these conditions were satisfied before entering the subroutine. The routine could be made more general, by allowing the main program to supply as parameters, details of the size and location of the field to be converted and possibly also the location of the output field. Subroutines in a program may be written by the programmer or may be inserted from a library of common routines supplied as manufacturer's software or forming part of a subroutine library built up by the user. Compiler systems usually allow the user to store subroutines of all types on a special subroutine file, and these can then be called into a program when it is compiled.

Switches

A program switch is a programming technique which enables the program to follow different paths according to given conditions. A switch may be a particular program location which can have two or more different values. At various stages in the program the value of the switch can be tested and specific paths of action taken according to the value of the switch. The value of a switch will vary during the operation of a program. A switch may be altered by a variation on input data, for example, a designation on a card may switch the program to different input validation routines according to the type of card being read. Switches may also be set by external action, such as by inputting a switch-setting message from

149

the processor console. Switches may also be set as a result of different conditions on input or output peripheral devices: for example, when the paper in a printer is running low, a hardware switch is set which can be tested by program, perhaps causing the program to wait until more paper is loaded.

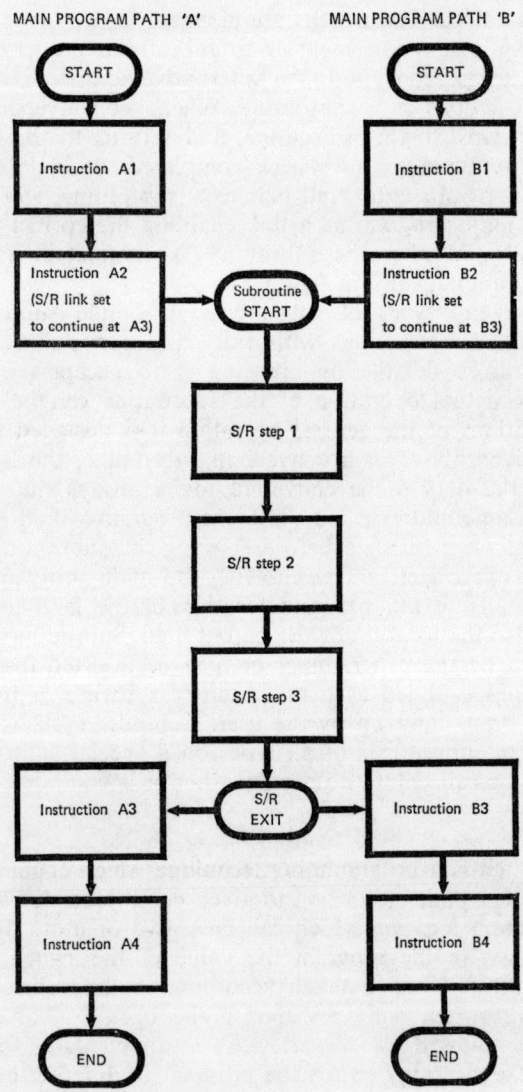

53. The use of a sub-routine

Table look-up

Table look-up is a technique for simple data retrieval within a program. Frequently a program has to store a large amount of information, part of which may be required at any given time. For example, an invoice program may require to 'remember' the prices of say 100 items, each identified by a product code. When a detail is read, the program is required to match a price to the item code. A quick way for a particular item code to be located in store is to relate the code to the address of the location in which it is stored by some simple rule or algorithm. The program then calculates the address from the code and is able immediately to find the required information.

Buffering

Buffering is a technique used to speed up input and output routines, in particular with backing store input and output. Data is normally organised in a backing store in the form of blocks, each of which contains several records, which are the logical units of data to be processed. The unit of input and output to the store, however, is the block of records. Buffering input or output means allocating two input/output areas, each the maximum size of a block of data. Blocks are written or read alternately to each area, and while one area, is being 'filled' or 'emptied', the program can be accessing the other area. This keeps the storage device working at virtually full speed, rather than waiting for processing for one block of data to be completed before starting on the next. While buffering speeds up the operation of a peripheral, it uses up more program storage space, so the technique has to be used with caution (see section 4.5.3).

3.5 UTILITY SOFTWARE

A computer can do nothing without a program of instructions, and each job required must have its own special program. However there are many tasks of a routine nature that all computer users require their machine to perform from time to time. It would clearly be wasteful if each user spent a lot of time writing programs for these tasks and it is normal practice for the computer manufacturer or other programming specialists to write programs for these tasks. These programs are then supplied with the equipment to all users (although in some cases users who operate large numbers of computers may write their own programs). These programs are known collectively as utility software, and this section outlines some of the basic programs available from most manufacturers.

3.5.1   Housekeeping packages

General

A housekeeping package is a routine which performs functions associated with the control of input and output of data. At the basic machine code level a program may need large numbers of separate instructions to a peripheral unit in order to control its performance. Such instructions may order the unit to begin a transfer of data, may check the parity of such a transfer, may control the area of storage from or to which transfer takes place and perform conversion of data from one format to another. The sequence of such instructions for any particular type of input or output will tend to follow a common pattern, so in order to simplify the task of programming, standard routines will be supplied either by the user or the manufacturer to perform these standard functions. All the user need do when using these facilities is to supply information to the housekeeping routine giving specific values to any necessary variables, such as file labels, record sizes, etc.

Basic peripheral control software

Housekeeping packages for basic peripherals such as paper tape and card reading and punching devices do not normally need to be very sophisticated.

Input is in the form of unit records (cards or blocks of paper tape) which are placed in record image format in store as a result of the normal read instruction, or output as a result of an output instruction.

On some systems, control data has to be preset in memory to enable the peripheral device to function: for example, an area may be required for communication of error conditions by the peripheral to the program. Software routines can be provided to simplify this process. Normally, however, packages simplifying the use of basic peripherals form part of a general data editing package as described below.

Data editing software

Data editing software performs the following functions:

1. Activating a slow input peripheral.
2. Distributing the data from an input record into predetermined memory locations.
3. Converting specified fields from external to internal format (e.g. character fields may be converted to binary representation).
4. Distributing data required for output into predetermined memory locations.
5. Converting specified fields from internal to external format.
6. Activating a slow output peripheral.

Each of these functions may be supplied and utilised separately as individual routines, or may form part of a general package. Functions 2 to 5 may be utilised on data which has been input from or is to be output to a fast peripheral or storage device.

Storage device control
The use of storage devices such as magnetic tapes, disc stores or drums involve the control of writing and reading operations, and of organisation of data within the device. Housekeeping packages provide the user with the ability to check automatically the labelling of files, both by checking labels of files open for reading, and writing labels to output files. In the case of serially organised files, the batching of records into blocks may be controlled by the package, and double buffering techniques catered for automatically if required. Direct access devices require housekeeping routines to control the storing of data according to the particular methods chosen (see 4.8.1) and for retrieving data from store. Where direct access files are treated identically with magnetic tape file storage, that is by storing records serially within the file, the housekeeping functions may be identified with the corresponding magnetic tape housekeeping routines. In this case, a program using the housekeeping package may be recompiled to accept different storage devices without any program modification, the particular device required being specified by a compile-time parameter.

Interrogating and display unit control
Housekeeping packages for units such as interrogating typewriters, visual display units and digital incremental plotters provide the programmer with routines which perform the functions peculiar to the unit in question. For example, interrogating typewriters require programming for access by the unit to the program, the input of data from the unit, and the output of data to the unit. Visual display units require packages to control the creation of various output symbols, and their combination into more complex symbols; the ability to alter the scale, axes, and three dimensional aspects of the displayed image; and the ability to access and interpret signs from an input light pen device. Digital incremental plotters will have routines to control the x-axis and y-axis movement of the plotting pen, the speed of rotation of the paper, and scaling operations to modify the significance of the digital increments.

3.5.2 Generators

General principles
A generator is a program similar to a compiler or assembler (see 3.4.3) in that it produces an object program in machine code from a series of source language instructions. However, a generator will produce programs of a restricted type only, rather than general purpose programs. Since the type of program produced by a generator is predetermined, the language is necessarily restricted to specifying a particular version of the type of program required.

153

For some types of generator the restrictions are so severe that the language becomes virtually a set of parameters, while in other cases considerable flexibility is reflected in a language approaching general purpose high level languages. The two main types of generator provided by most manufacturers are report generators and sort/merge generators.

Report generators

As the name implies, a report generator is a generator whose output is a report program. A report in this context is defined as an analysis of a data file which has been sorted to some predetermined sequence. Processing takes place at breaks in the file sequence specified in the report source program, and the results of this processing together with details from the records on the file are printed out. Sophisticated generators enable the user to specify complex processing and editing requirements and approach high level languages in their flexibility; however report programs can usually only process input files, and are limited to printing or listing results to an output file.

Sort/merge generators

The next section (3.5.3) describes the general principles of sorting files. Programs to sort files are provided by manufacturers either as fixed packages, when the user is limited in his options to specifying the location and priority of record keys, or in the form of generators. A sort generator provides the user with greater flexibility in output formats, perhaps including the facility to select and edit records at the same time as sequencing or merging them.

3.5.3 Sorts

Principles of sorting

In order to extract information from a mass of raw data, some ordering of the data into a meaningful sequence is required. Where processing involves the presentation of data in sequence to an analysis program, the data within the file must be sorted into the required sequence before being analysed. Sorting may take place before raw data is read onto a computer storage medium, and electro-mechanical devices such as card sorters are available for this purpose. However, where large volumes of data are involved, or sequencing of lengthy or complex keys is required sorting of files on computer storage media is performed. Since sorting is a very common data processing requirement, manufacturers provide utility software which enables users to specify their particular sequencing requirements by means of simple parameters to a general purpose sort program. Software is usually available for sorting files held on all types of storage device. The user specifies the sort keys, and also details about the type of file such as storage device, file labels, record structure. The sort program reads the unsequenced input file, and by means of various copying

154

techniques ultimately produces as output a copy of the input file in the required sequence. Various sorting techniques, such as classical and poly-phase sorts, are used: these terms refer to methods by which the final sequenced file is built up from intermediate files produced in a series of 'passes'.

Sort keys

The purpose of sorting a file is to present records within a file in some defined sequence. This sequence is determined by the ordering of certain specified fields within the record. Fields whose ordering determine the sequence of a file are known as *keys*. The simplest case is an ordering of a single key: for example, a file of personnel records may be sequenced by ascending order of a numerical staff number. A more complex ordering may be produced by introducing a further key. Each record on the personnel file may also be allocated a department number and the required sequence may be 'staff number within department number'. This means that all records for the lowest department number are presented first, each in ascending sequence of staff number: then all records for the next department number, and so on. This is indicated in figure 54.

Ascending Staff Number Sequence (1 Key)		Staff number within department sequence (2 Keys)	
STAFF NUMBER	DEPARTMENT	STAFF NUMBER	DEPARTMENT
101	2	124	1
123	3	178	1
124	1	213	1
176	2	101	2
178	1	176	2
202	3	123	3
213	1	202	3

54. Sorting on different keys

According to the extent and sophistication of utility software available, the size and number of keys which can be specified, and the type of ordering (e.g. ascending, descending, alphabetic) will vary. If a complex sequence is required which cannot be achieved using standard software, special purpose sort routines may be necessary. However, the analyst should note that by judicious record design the number of keys specified to a

155

utility program can be reduced by compounding two or more key fields into a single field. In the example above the department number could be prefixed to the staff number. A single ordering of the resulting four digit field would give the same sequence as the ordering of the two fields separately.

Sort techniques (magnetic tape)

Sorting of files held on a computer storage medium is required when files are processed serially. Normally each record in a file is treated as a whole, that is the data and key fields are moved together. Some sort techniques separate data from key fields, sorting only keys together with a pointer to the associated data. However the most common form of sorting is of serial files held on magnetic tape, or *tape sorting*. Tape sorting techniques are all variations of *merge sorting*.

Merge sorting techniques consist of two phases:

1. *String creation*: Records are input from the unsorted file, and formed into groups which contain records in correct sequence. A group of records in sequence is known as a 'string', and these strings are written to other intermediate tapes.
2. *String merging*: The intermediate tapes containing strings created at the first phase are read, and strings from alternate tapes are merged to form longer strings, which are in turn written to further intermediate tapes. This phase continues until finally a string is created which contains all the original records: this is the sorted output file.

Each repetition of the string merging operation is known as a *pass;* the maximum number of strings which can be merged together in each pass is known as the *way* of the sort. In general, the more passes that have to take place, the longer the sort will take. The greater the way of the sort, the shorter the time taken will be.

The various techniques of tape sorting, such as classical, polyphase, oscillating and cascade sorting differ mainly in the methods adopted for determining the length of strings and their distribution between available tape decks. The particular technique used in any given application will depend on the number of tape decks available, the amount of storage available in the processor and the time allowable for sorting. In general, the larger the number of available decks and storage, the shorter is the time required.

Details of these factors, and of approximate sort times for various combinations of record size and file length are generally supplied with appropriate software produced by manufacturers. Normally two types of software package are supplied. Sort routines are complete programs, to which the user supplies details of keys, record length, file labels and other variable factors for each particular run. Sort generators require a more extensive specification of sorting requirements by the user, but result in the creation of an object program specifically for the required sort: this may be more efficient than the more general sort routine. Section 4.4.3 deals with the systems implications of magnetic tape sorting.

Sort techniques (other devices)
Sorting of files held on direct access devices follows generally the pattern of magnetic tape sorting, since sorting implies the serial ordering of a file. However, with large direct access files it is sometimes more efficient to sort an index file consisting of keys for each record together with the address of the appropriate record on the direct access device. The sorted index may then be used to access each record from the direct access store in the required sequence.

Where all the data to be sorted can be contained in internal storage, internal sort routines can be used. With these techniques, records in store 'swap places' until the final sequence is achieved.

3.5.4   File control

Introduction
File control software is concerned with routine operations in data files. File control software is distinct from data management software (3.5.5): the former deals with *complete* files of data regardless of contents, while the latter is concerned with the data contents of the files.

Labelling
In most computer systems files held on a storage device are identified by a special block of data held as the first block on the file. This block will contain certain control information enabling the file contents to be identified, and may also contain additional information about the storage unit itself, such as the date when the unit was last written to, the number of times written to, and the serial number of the unit. Labelling packages enable the user to write properly formed labels to files, to modify labels on existing files, and to read labels from files for checking purposes.

Scratching
In systems where file labelling is used, it is generally not possible to write to a file unless the control information in the label block indicates that the data on the file is no longer valid. One method for doing this is by the use of purge dates, that is a date recorded on the label at the time the file was written to, giving a date after which the data would cease to be valid. An alternative method is to give a special name or label to a file which may be used for recording data.

In any system, however, certain files may hold data which has become out of date, or is invalid for other reasons, and thus is available for writing to. A routine must thus be available to alter the label so as to indicate that this state has been reached, even though the purge date has not yet been reached, or the file label is unchanged. The normal labelling

157

routine is only available for giving a current data label to a file already available for writing to. The scratch routine will enable a new label to be written to a file with an already valid label, and must therefore be used with care, as it is possible to destroy valid data if wrongly used. Normally a scratch routine will print the existing label of the file for checking before the label is finally overwritten. Certain highly important data files may have additional security codes present in the label rendering it impossible to scratch them until the purge date is reached, or needing a special code to be supplied to the scratch routine before overwriting the label.

Copying

File copying routines are provided for producing an exact copy of a file, either from one unit of a storage device onto another similar unit, e.g. from one tape reel to another, or from one storage medium to another, e.g. copying a card file onto tape. Normally copying software will not involve any modification of data, although editing from one format to another may occur, e.g. when transcribing from cards to tape, character punching may be converted to binary representation. File labels may also be changed, so that the copied file may be given a distinct label if required. More than one input file may be copied onto the same output file at the same time.

Printing

File printing routines are a special case of the general copying routine, where file contents are transcribed to a line printer. File contents may be reproduced in different formats on the printer if there is a difference in the internal and external data formats, e.g. character to binary conversion. Special printing may be provided if the file contains program instructions rather than data. Some selection and editing facilities are normally provided with printing routines to enable parts of files to be output, e.g. specified numbers of records or blocks and certain portions only of records to be printed. However, general file printing routines will not provide any data analysis function: these will be found in data management software (3.5.5) or report generators (3.5.2).

Merging

Merging of files involves the combining of records from two or more sequenced files into a single sequenced file. Each of the constituent files must be in the same sequence, although the record layout of files need not be identified. The output file will be in the same sequence as the input files, placing records from each in their correct relative order.

Maintenance

File maintenance is the term given to any system of reorganisation of data items within a file, where the reorganisation is independent of the information content of the file. It is thus distinct from file updating, which it concerned with the routine modification of variable information in the file in response to transactions. File maintenance software in effect is a form of selective copying. Facilities include the combining of data from more than one file, the deletion of items by exclusion from the copying function of records identified by record key or record count within a file, and the selection of specific portions of records to be copied. The systems implications of file updating and file maintenance are discussed further in sections 4.4.4 and 4.4.5.

Loading
With files stored on direct access devices (see section 4.6) software is generally provided to enable data to be loaded into the device in accordance with the various sytems of file storage available. Software is also available for the function of file reorganisation. These topics are discussed more fully in sections 4.6.1 and 4.6.4.

3.5.5   Data management software

General
Data management software is a general term given to routines which perform a variety of generalised operations on data files. The scope and power of data management software varies considerably between different manufacturers. However the common principle for all data management software is that the routines are data-independent. This means that the routines are written quite generally, and will operate on any data formats and even on data held on different types of storage medium. Usually the parameters supplied to a data management routine will consist of some form of data specification, describing record structures and file layouts, and a set of procedure steps, specifying the particular details of the generalised operation required for the purposes of a given run. This 'parameter set' may be more or less complex, according to the complexity of the operation and the requirements of the user. In some cases parameters are supplied in the form of answers to a fixed format questionnaire setting out the options available to the user. Other forms of specifications resemble procedure statements in some programming languages, or the specification of generator programs. There is in fact no hard and fast line between data management software generators and some high level languages.

Data management operations
Common operations for which data management software is supplied include data validation and editing, file updating, comparing files, collating files, and reporting from files. The method by which different types of software achieve these operations differ greatly between manufacturers. However in most cases the user will specify the structure of the records and

files he is using, and then select the particular form of the operation he requires. If a sequence of several operations is required, the user may be able to specify all his requirements at the same time, with the software system selecting appropriate programs, or the user may specify each operation as a separate run.

Systems applications

The use of good data management software can in many circumstances be of great assistance in systems design. The operations covered by comprehensive data management systems will tend to be common to a great many data processing requirements. Since using software avoids programming bottle-necks, once the overall systems design is completed it can be rapidly implemented as a series of software specifications. Use of software may be less efficient than specially written once-off programs, since the system may have to be broken down into a series of separate runs each making use of one data management routine. However, the gain to the analyst is the ability to test the validity of a system structure, modifying the fundamental parts of those systems, such as record formats and data elements, and output layouts, without having to waste slow and costly programming time. Once a system has been proved, and if greater efficiency is required, some sequences of data management software can be replaced by special-purpose programs. The prior use of software will have enabled the analyst to specify programs with greater precision, and will also provide an extremely useful source of test data and expected results against which to prove the resulting programs.

3.5.6 Program handling software

General

Program handling software is concerned with the functions necessary to insert programs into memory and to control their operation while in memory. Many of these functions are also performed automatically when programs are under the control of operating systems (see 3.6). The functions described in this section are basic routines, usually available in all computer systems.

Loading

Loading is the term given to inserting a program into memory from some external medium, such as cards, paper tape or magnetic tape.

The method by which this is achieved will vary according to the hardware and software philosophy of the particular computer system.

The most basic method for loading is to employ a 'bootstrap' technique. In essence, this means forming a few instructions in the processor, perhaps by setting manually a series of switches. These instructions then themselves read in a more comprehensive set of instructions, known sometimes as 'initial orders' or 'initial instructions'. This set of instructions is in effect a small program designed to read and store the programs required for running. The functions performed by this loading program may include the checking of the validity of the input program through parity checks or a 'checksum' technique. If the input program is stored on a program library, the loading routines will search the library medium for the required program.

In some cases, the user may modify the address of the start of the program, thus exercising control over the location of the program in store. The loading program may then adjust all other addresses in the program whose value depends on the value of the address of the beginning of the program.

Where some form of operating system is in use, the functions of program loading form an essential part of the operating system. In addition to loading required programs into store, the operating system organises the allocation of peripheral units to the program, and in the case of multi-programming, organises the location and operation of several programs together. Full details of the handling of programs in an operating system will be found in 3.6.

Run-time amendment
In addition to inserting a program into store, the program handling software may perform the function of modifying the program after it is loaded. The changes to be made to the program will be specified in machine code and may be input from the same or a different medium to that of the loaded program, or directly through the console. In some cases this run-time program amendment is performed by a separate program, such as the program used for loading. In other cases, the routine for reading and implementing run-time amendments will form part of the operational program. The facility for making run-time amendments may be used both for input of run-time parameters, such as file generation numbers, dates, etc., or for changing instructions in the program. Modification of instructions at run-time is one technique used in program testing. This is discussed more fully in 3.4.5.

Dumping
An additional facility which sometimes forms part of program handling software is the facility for outputting a program from memory onto another storage medium from which it can be subsequently loaded again. For example, a program may be stored on a magnetic tape library, loaded into memory, then dumped onto cards or paper tape, giving a version of the program which can be read from a slow peripheral, perhaps on a system without magnetic tape storage. Dumping a program to a printer is a special case used in program testing (see 3.4.5).

3.5.7 Program maintenance

General
Once a program has been completed and listed, it might be imagined that

161

F

the story is finished: the program will continue indefinitely to be used for the purpose for which it was written. However in practice this never happens: changes will always be required in a program, either to correct previously undetected errors, or to enable it to perform functions not originally specified. The purpose of program maintenance is to simplify the task of keeping programs up to date with amendments. The importance of this activity is increased when a large number of programs are in use. In these circumstances programs are held in a program library. This is another term for a specialised form of file held on a storage medium such as magnetic tape or direct access device, whose contents are individual programs. Program maintenance software is in effect an updating system for such specialised files.

Program libraries

Program libraries hold all the programs in use in a system. Libraries may hold programs in machine code, or in source form. In the latter case, individual programs will need to be compiled before running. Normally, all programs used in any system will be held in machine-code form on a library file: this will include both manufacturer's software, and the user's own programs. Software is not normally supplied in source form by manufacturers, so the source library tape will hold the users own programs only. Program maintenance software performs several functions. It enables new programs in machine code format to be added to a machine-code library, and enables programs on the library to be removed. In some cases, the coding of individual programs on the library may be modified directly, as in run-time modification discussed in 3.5.6. The function of adding and deleting complete programs is also available with libraries holding programs in source form. It is in this form that program amendments are also normally applied. The user specifies lines of source coding needing replacement, addition or alteration, and supplies corrected versions. Once the source coding has been modified the program can be re-compiled to give a new machine code version which can be run and tested in the normal way, and finally included on the machine code library tape.

In some systems programs are identified by version numbers indicating which modifications have been applied. The maintenance routines may either update these version numbers automatically or enable the user to specify which number to insert.

Documentation

The use of maintenance software to update program libraries provides a useful source of documentation for program amendments. The updating action requires the user to specify precisely the changes required and the software normally provides a printed record of all changes applied, showing the state of the program before and after modification. This output from the system enables later programmers to understand changes made to original versions, and also enables efficient auditing of programs to be carried out where necessary.

3.5.8 Dump and restart

General

If something goes wrong during the running of a very long program, such as the failure of a peripheral unit, or if the run needs to be stopped for any reason, it is clearly wasteful to have to start again from the beginning. One method for avoiding this is to record the state of the program at regular intervals during its operation. The program can be restarted at a later date by resetting the state of the processor and peripheral units at their condition at the time the last record was made of these states. This process is known as dumping and restarting, and standard program handling software is normally available to cater for the various procedures necessary in order to make the operation automatic and simple to use. (Particular systems considerations are given in section 4.4.9).

Dumping

Dumping is the term given to the recording of the state of a program at given intervals. This entails outputting the state of the program in the processor onto some backing store such as magnetic tape or discs, or on to punched cards or tapes; the recording of the positions reached by any serial storage medium such as magnetic tape; the recording, usually by the operator, of the position reached on card or paper tape input and output files, and the recording of any other necessary run conditions. The points at which dumping occurs may be determined by physical events, such as the completion of reading or writing to successive reels of tape or the printing of a specified number of pages; or the routine may be entered after fixed time intervals or by the manual intervention of an operator. Normally, a record of the serial number of each dump made is recorded on a line printer or console typewriter, and the run continues its normal operation automatically.

Restarting

Restarting a run from a given dump point is also performed by a program handling software routine. This will look for and read in the state of the program at the given dump point, as recorded on the dump storage medium, and will set up data files at the last location reached. A restart routine will also instruct the operator to load any other peripheral units in operation at the time of dumping, and then recommence operating normally from the given dump point.

3.6 OPERATING SYSTEMS

As computer hardware has developed, providing users with greater and greater power, enabling more and more programs to be run simultaneously with an ever increasing number and range of peripheral equipment, so the task of utilising a computer efficiently becomes more complex. Each program being run requires many operations: the loading of the program; loading of input data and data held on backing store; allocating peripheral devices to the program; monitoring the programs actions and obeying any requests from the program; loading storage units for output. When all these factors, and more, have to be considered for perhaps 16 or more programs running simultaneously, each requiring attention of a different sort at the same time, it can be seen that the task of computer operators can soon become impossible. Inevitably this confusion leads to inefficient utilisation of the computer processor and peripherals. Programs will remain idle waiting for attention: devices will be overloaded with programs queuing up to use them, or will remain unused for long periods. Operating instructions, however carefully written, will be overlooked by harassed operators.

The purpose of an operating system is to use the computer's own ability to carry out routine tasks of a complex but precisely defined nature to relieve the operator of much of this burden. Most computer systems, especially multiprogramming systems, have a special program always resident in core storage, to carry out certain routine tasks such as program loading and dumping. An operating system is an extension of this principle. In effect, an operating system is a program, or series of program modules, designed to control all the resources available to a computer system in the operating of programs. These resources may be defined as core storage, peripherals, the processor time, together with operator intervention. The operating system is informed of all the requirements which it has to meet, in the way of programs to be run, the input data and output media they require, and their relative importance. The system then allocates resources to meet these requirements in the most efficient way possible, and procedes to implement its requirements as far as possible without any manual intervention.

3.6.1 General principles

Foreground and background working
The work load on a computer system can be classified into two main types: regularly recurring jobs of known duration and predictable operational requirements, such as a weekly payroll run; and work of an ad-hoc nature in which either the operating requirements may be unpredictable, such as in program testing, or where the load and duration may be unpredictable such as the utilisation of a multi-access system in

164

conversational mode. One of the functions of an operating system is to so utilise the resources of the computer system that both types of work load can be dealt with together. This is done by organising the work in two modes, *background* working, which consists of the regular and known work load, and *foreground* working in which the ad-hoc and unpredictable type of work is dealt with.

Hardware considerations

An operating system is, basically, a software package, in that it is a special series of programs designed to carry out certain defined tasks. However, in practice for an operating system to be an economic proposition the hardware of the computer system must be of sufficient capacity both to hold the operating system itself and allow for the user's programs to be run. This means that an operating system is only viable in a multiprogramming system, with fairly extensive backing storage. Backing storage—magnetic tape, discs and drums, used both to contain the operating system modules themselves, and as a store for user programs, and intermediate data files. Since most operating systems are able to off-line output (see section 3.6.3) it is possible to reduce the number of slow output peripherals such as printers and punches to the minimum necessary for continuous operation. For example, in a situation where four programs are being run simultaneously, each requiring a line printer for 25 per cent of its time, instead of four printers operating 25 per cent, an operating system will enable all output to be produced from one printer operating continuously. An operating system enables programs to be written without direct reference to specific types of backing store or peripheral. Programs refer to files as the basic level of data, and the operating system itself informs the program of the particular medium on which any given file is recorded. This enables programs using the same file to be run simultaneously, provided the file is on a direct access device. Thus direct access storage is of particular importance for the efficient operation of a computer using an operating system. Operating systems are also particularly important in a multi-access system, enabling processor time to be shared efficiency by many users of terminal links, both where the users are all using the same routines, as in a reservation system, or where each user is developing and running his own programs.

3.6.2 Command language

Job description

The basis of an operating system is the description by the user of the operating requirements of the job, together with likely events during the running of the job and the resulting actions required. The description of a job is generally prepared by the user or the computer operations staff. They will use a specialised form of language, resembling in some respects a high level programming language, to specify the job requirements to the operating system: this language is known as the operating systems command language. The types of statement can be divided into four main categories: statements concerned with job set-up; statements concerned with the running of the job; statements concerned with the use of the operating system in conjunction with standard software and statements available to the user in connection with the control of the operating system itself.

Job set-up

In order to allow the operating system to accept a job, it must be provided with information about the data files required by the job both as input and output, and the peripheral requirements for input and output. Peripherals specified by the user will be those for which the program was originally written. The operating system may store required files on different storage media, and may off-line output data prior to outputting onto the required device: in all these cases the system itself interprets user program requirements into the operating system's allocation of peripheral units and storage devices.

Job running

During the running of a program, a human operator may be required to perform many actions. Initially, switches may need to be set to inform the program of certain user options available at run time. At stages in the program the operator may need to take actions, such as giving a core dump when errors occur, or choosing one of several alternative programs to follow a job depending on the results of the first job. Operating instructions have to be prepared indicating to the operator the action to be taken when recognisable program events occur. With the job running control statements available to the user of the operating system, a description of these events can be specified together with the required action or actions. In some cases it will not prove possible to provide an automatic continuation from every part of a program, and manual intervention by an operator may be necessary. However, most operational situations will be covered by an operating system statement.

Software utilisation

The use of an operating system can naturally be extended to use of standard software routines. In some cases specialised operating system statements are available for the more common utility packages. These will generally include compilers and generators, debugging aids such as storage printout routines, sorting programs and other housekeeping packages.

Operating system control

An operating system is capable of handling the operation of many jobs, and will normally be given control statements for large numbers of jobs at the same time. Statements are available which enable the user to categorise jobs, identify them with codes which can subsequently be analysed for accounting and statistical purposes, and allocate relative priorities between jobs, either by setting deadlines or by specifying an order of importance. Jobs may be further categorised as requiring all input and output on-line (for example when specialised peripheral units have to be allocated to specific jobs) or jobs may be described as available for off-lining of input and/or output.

The operating system can be required to give messages to the human operator when certain conditions are reached, or to halt operating at specified intervals. These various requirements are implemented by operating system statements defined for the control of the system.

3.6.3 On-line and off-line working

General

A peripheral unit is said to be on-line when it is operating under the direct control of a program. When an operating system is in use, peripheral units may be under the direct control of a particular program, in which case they are said to be on-line to that program. The operating system may however substitute different units for those actually requested by the program. The units for which the program was written are then said to be operated off-line. The purpose of off-line working with an operating system is to utilise slow peripherals with maximum efficiency. When a program reads cards directly from a card reader, or prints directly from a line printer, the difference in speed with which the electromechanical devices handle data and the program itself operates on the data means that valuable processor time will be wasted while the program waits for the peripheral transfers to be completed. Multiprogramming processors go some way to solving the problem, since while one program is waiting for a peripheral transfer to be completed, another program can make use of the processor. However even this arrangement leaves much to be desired. Peripheral units may not be working at their full speed, or there may not be sufficient peripherals of a given type to satisfy the needs of all programs being run at the same time. By off-lining the use of slow peripherals, an operating system ensures that programs run with maximum efficiency. All slow peripheral output is stored on a backing store, and separate operating system routines output the contents of the backing store onto the required peripherals simultaneously with the running of other programs. All available output peripherals can thus be operated at full speed, outputting the results of programs completed earlier. The same applies to input from slow peripherals: input data is read by the operating system *before* the program requiring it is started: thus all input data from slow peripherals which will be required by later programs is read first, utilising slow peripherals working at full speed and capacity.

167

The document concept

In order to enable the operating system to utilise the computer resources efficiently, files of data are treated as units independent of the medium on which they are originally recorded, or to which they will be finally output. Each file in the system is given a name, sometimes known as the document name. A user will specify the 'documents' required by his program, either as input files or output files.

In the case of off-lining of input files from slow peripherals, the operating system will look for files bearing the document names, and read these into backing store before the user program is loaded. The store into which files are loaded is also known as a 'well'. The well contains all 'documents' required by programs being run under the operating system, and all output files are also placed into the well. These wells thus act as large buffer areas shared by all programs under the control of the operating system. Three activities can thus take place in parallel: the operating system fills the input well with documents input from slow peripherals, keeping all available units busy at their optimal running rate; programs in the system extract relevant documents or files from the input well, access any files required on backing store, and output results for final output to a slow peripheral into the output well; finally, the operating system will be extracting documents from the output well and outputting them on the appropriate peripheral device. Figure 55 illustrates the function of the well diagramatically.

3.6.4 File storage

Types of file

As explained in section 3.6.3, an operating system treats a file of data as being independent of the storage device on which it is held. A user program requiring to read or write data is provided with the appropriate file by the operating system, which controls the medium on which the file is stored. However, there are certain restrictions on the generalised concept, since files may be of two basic types, serial files and direct access files.

With a serial file, data is presented to a program in the sequence in which the records are stored on the file, and in order to examine a particular record within the file, all preceding records must be read. With a direct access file individual records have unique addresses within the file enabling particular records to be extracted directly without the necessity for reading all preceding records. Since direct access files must necessarily be stored on a direct access storage device such as magnetic disc or drum, the operating system must be informed of the *type* of files it is handling.

File creation

An operating system has the power of creating new files when required by

168

Documents input from slow peripherals

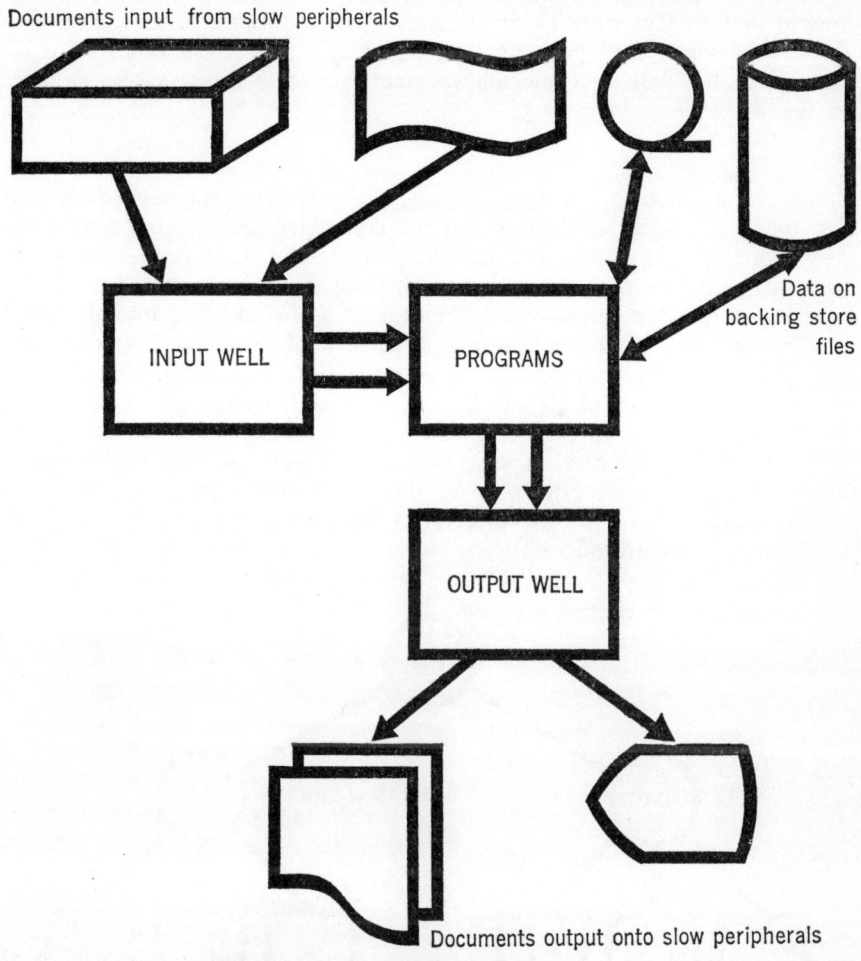

INPUT WELL

PROGRAMS

Data on
backing store
files

OUTPUT WELL

Documents output onto slow peripherals

55. Function of the well

user programs. When off-line operation of slow peripherals is required, the operating system itself creates input files by reading data directly to a backing store as described in section 3.6.3. Programs themselves may create files when data is output, either as direct output to a backing store of intermediate results, or when output is off-line. When off-line operation is requested the files created are always serial files, since data is read in from or written to slow peripherals in serial form. The actual medium used may be either magnetic tape, or a direct access store used serially: in some cases where the file is relatively small it may be stored completely in core storage. Similarly, all magnetic tape files used by programs as

F*

input or output are treated as serial files, even though the operating system may in fact store them on direct access devices. When creating a direct access file, the operating system has to be informed of the size of the file, and it will then allocate so much available direct access storage to the file.

File storage utilisation

An operating system will have to handle a very large number of files at any one time. Some of them it will be creating or storing under off-line operation; some will be more or less permanently in storage for use by user programs when required; others will be temporary storage of inter-mediate results for run-to-run communication. In addition to data files, the system will also be controlling large files of programs and software routines. One method by which an operating system can keep control over this complex pattern of files is by means of a hierarchial tree structure. In this system the operating system creates specialised files known as directories or index files. A directory will contain information about a series of related files. Higher level directories will contain information about related lower level directories, giving a structure which is represented in figure 56.

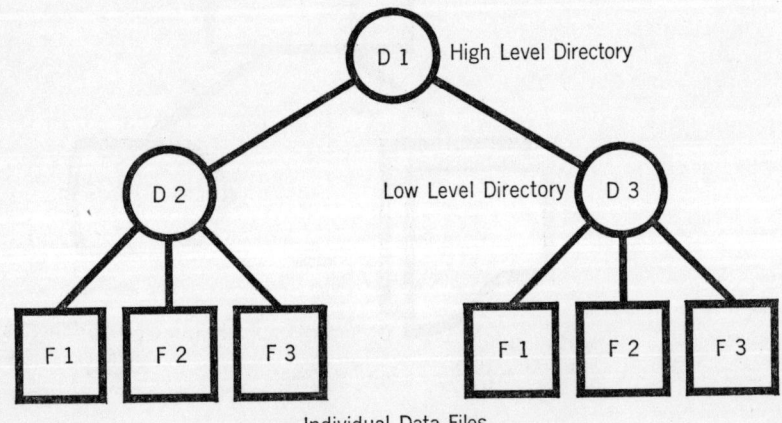

56. Hierarchical file structure

The sort of information held in a file directory will be the names by which files it controls are known, and housekeeping information, such as the file generation number, the date the file was created, etc.; information about the type and structure of the file; information about the user of the file and any information required to preserve the security of the information. Most important of all will be information enabling the operating system to 'find' the file among all the various storage devices it controls.

When a user wishes to refer to a file he can do so by specifying the names of the directories making a 'chain' leading to his file.

For example by referring to a file as file F_1 of directory D_2 of directory D_1 a unique file is located by the system. The methods by which files are organised within directories will depend on the requirements of the user of an operating system.

One common method of utilising this hierarchial system of file organisation is to make the file structure reflect the 'authorisation' structure of those entitled to access to information in particular files. Thus individuals may be limited to accessing data in a specific lowest level file only, or to all files held in a specific higher level directory, according to their position of authority within an organisation's management structure. The system also enables different users to be charged for file usage where separate accounting is required either between parts of the organisation or when a computer is being shared by different organisations.

File security

Every user of an operating system is allocated a 'user name'. The system uses this name both for accounting and statistical purposes, so that a record is kept of how much of the system's resources are utilised by each user, and also as an authorisation code, indicating which programs and files are available to the user.

A hierarchial file structure as outlined above enables each file directory to be associated with one or more user names, indicating that the specified user is allowed access to all files under the control of that particular directory. The system will not normally allow any unauthorised user access to any files unless some specific action is taken to inform the operating system that the user is permitted to use a given file.

In some cases it may be necessary to have files which do not come under the control of the operating system at all. Such files, sometimes known as 'private files' are accessed directly on-line to a user's program and thus come outside the hierarchial structure and system of user names. The purpose of private files may be to use particular magnetic media such as specific reels of tape or discs, which may be required for communication between different computer installations, or to enable the user to control the exact layout of data on a particular storage device, for example when developing direct access storage systems with minimum head movement.

3.6.5 Job scheduling

Principles of scheduling

In section 3.6.3 the concept of off-line working was described. This concept enables an operating system to share the use of slow peripheral devices so that no program is held up by the non-availability of a unit, and means that all units will be working at their full rate. Another facet of computer utilisation which an operating system will take care of is the

171

sharing of processor capacity between programs. Every program being run on a computer takes up *space,* or core storage within the processor, and uses *time* by requiring the processor's control logic for its operation. Under manual operation, a computer operator loads jobs on a computer. Each job will normally state on its operating instructions the data files it requires, the amount of core storage it occupies, and the approximate amount of time it will take. Faced with a large number of separate jobs to run over a period of time, the operator must work out the best running schedule so that all the machine's resources are fully utilised. At the same time some jobs may be more important than others: high priority work must be completed before work of lower priority. The situation is further complicated if there is a multi-access system in which several different users have immediate on-line access to the computer: the work load must be shared effectively between foreground and background work.

Job concept

To an operating system, a job consists of programs and associated data files. Each job is described to the system, and the system stores the descriptions of all jobs it is to be required to run in a specified period. Each job will have as part of its description, in addition to the list of programs and files it requires, a description of the time it is expected to take and an indication of the priority to be allocated to the job. The operating system then arranges the running of jobs in turn by extracting the job descriptions in sequence from backing store, and calling in the required programs and files.

The operating system can keep track of progress of all jobs, and may interrupt low priority jobs and even suspend them altogether to allow higher priority jobs to be completed. This also enables the operating system to balance the use of the computer with multi-access facilities, allowing different users to have access to the processor when time and space is available, according to the user's priority rating.

3.6.6 Conversational mode

General

In everyday life, carrying on a conversation implies the exchange of ideas or information between two or more people. A computer is said to be used in conversational mode when the user is able to exchange information or develop ideas with the aid of a computer, where the particular information or pattern of ideas is not predetermined but evolves as a result of the interaction between the user and the computer. When used in normal mode, a computer is presented by the user with a pre-defined program of instructions to be carried out on a file of data whose contents are of a predictable type. The user knows what sort of data is being fed in and expects to get a certain type of answer out. When a computer is used conversationally the situation is not so clear cut: the user approaches the computer without a precise definition of his requirements, whether this is for information or for the development of a computer program. By using the facilities provided by the system he is able gradually to approach the solution he seeks until he has achieved a satisfactory result.

Types of conversational mode
A computer is used conversationally for two main purposes: information retrieval and program creation. When used for information retrieval, the user access a file of data. The system informs the user of the broad categories of data on the file, and the user is able to make a choice: the system further breaks down the categories, each time enabling the user to make a further choice or revise an earlier one, until the required item of information is retrieved. A similar process is used when writing program using a conversational compiler. The system offers the user a selection of instructions and gives him rules for forming a correct program. The user selects instructions, and is informed if at any stage his choice is illogical or incorrect. He is able to test his program as he goes along and, as parts are completed, these can be stored for further use.

Function of operating system in conversational mode
The use of a computer in conversational mode is, superficially, extremely wasteful of processor resources. While the user is thinking, making his choice between offered alternatives, the system has to wait for him and is thus doing nothing. When used with an operating system, this situation is avoided. The operating system can utilise the thinking time to activate other programs, or share resources with other conversational mode users.

If a user takes over-long in thinking, the system can dump his section of store and retrieve it into core store when the user wants to make another choice. Users can also be given a time limit: when this is exhausted the operating system will not allow them further access to the processor until they are allocated more time.

3.6.7 Multi-access

General
In section 3.6.6 we considered the use of an operating system with the conversational use of a computer. Using a computer conversationally implies several users each having access to the processor so that each can ask questions and receive replies. This is in fact a form of multi-access working. Multi-access means the sharing of computer time simultaneously between several users, the numbers of independent users being anything from two or three to several hundreds. Multi-access users communicate with the computer via remote terminals and data links. These terminals can take the form of typewriters or may be a form of visual display device, and may be used both for data input and as output devices. Multi-access use of a computer system is more general than conversational mode

working, since not all multi-access applications involve interaction with the computer in the same way as in conversational mode. Among examples of multi-access systems are airline reservation systems, in which thousands of terminals may be remotely connected to a control system which organise flight bookings. Another application of multi-access usage is the sharing of a central computer by numbers of research scientists in a university or laboratory: each user can write and test his own programs, having available to him the apparent resources of the whole machine. Multi-access working is also utilised by large computer bureaux. Customers store their data files on a central computer, and access their files for updating and information retrieval through their own terminals.

Operating systems and multi-access
Effective use of a computer working in a multi-access mode requires sharing the computer's resources between users. For this an operating system is an essential part of multi-access working. The system enables users to set up jobs from remote terminals, operate them as background work, or use the computer in a conversational mode.

3.6.8   Systems design for operating systems

Systems and programs not specifically designed for an operating system can normally be run satisfactorily within an operating system. However the effective power of a computer system can be considerably extended if systems and programs are designed to make sensible use of operating system facilities. Those parts of system design most affected will be input and output functions, file structures, and the specification of individual programs making up the various runs in the system.

Input and output functions
The main advantage of an operating system when considering input and output is the ability to use off-line operation of slow peripherals. When batch processing of large volumes of input data from cards or paper tape is involved, the standard method for maximising the efficiency of slow peripheral usage without an operating system is to accumulate as large a batch of data as possible before input takes place. An input program will read and possibly validate this raw data and store it on a backing store, from which it can subsequently be used for analysis and updating purposes. With an operating system, relatively small batches of data can be presented to the system at any time: these are automatically placed into backing store, from which the validation, analysis and updating programs can retrieve successive batches of input data when required. This improves efficiency, in that the frequency of updating or analysis need not wait for the accumulation of economically large batches of input data. To make full use of this approach, care should be taken in the design of validation routines. If possible, the system should be designed to accept

and use valid input without having to wait for errors to be corrected and resubmitted. Error lists can be accumulated in the output well, and output off-line at regular intervals rather than being output on line each time the validation program is used to validate a batch of input documents. If this is not possible some of the advantage of off-lining input batches will be lost, since the system will again have to wait for the accumulation of sufficiently large amounts of input data before the data can be used.

Advantage can also be taken of the off-line operation of output peripherals. For example, in many systems analyses are required from data files sorted into different sequences. More than one analysis may be required from the same file sequence, and it is clearly sensible to produce all required analyses from the given sequence before re-sorting to another sequence. This can be achieved using an operating system by nominating the sorted file to be held as a direct access file, then allowing all output programs to access the file together, storing all output documents in the output well. Some output may be of higher priority than others, and the subsequent off-line production of the output can thus be done in order of priority. Points to be considered in a system off-lining printer output include the use of pre-printed stationery. If this requires careful initial set up, or special control loops on the printer to allow for non-standard paper throwing, it is not possible to use off-line output: a printer must be allotted to the output program on-line to allow for set up and special loops. One method of overcoming this problem is to output without any headings on plain stationery and add the heading information afterwards by means of one of the various electrostatic copying devices available.

File structures
File storage in an operating system has been discussed in section 3.6.4. The main advantage of an operating system is that the user or systems designer need only specify the type of file he is dealing with: the system takes care of the actual storage device allotted to a file, and makes sure that it becomes available to programs requesting access to the file. Access to files can be controlled by the design of a suitable document indexing system such as the hierarchical system described in section 3.6.4. When deciding between nominating files as serial files or direct access files consideration must be given to the use of the file for analysis and updating, and the relative frequency at which individual records within the file are required. In cases where the majority of records receive attention either for updating or for analysis a serial file is adequate. With files where some records have considerably higher hit rates than others when updating takes place, or where records are required by large numbers of different analyses, economical use of resources will be achieved by making the files direct access.

Program specification
Specifications for programs for use within an operating system should state

175

whether or not peripheral units will be operated on- or off-line. When off-line working is specified, the program will be written as for a particular type of unit, but the data will in fact be presented to the program from the well, or will be output to the well. This means that the programmer need not consider the programming of input/output device control routines. For example there is no point in catering for 'paper-low' conditions on printing if the output is in fact written to a backing store. Similarly, where direct access files are opened or read by a program, the operating system will carry out all the data organisation functions required. If, however, it is necessary to include specialised file organisation techniques then the program will require the storage device on-line.

PART 3C : MACHINE ROOM EQUIPMENT

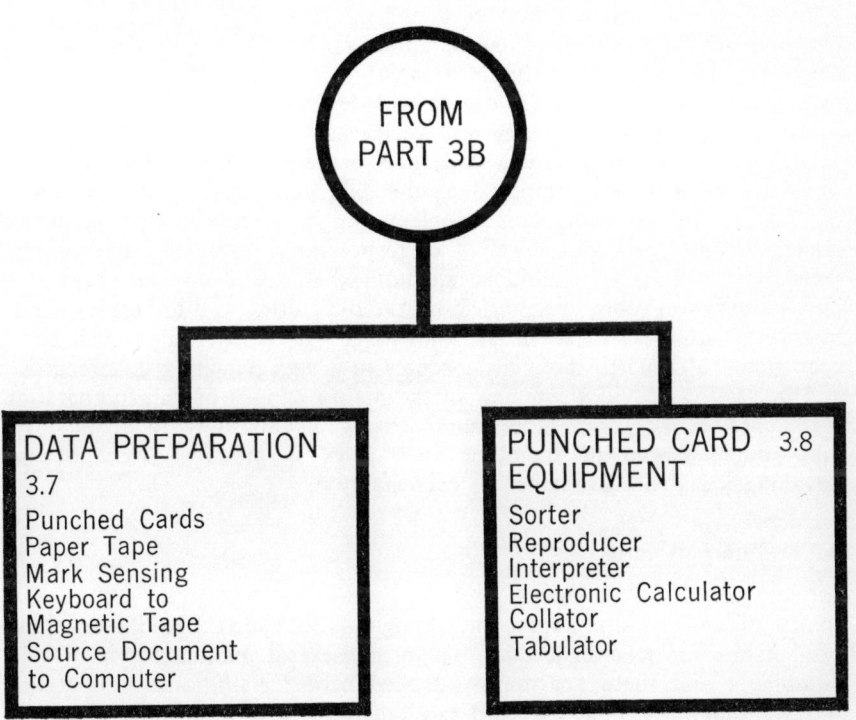

FROM
PART 3B

DATA PREPARATION
3.7
Punched Cards
Paper Tape
Mark Sensing
Keyboard to
Magnetic Tape
Source Document
to Computer

PUNCHED CARD 3.8
EQUIPMENT
Sorter
Reproducer
Interpreter
Electronic Calculator
Collator
Tabulator

3.7 DATA PREPARATION

Input to a computer may be prepared in a number of ways, and in section 3.2 we have examined the types of equipment used to accept data in such different media as punched cards, punched paper tape, magnetic ink, and magnetic tape. Considerable emphasis has been placed on the need to ensure that data is correct when it enters a computer system, and we have seen how operations on incorrect data and correction after input waste valuable time and can spoil an apparently well-designed system. Sometimes it will appear to the systems analyst that a system is being jeopardised by the unaccountable failure of punch operators to carry out the seemingly simple task of transferring data from source documents to punched cards, and he will demand with increasing urgency a higher standard of punching and verification. But a closer look at the punching documents and data collection procedures may reveal that too great a burden is being imposed on the data preparation staff and that a greater familiarity with their problems might have resulted in better documents and procedures and a correspondingly more efficient system. Every systems analyst should be encouraged to spend a week or so in a data preparation room, learning about the difficulties, and the errors which may result from such difficulties; knowledge gained in this way can be of very great value when the analyst comes to design punching documents.

This section provides a review of the main methods of data preparation —punched cards, paper tape, mark sensing, keyboard to magnetic tape, and source document handling—and gives a practical view of the advantages and disadvantages of each method.

3.7.1 Punched cards

Introduction
Punched card machines, sometimes known as unit record equipment, have been in use for several decades. There are several manufacturers of such equipment and there are many punched card installations in operation today. Sometimes punched card equipment is used alongside electronic computing systems to perform auxillary processing operations, but more commonly punched cards are used nowadays as an input/output media. The basic element in punched card processing is a card which contains data recorded as holes punched in columns; these holes are sensed electronically (or mechanically) to activate machines which process the data to produce printed results or intermediate files of cards for further processing.

The operations performed by punched card machines include reading information, comparing, classifying, summarising, printing and punching information, and in a later section there is a discussion of the facilities provided by these machines (see 3.8). In the present section consideration is given to the punched card and its initial preparation from source

178

documents; this topic is discussed in relation to the standard 80 column card, other types of card being mentioned in the later section.

The 80 column card
An 80-column punched card measures $7\frac{3}{8}$in. by $3\frac{1}{4}$in. and may be .007in. or .009in. in thickness. The card is divided into 80 vertical columns numbered 1 to 80 from left to right as shown in figure 58. Each column is divided into 12 index positions designated as 10, 11, 0, 1, 2, 3, through to 9 from the top of the column downwards. The three topmost positions are referred to as zone positions or the upper curtate, and the significance of these positions is made apparent when the punching code is explained below.

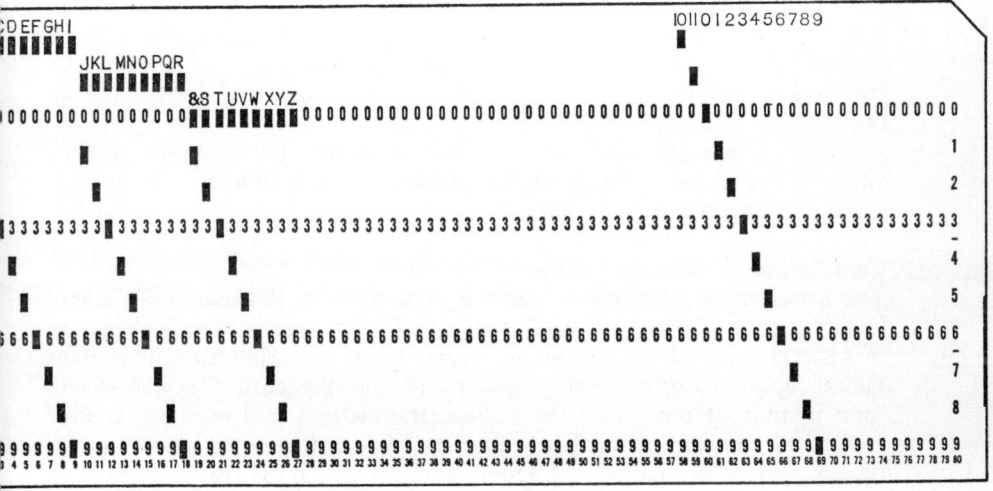

58. An 80-column punched card

The positions 1 to 9 are known as the numeric positions or the lower curtate. The 10 and 11 positions are sometimes given different names and may be known as 12 and 11 positions or the Y and X positions respectively, and just to add to the confusion Y and X are sometimes given the reverse significance. Throughout this book we shall use 10 and 11 for the top and second positions.

Each column of the card is generally used to accommodate a single numeric or alphabetic character, this information being represented by holes punched in the index positions on the card. The columns of the card are grouped to form fields and the fields may be from 1 to 80 columns in size according to the requirements of the data that the user wishes to process.

179

Punching codes

There are a variety of punching codes in use, introduced by different manufacturers at various stages in the development of punched card techniques. A systems analyst must be aware of these codes when developing and planning new procedures—some coding systems are not compatible with others and special action may be needed to modify equipment, or reproduce card files into new formats. During the creation of magnetic tape files or disc files from existing punched card files special conversion routines may be needed in order to effect code conversion (see section 3.2.2).

Figure 59 shows a typical 4 Zone card code as used in some punched card systems. It allows for 48 separate characters to be represented— numeric digits are recorded by punching in the digit positions from 0 to 9, and the twenty-six alphabetic characters are represented by using combinations of a zone and a numeric punching. Additional special symbols are created by using combinations of numeric punchings. It will be seen that the card shown in figure 58 is punched in this code and interpreted on the top line.

The 64-character card code shown in figure 60 is more in line with the requirements for input to modern digital computers; it provides a greater range of special characters.

Card layout

The arrangement of fields on cards is controlled by the user; card manufacturers generally provide cards preprinted to the user's specification and two samples are shown in figure 61 and 62. One of the most important factors about card processing systems is that the card files themselves form permanent records of the various transactions and events that they record. Reference can be made directly to files for visual inspection of records, and for manual extraction and correction where necessary. In punched card processing each transaction is represented by a single card; individual print lines on an output statement printed by a punched card tabulator are represented by individual cards. Thus in an order processing system each item ordered may be represented by a card giving customer number, customer order number, item number, description, quantity, unit cost, extended price (see figure 61).

Packing

Where punched cards are used solely as an input medium for an electronic computer, the cards are not generally retained as a permanent record, and input formats can be designed to take advantage of the data stored within the computer's backing store. For example, in figure 62 a card as used in an order processing system is shown, and here several items relating to a particular customer's order have been packed onto the same card. This technique reduces the time taken in reading data

180

into the system and considerably reduces the effort initially expended in data preparation.

PUNCHING CODE	CHARACTER or SYMBOL	PUNCHING CODE	CHARACTER or SYMBOL
0	0	11/3	L
1	1	11/4	M
2	2	11/5	N
3	3	11/6	O
4	4	11/7	P
5	5	11/8	Q
6	6	11/9	R
7	7	0/1	&
8	8	0/2	S
9	9	0/3	T
NONE	SPACE	0/4	U
10	10	0/5	V
10/1	A	0/6	W
10/2	B	0/7	X
10/3	C	0/8	Y
10/4	D	0/9	Z
10/5	E	1/2	%
10/6	F	1/3	¼
10/7	G	1/4	−
10/8	H	1/5	/
10/9	I	1/6	½
11	11	1/7	=
11/1	J	1/8	@
11/2	K	1/9	¾

This card code uses combinations of holes employing 4 Zones (10, 11, 0 and 1) to represent figures 0 to 9, letters A to Z, and a range of other symbols.

59. A 48-character card code

Dual-purpose cards
Punched cards are also sometimes used as source documents in clerical procedures, and thus serve a dual purpose although punching usually takes place in a part of the card especially reserved for it. The card circulates in a clerical routine to be annotated with data and is eventually punched to inject data into the mechanised processing procedures.

181

PUNCHING CODE	CHARACTER OR SYMBOL		PUNCHING CODE	CHARACTER OR SYMBOL
0	0		11/7	P
1	1		11/8	Q
2	2		11/9	R
3	3		0/2	S
4	4		0/3	T
5	5		0/4	U
6	6		0/5	V
7	7		0/6	W
8	8		0/7	X
9	9		0/8	Y
NONE	space		0/9	Z
10 or 10/0	&		11	-
3/8	#		11/0	"
4/8	@		0/1	/
5/8	(		10/2/8	+
6/8	)		10/3/8	.
7/8	]		10/4/8	;
10/1	A		10/5/8	:
10/2	B		10/6/8	'
10/3	C		10/7/8	!
10/4	D		11/2/8	[
10/5	E		11/3/8	$
10/6	F		11/4/8	*
10/7	G		11/5/8	>
10/8	H		11/6/8	<
10/9	I		11/7/8	↑
11/1	J		0/2/8	£
11/2	K		0/3/8	,
11/3	L		0/4/8	%
11/4	M		0/5/8	?
11/5	N		0/6/8	=
11/6	O		0/7/8	←

This card code has been designed to incorporate the figures 0 to 9, the letters A to Z and all the usual punctuation marks as well as the more common scientific and mathematical symbols.

60. A 64-character card code

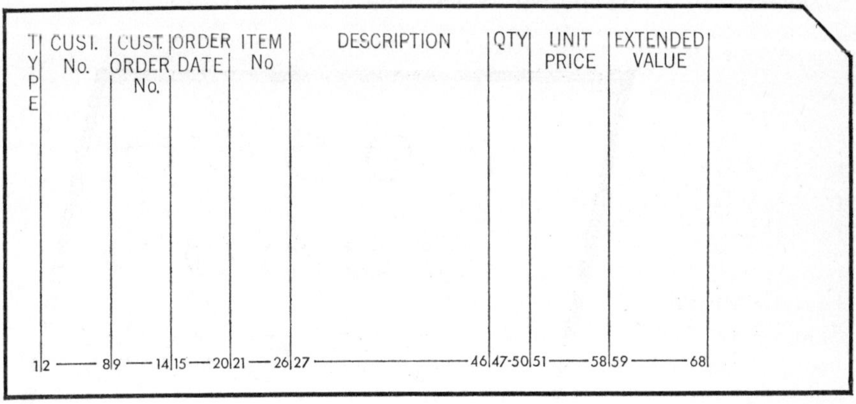

T Y P E	CUS I. No.	CUST. ORDER No.	ORDER DATE	ITEM No	DESCRIPTION	QTY	UNIT PRICE	EXTENDED VALUE

1 2 —— 8 9 —— 14 15 —— 20 21 —— 26 27 ———————————————— 46 47-50 51 —— 58 59 ——— 68

61. Detail card

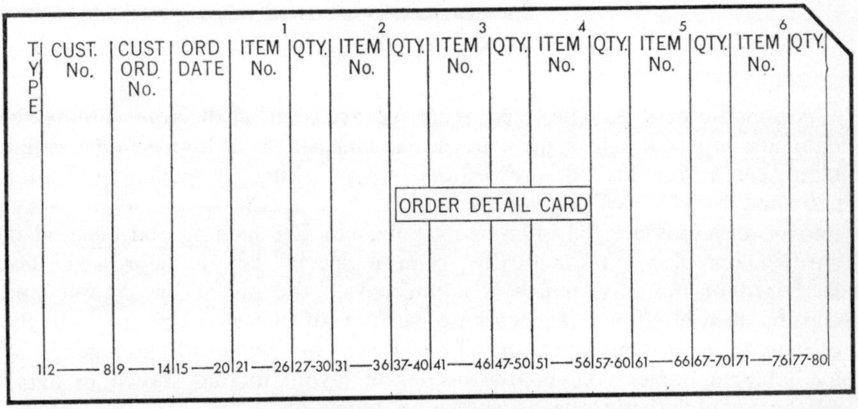

T Y P E	CUST. No.	CUST ORD No.	ORD DATE	ITEM No. 1	QTY.	ITEM No. 2	QTY.	ITEM No. 3	QTY.	ITEM No. 4	QTY.	ITEM No. 5	QTY.	ITEM No. 6	QTY.

ORDER DETAIL CARD

1 2 —— 8 9 —— 14 15 —— 20 21 —— 26 27-30 31 —— 36 37-40 41 —— 46 47-50 51 —— 56 57-60 61 —— 66 67-70 71 —— 76 77-80

62. A spread card for computer input

Punching cards

Punched cards may be prepared by the use of simple hand punches, which consist of a card rack which transports the cards column by column past a row of 12 punch knives, one for each card index position. The rack is moved continuously as keys are pressed on a simple numeric keyboard similar to that shown in figure 63; the keyboard is linked by direct mechanical connections to the punch knives and holes are punched according to the keys selected. With hand punches the cards must be loaded and extracted from the card track manually. Hand punches are suitable only for small volumes of data or for punching on an ad-hoc basis. They

183

are often found in or adjacent to computer rooms to enable operators and programmers to prepare data or parameter cards at run time.

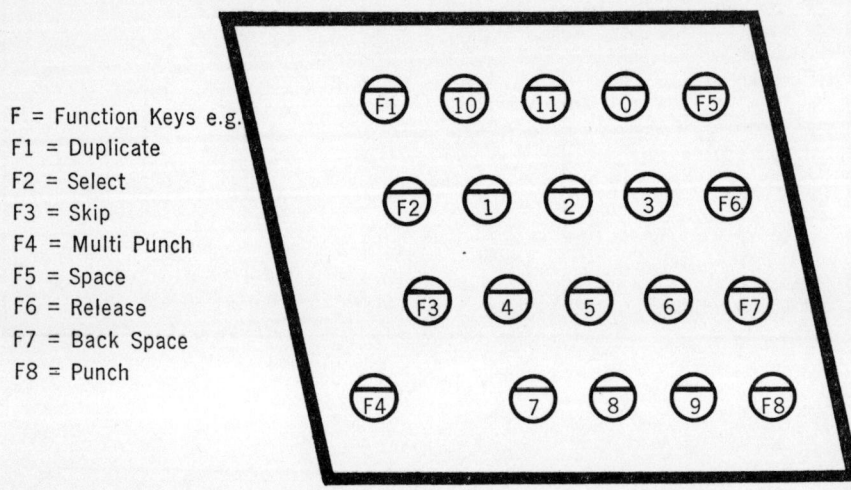

F = Function Keys e.g.
F1 = Duplicate
F2 = Select
F3 = Skip
F4 = Multi Punch
F5 = Space
F6 = Release
F7 = Back Space
F8 = Punch

63. Numeric keyboard

Automatic card punches (see figure 64) are used when large volumes of data are required. Here the card track consists of a hopper into which blank cards are loaded preparatory to punching, a punching station activated by electrical signals generated from a keyboard, and a stacker into which cards are fed after being punched. The feeding and ejection of cards takes place automatically when a special key is depressed. The keyboard of the card punch is connected to the punch by a cable and can be moved about the working surface of the machine to suit the convenience of the operator. Two types of keyboard are generally available: a numeric keyboard, similar in layout to that shown in figure 63; or an alphabetic one as shown in figure 65.

Numeric keyboards are operated using one hand, whereas the alphabetic keyboard resembles a typewriter in layout and can be operated two handed.

With a numeric keyboard alphabetic characters and symbols must be punched by depressing two or three keys simultaneously, and therefore the potential operating speed is reduced as compared with the full alphabetic keyboard. However, for punching numeric data the single handed method of operation is usually faster, and for this reason alphabetic keyboards are designed to include ten numeric keys grouped for single handed operation as on a numeric keyboard.

Other features provided as standard on automatic punches are a column indicator, which tells the operator the number of the next column to be punched, a backspace key which back spaces the card rack as required, and a key to cause cards to be ejected from any column to the end of the card track.

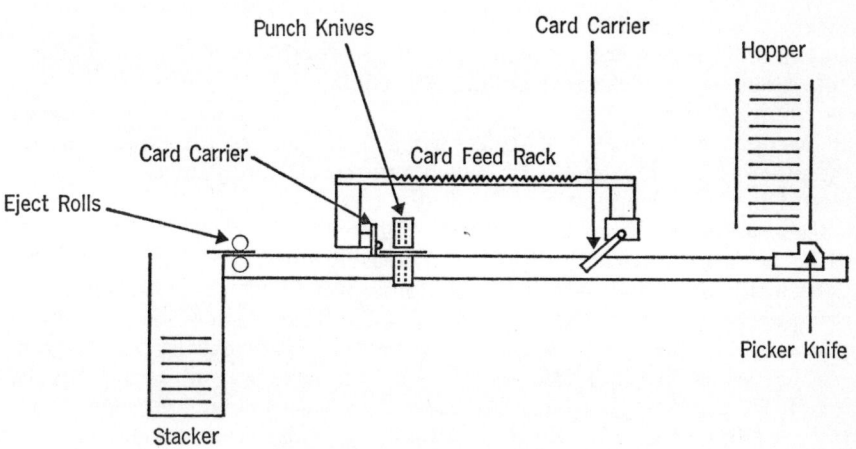

64. Outline of automatic card punch

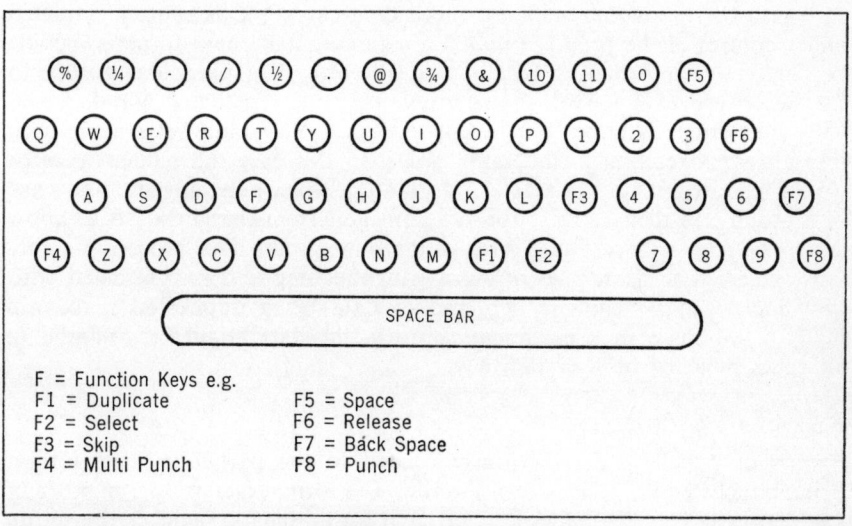

65. Alphabetic keyboard

185

Another feature which may be included is a reading head (either photo-electric, mechanical, or brush sensing) which detects holes in card columns after they have been punched. This feature may be used to effect gang-punching (see below) or to activate a print unit which prints characters along the top edge of the card above the corresponding column positions.

Programming a card punch

Automatic card punches can be programmed to permit automatic operations to be performed at specified columns of a card. The user has the ability to control these functions on some machines by making plugged connections on a small plugpanel, wired in as part of the punch, or (on the other types of punches) by punching a special program card which is loaded into a sensing device known as a program drum.

The object of these features is to allow certain repetitive operations to be activated automatically, thus increasing operator efficiency. Operations available include automatic skipping over blank columns, automatic gangpunching (see below), and the shifting from numeric to alphabetic punching mode and vice versa. The program unit is usually set up from one job to another providing a completely flexible way of setting up the automatic punch for all the jobs required by a particular user. Sometimes these units permit operator discretion even within the framework of a particular program when different punching formats are needed in the same run.

Gangpunching

This operation entails the duplication of punching from fields on one card into corresponding fields on succeeding cards and is usually activated under control of the program unit. For example most card formats include key indicative information common to all cards of a particular group. In the order processing card shown previously the Customer Number and Customer Order Number fall in this category and would be gangpunched into all cards relating to the same order. In this case the punch operator would have to set up the master information when punching the first card of a group and then activate the program unit from the keyboard to allow gangpunching for succeeding cards of the same group. This process is some-times referred to as *semi-permanent* gangpunching and is contrasted with *permanent* gangpunching which implies data being duplicated in certain fields of all cards in a particular run, e.g. the date might be common to all cards punched on a certain day.

Speeds of card punching

Speeds of punching are obviously related to the particular card format being punched and the length of the run. The format dictates the number of key depressions to be made per card, and for automatic punches the set up times must be included within the overall time for a job. Obviously where an operator has the opportunity to settle into a long run using the same format the output rate can be higher than with many small jobs.

In addition one has to consider the speeds permitted by the card punch itself in relation to the automatic facilities available — some sample speeds are quoted below: —

feed and eject	20 columns per second
automatic gangpunching	20 columns per second
automatic skipping	20 columns per second
automatic spacing	10 columns per second

The terms *spacing* and *skipping* are used to signify the automatic movement of the carriage over columns where punching is not required. *Spacing* is used to describe the movement over, say, 4 or less columns, whereas *skipping* is used to describe automatic movement over more than 4 columns. The distinction is made because mechanical limitations impose slower speeds where only a few consecutive columns are to be ignored.

The overall speed of punching depends upon many of the factors already mentioned, e.g.:

> type of punch
> type of keyboard
> source document design
> card format
> nature of the data (alphabetic or numeric)

The speed is also affected by general operating conditions in the machine room and by the level of training and development of the operators.

Generally speaking an operator is capable of maintaining speeds between 10,000 and 15,000 key strokes per hour. Within any particular installation, standards are usually set for estimating the time required for jobs, and for measuring operator performance. If the systems analyst is working in an organisation where such standards are established he should familiarise himself with them and ensure that he adopts them in evaluating the time required for data preparation.

Verifying punched cards
When cards have been punched they must usually be checked using a device known as a *verifier,* to ensure that the punching operation has been completed accurately. Some verifiers are similar to the handpunch previously described, whereas others are the counterpart of the automatic card punch. The principle of a verifier is very simple: it is equipped with a reading unit which senses holes in the cards and passes the signals thus generated to a comparator unit to compare with entries made manually by the verifier operator on the keyboard of the device. The verifier operator works from the original source documents repeating the keystrokes made by the original punch operator. This is an independent check and the possibility of the second operator making the same mistake is remote.

187

If the comparator unit encounters a discrepancy the keyboard locks and an error signal lamp glows: the verifier operator then repeats the keystroke to see whether she herself has caused the error. If the original card is incorrect it is rejected and returned with that batch and the original document to be repunched.

All correctly verified cards are automatically notched on one edge as they are ejected into the stacker of the verifier, cards which fail verification are stacked separately or are offset to permit easy identification and correction.

3.7.2 Paper tape

Telegraph systems had been using five-track punched paper tape for some time before its potential as a medium for computer input and output was recognised. In telegraphy it had been seen that, while it was possible to transmit telegraph signals directly to a receiving device, it was often preferable to store the information on paper tape as an intermediate medium so that efficient use of the telegraph line could be allowed by the high-speed transmission between paper tape devices. Similarly, paper tape could be used as a means of storing data converted into a form readable by a computer, and it was seen that it was a compact medium, less bulky than punched cards, and not so subject to the restrictions imposed by a rigid card layout.

Characteristics of paper tape

Paper tape is supplied in rolls approximately 1,000 feet long and widths of either $\frac{11}{16}$in., $\frac{7}{8}$in. or 1in., depending, as we shall see, on the number of channels punched. In the same way that a character is represented in a punched card by the punching of certain holes in a column, so characters are represented in paper tape by a single row of holes across the width of the tape. There are 10 such rows to the inch, allowing ten characters to be represented in each inch of paper tape, and figure 66 shows two examples of punched paper tape with the decoded representation shown at the side of the tape. It will be seen that many of the codes in the five-channel tape are doing double duty, while the eight-channel tape, with its greater range, can afford a different code for each character in the set. Several different codes are in use, although much co-operation among various standards organisations has resulted in recommendations of standard codes. The five-channel version is now less common for data processing uses as there is not enough capacity for a full range of characters as well as the special symbols needed and error checking facilities, and it is now unusual to find the $\frac{11}{16}$in. width paper tape as a computer input medium. The 6-channel tape is $\frac{7}{8}$in. and the 7-channel and 8-channel are both 1in. wide.

In addition to the channels used for the representation of codes there is a further channel containing a small hole punched in each row. These holes are sprocket holes and, on slow readers, are used for driving the tape mechanically past the reading station; on fast readers which use a friction driving technique, their purpose is to provide a clocking pulse when the tape is read by a photoelectric head. In either case, the position of the sprocket hole channel can be used as an indication of the direction of the tape, although this is often supplemented by arrows printed on the tape.

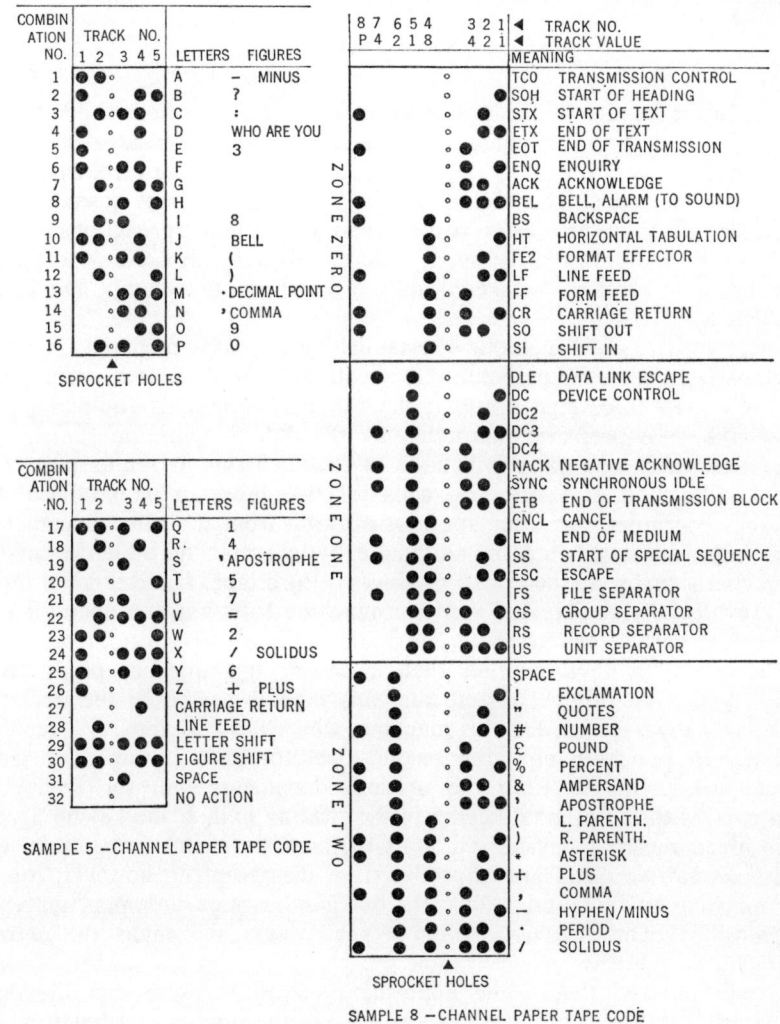

66. Paper tape codes (The complete codes are not shown)

Because there is little or no wasted space on a strip of paper tape and the medium is considerably less bulky than punched cards (which may well contain unused card columns) paper tape can be easily transported and is an excellent medium when large quantities of data have to be moved from outlying stations to a central area. For this reason it is often produced as a by-product of some other operation such as the use of a cash register or a Flexowriter so that considerable quantities of data can be generated and moved without severe transport problems. Once the tape reaches the computer, it can be read at speeds of around 1,000 characters a minute.

Error detection
A significant feature of the use of paper tape is that where a completed tape is found later to contain an error character the tape must be repunched—it is not, of course, possible to 'overwrite' the error character with the correct one. This means that an important systems consideration in the design of systems using paper tape input is that transactions should be batched in appropriately small batches, with batch control totals so that, when an error is discovered, only a small amount of tape has to be repunched.

There are two main methods of establishing that data prepared on paper tape has been correctly prepared. (This will still not, of course, ensure that the data itself is correct, nor that the original documents containing the data have been properly interpreted by the data preparation staff—later validation checks on entry will have to establish this as far as possible.) The first method of verification calls for two tapes to be produced by different operators, one after another, working from the same documents. These tapes are then compared automatically by a machine which compares them character by character and produces a third tape. If the original tapes are not precisely similar, the verifying machine halts and the operator can key in the correct character.

The second method requires that, after one operator has produced a paper tape from original documents, this tape is passed to the operator of a paper tape verifier. On this machine, which has a keyboard, a reading unit and a punching unit, the second operator goes through the same process of keying in from the original documents, but the keying is compared with the original keying in the reading unit. If the keying agrees with the characters punched in the original tape, the punch unit is activated and a second, verified, tape is produced. A disagreement, however, causes the machine to halt and allows the second operator an opportunity of examining the original documents to ensure that she enters the correct character on the new tape.

It will be seen that, while both these methods ensure that the data preparation staff will produce tapes which are the result of an independent checking operation, there is still the very real danger that two operators have both misinterpreted a character on the original document. A careful study of a number of documents in any data preparation room will convince a systems analyst of the need to design punching documents which give the maximum help to those who must punch from them. The greatest benefits will derive from forms which give the minimum scanning requirement (i.e. that data is picked up from the form in a regular flow); which have the best use of space so that the form contains the maximum amount of data without being so condensed that characters are squashed and therefore unrecognisable; and which are designed with the data preparation staff in mind as well as the requirements of the data.

Further systems considerations
An indication has already been given that batches should be of a suitable size to allow re-punching when an error is established. This should be balanced against the overall consideration that batch sizes should be of size suitable for one operator to punch in a reasonable period. This means a careful consideration of the difficulty in interpreting the document and whether or not it can be classed as sufficiently interesting to maintain the operator's interest. A bad estimate of a suitable batch size can result in a number of errors towards the end of the batch, invalidating much good work in the design of the punching document itself.

It should be remembered that, while it is fairly easy for a skilled punched card operator to read the data contained in a punched card, this is extremely difficult with paper tape and, if a piece of tape needs interpreting then a tape print will be called for. This means that, as a regular part of the system each tape should be marked with the contents, and an indexing system should be instituted so that any piece of tape can be quickly found. Such a procedure will probably be already in being in a data preparation room, but if the systems analyst is designing a system which calls for the acceptance of already prepared paper tape from outlying units, then a rigid control on the method of classifying and marking tape will be necessary. For example, pins and paper clips must never be used on paper tape, and any large reels of prepared tape should be kept in the boxes supplied by manufacturers (small reels can be secured by an elastic band around the circumference of the reel). Storage conditions of paper tape are important, as the machines to read them are set up with fine limits in order to avoid the possibility of misreadings and tape wrecks. It is also worth making a rule that, even when only a few characters are being prepared for onward transmission, the operator punches leader and trailer characters so that the piece of tape is at least three feet long.

3.7.3 Mark sensing

General
In preparing punched cards and paper tape, it is necessary to transcribe data from original documents to another medium. One method of allowing original data to be translated into machine language without such a

191

potentially error-prone transcription process, is to use mark sensing devices. A typical mark sensing card will look like an ordinary punched card except that its pre-printed format will contain a number of windows or boxes 2 or 3 columns wide and one or two rows deep. It will be possible for these boxes to be marked in a pre-arranged code so that a number of straight lines in different columns can be used to represent, for example, an account number and a quantity.

Method of operation
Cards marked in this way can now be fed direct to a mark sensing reader and the graphite content of the original mark will be used to complete a circuit causing appropriate holes to be punched in the same card. As a result of this action, a pack of punched cards will be produced and a mechanical transcription process will have taken the place of the manual one called for in an ordinary punched card or paper tape data preparation process.

Characteristics of mark sensing systems
It will be seen that the capacity of a conventional 80-column punched card is reduced by three if each window is to take three columns, and mark-sense cards are therefore suitable only for those applications requiring a limited number of columns-worth of data. A further restriction is that if the cards are to be completed at the source of data, then this source should be in a clerical environment rather than one where grease, oil or dirt might make the card impossible to read.

3.7.4 Keyboard to magnetic tape equipment

General
Magnetic keyboard equipment enables data to be recorded directly onto magnetic tape by means of a keyboard operation. The magnetic tape produced may be read directly by computer tape units.

Description of unit
A magnetic keyboard recording device consists of a keyboard and a spool on which the recorded tape is wound. The keyboard resembles an ordinary typewriter keyboard, together with a special numerical keyboard, similar to the keyboard attached to automatic card punches and paper tape punches. In addition the operator has other controls enabling various feature such as automatic skipping, block size control, and field justification to be controlled. In effect, the keyboard operation creates blocks on tape which consist of one character for each key depression. The total number of characters in a block may be fixed or variable depending on the mode of operation selected, and there is usually an upper limit of 80 or 160 characters to a block. Verification can also be performed on the recorder in the same way as punched cards are verified, that is to say by reading

an already prepared tape and comparing the block on tape with the same block as keyed by a separate operation. In addition two or three recorders may combine data from several tapes onto a single tape.

Error checking
Data recorded on tape is automatically check-read after the writing operation, and a bit-by-bit comparison of the character is made with the data previously keyed in. Any discrepancy is signalled, and the faulty record may be re-entered.

Systems advantages
The main advantage of a magnetic tape recorder as an input medium is that it eliminates an intermediate stage of off-line storage such as cards or paper tape. Data on the magnetic tape may be read immediately by computer, and normal validation procedure performed on it. Cards and paper tape may be damaged or misread by card and tape readers: such problems are eliminated by keying data directly to magnetic tape, and deterioration through storage is also avoided.

3.7.5 Source documents direct to computer

Under character recognition equipment (section 3.2.10) we have discussed the equipment available for handling original documents, and recognising the information contained on them by use of MICR or OCR. It will be readily appreciated that the main significance of the use of such equipment is that in a more general and more sophisticated way than by mark sense equipment, the transcription process in data preparation can be avoided. Currently there are limitations to the varieties of characters which can be recognised and the varieties of document which can be handled, and the speed of advance in this field has been disappointing to those used to significant leaps forward each month in the technology of data processing. This has tended to mean a holding back in the use of document handling and character recognition equipment, but there is no doubt that this equipment will eventually allow considerable changes in data collection and data preparation habits, and systems analysts should make themselves particularly aware of additions to manufacturers' ranges in this field. All changes which lead to an accurate input to a computer will improve computer systems and, if a user is aware that the characters he writes or the marks he makes will become direct input to the computer, he may cease to expect the double standard which requires a machine not only to detect what is wrong but to put it right on entry.

G

3.8 PUNCHED CARD EQUIPMENT

A description of punched cards and their methods of preparation has been given previously in section 3.7.1; the present section deals with the equipment used in processing cards to create files of information and to generate listings and tabulations from card files.

This subject provides useful background information to the systems analyst, since in many cases he is asked to investigate existing punched card applications in developing a new computer system. In many instances an analyst may find it necessary to utilise the resources of a punched card machine room in creating computer files from existing card files, or in supporting a computer based system during parallel running.

3.8.1 Punched card sorting machines

Why sorters are used
The end result of a punched card system may be the preparation of a list of information from all cards in a particular file, perhaps with totals at intermediate stages. On the other hand a printed tabulation giving totals from only predetermined groups of cards may be needed.

To illustrate these cases consider a sales analysis application where individual cards are punched from orders submitted by salesman; each card representing one line item, and containing the following fields of information:—

Field	Number of Digits	Column
Customer number	5	1– 5
Order number	5	6–10
Sales area code (assigned codes 08–15)	2	11–12
Salesman number	6	13–18
Product code	3	19–21
Quantity	4	22–25

The cards initially are punched from batches of orders arriving daily in the head office from the various sales office, and after being punched and verified are not in any particular sequence. Assume further that the following characteristics apply:

80	Salesmen
8	Sales areas
50	Product lines
16,000	Customers (2,000 per sales area)
8,000	Line items ordered per day

Each week some 40,000 cards have to be sorted to provide various analyses of information contained in the file:

1. A sort of product code would provide an analysis of demands for various products. Perhaps within this sequence cards could be further sorted by sales area code to show the popularity of different items in the various sales areas.
2. A further sort to re-arrange the cards by customer number might be needed to show buying trends of various customers.
3. A further re-arrangement of the file could be carried out to show sales by salesman within sales areas.

These sorting operations would be an awesome task if conducted manually, but punched card sorters are designed to carry out this work with considerable speed.

Sorting numeric data
A sorter consists basically of a card track that transports cards at high speed over 13 pockets, which correspond to the 12 index positions of a card column plus a further pocket for receiving cards that are blank in the column being examined. A diagram is shown in figure 67.

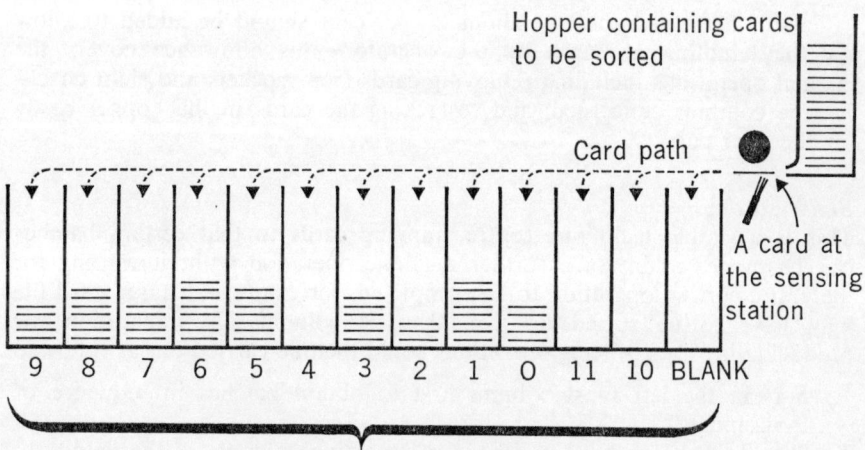

67. Punched card sorter operation

To sort on a field containing numeric data the cards are examined one by one in each column of the field concerned. Sorting consists of separate passes, column by column, from right to left across the field, the cards being directed to appropriate pockets according to the punching detected on each occasion in the particular column. A simple device is used to detect the presence of holes in particular index positions; it consists of a carbon brush which makes electrical contact with a contact roll whenever a hole is present. This action activates a flap above the relevant pocket to ensure that the card is stacked in the correct position.

Numeric sorting requires cards to be sorted into pockets 0 to 9 only; punchings in other positions would generally be considered as errors in any column of a numeric field, although a blank column may be acceptable in certain cases, and positions 10 and 11 are used for tenpence and elevenpence in some card files.

To sort by product code in the previous example would necessitate three passes of the complete card file though the sorter: on the first pass the cards are arranged by sequence of digits in column 21, then a second pass arranging by column 20, and finally a pass to arrange by column 19.

To sequence further by sales area code would require additional passes to examine column 12 and then column 11, a total of five passes in all.

The speeds of sorters vary from about 600 to 1,000 cards per minute, thus to sort 40,000 cards on five columns as suggested in the previous paragraph would require:

$$\frac{40,000 \times 5}{1,000} = 200 \text{ minutes}$$

To this figure an uplift of about 20 per cent should be added to allow for the handling of cards by the operator—this allowance covers the manual operations including removing cards from pockets and sight checking the columns concerned, and restacking the cards in the hopper ready for the next pass.

Block sorting

This is a simple technique for rearranging cards to that certain batches can be processed in some further machine operation without waiting for the entire sorting operation to be completed, for example, a large card file may have to be sorted on a field of 4 columns in preparation for producing invoices, sorting operations could then be carried out as follows:

1. Sort in the left most column first to obtain batches in sequence of thousands (0, 1, 2, 3, etc.).
2. Sort the 0 batch first on all three remaining columns of the field.
3. Then sort the 1 batch, the 2 batch and so on.

In this way the sorting operation takes the same length of time but some batches can be ready before others.

Alphabetic sorting

Sorting fields containing alphabetic data requires two passes on each column; cards are first sorted as for a numeric field and then a further pass is needed to sort on the zone or upper curtate positions. A switch is

196

provided on the sorter so that the mechanism can be set to distinguish only the top three positions of each column 0, 11 and 10.

The principles of block sorting can be applied to alphabetic fields, by alphabetic sorting the file initially into separate groups according to the first (or leftmost) column of the field.

3.8.2 Reproducing and gangpunching

Reproducers are machines which can copy fields from a card file into the same or different columns of blank cards in order to create a new file having data in a different format.

A reproducer as shown in figure 68, has two card feeds each having an input hopper and output stacker, one feed is known as the read unit and the other as the punch unit. The cards containing the data to be reproduced are placed in the read unit, and are fed in synchrony with blank cards through the punch unit. Selected fields are plugged across from the reading brushes in the read unit to desired columns of the cards at the punching station in the punch unit. The punch dies are thus activated to copy fields into the blank cards.

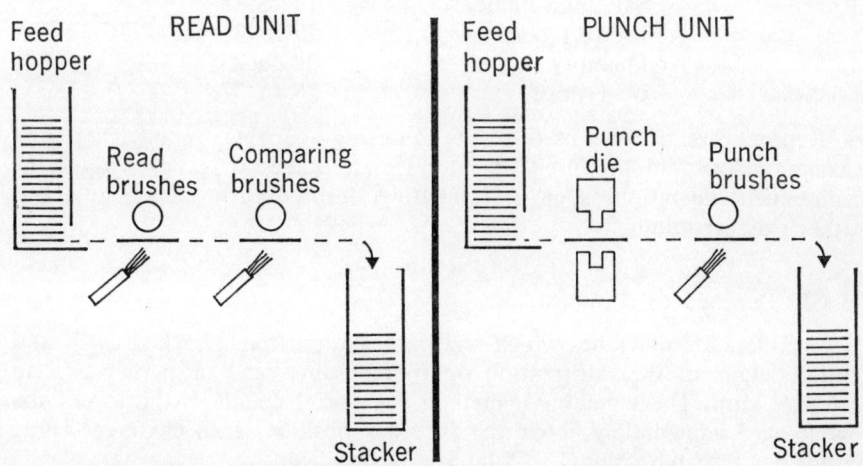

68. Punched card reproducer operation

After each new card is punched it passes to the punch brushes which are used to read data so that it can be compared with the corresponding fields in the original cards. If an exact match is not made the machine stops and an error condition is indicated.

A reproducer can also be used to gangpunch selected information from master cards into any number of related detail cards; this function can be achieved as a separate operation using the punch unit only. The cards to be gangpunched are placed in the feed hopper, the file concerned having been already arranged so that each batch of cards is preceded by a master card bearing the data to be gangpunched. The master cards bear a particular punching (usually in the 10 position of a predetermined column) which inhibits the punch drives when the master cards are at the punch station. However, when a master card arrives at the punch brushes data is read from it and gangpunched into the following detail card. Gangpunching of selected data then continues from one card to the next until a new master card is detected.

Returning to the sales analysis example introduced in 3.8.1, we can see that gangpunching could be used to get prices into the individual transaction cards. The transaction file would first have to be sorted by product code, and then be merged with a price file containing a separate master price card for each product type. Assuming that prices require 7 column positions a new detail card could be generated as follows:

Field	Columns
Customer number	1 to 5
Order number	6 to 10
Sales area	11 to 12
Salesman number	13 to 18
Product code	19 to 21
Quantity	22 to 25
Unit price	26 to 32

Reproducers can be used for reproducing (copying), comparing, and gangpunching and are used at intermediate stages to prepare data for subsequent operations such as tabulating. Reproducers operate at about 100 cards per minute.

3.8.3 Punched card interpreters

Interpreters are machines which read information from punched cards and print details of the information on to the same card or perhaps on to another card. The commonest method is to print details at the top of the same card immediately above the 10 index position; each character being printed above the column in which it is punched.

Interpreters are very useful where cards have to circulate in some clerical routine, and provide, in some situations, a transaction document in a machine readable form. The operating speeds of interpreters vary according to their method of operation but are generally between 60 and 100 cards per minute.

3.8.4 Electronic calculators

Electronic calculators were in many respects the forerunners of modern commercial computers. The earliest machines consisted principally of an

198

electronic unit containing a few storage registers, arithmetic and logical circuits. Input and output was via a punched card input/output device based on the concepts of the reproducer and is shown in figure 69. Cards containing punched data are fed through three stations in this unit. The first of these is a reading station which reads selected fields from each card into the storage registers of the electronic unit. While the card is travelling to the second station (punching station) the relevant data is processed in the electronic unit and results are assembled for punching into the same card on its arrival at the punching station. Finally each card is read at the third station (checking station) to compare the results actually punched in the card with results held over from the processing stage. Error cards may be diverted to a special stacker or be offset in the pack of stacked cards.

Input and output speeds in all these systems are limited by the card punch to about 100 cards per minute.

These machines are used principally for carrying out arithmetic functions not easily conducted with other punched card devices, i.e. multiplication and division. Each job can have a number of program steps, varying for different models between say 20 or 100 steps, which are used to manipulate data in the storage registers. Returning as a very simple example to the sales analysis cards previously described: the calculator could be used to extend the price by quantity to generate a new field in the detail cards as follows:

Field	Columns
Customer number	1 to 5
Order number	6 to 10
Sales area	11 to 12
Salesman number	13 to 18
Product code	19 to 21
Quantity	22 to 25
Unit price	26 to 32
Extended price	33 to 40

3.8.5 Punched card collator

Purpose of collators

Collators are used mainly for creating a single file by bringing two card files together prior to some further processing operation, or after having previously split a file to process some subset of the cards within it.

In straightforward situations this could be done by sorting as previously described, but a collator may be able to perform the operation more quickly since it looks at all columns in a particular field simultaneously. As well as this, a collator is able to identify situations where one of the two files that it is operating upon is incomplete because it does not contain cards corresponding to keys found in the other file.

Let us develop this further by considering the sales analysis file described in previous sections. Each day it may be necessary to produce despatch documents from the individual cards in this file; thus these transactions have to be sorted to customer number to be matched with name and address cards in a separate file also in customer number sequence

A collator can perform this merging operation by reading the two files and merging them into a new file in which name and address cards are placed in front of related transaction cards. At the same time the collator will outsort all transactions that do not have corresponding address cards in the name and address file. These transactions appear in a separate output stacker so that action may be taken. The name and address cards not needed in this particular run will also be stacked separately.

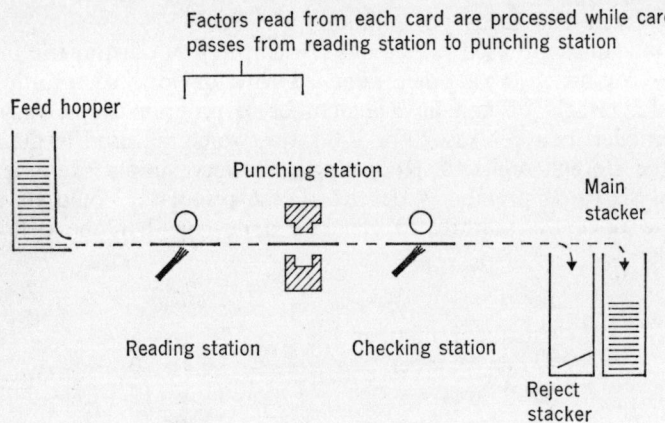

69. Input and output for an electronic calculator

Thus in this case, from two input files the collator will create three output files and at the same time will check the sequence of cards in the input files, and will stop if a sequence error is detected in either file.

Later, after the despatch documents have been produced, it will be necessary to select the name and address cards and merge them once more into the original name and address file whilst separating the transaction cards for use in subsequent analyses—this operation can be conducted using the collator once more. An outline of a merging operation using a collator is shown in figure 70.

3.8.6 Punched card tabulators and summary punches

Purpose of tabulators
Punched card tabulators or punched card accounting machines are devices which produce the final output from a punched card system. They are

200

able to read punched cards and produce printed reports and documents which are derived from information read from the cards. Such reports can consist of separate lines of information derived from individual cards with totals at intermediate stages corresponding to sub-groups of cards within the particular card file.

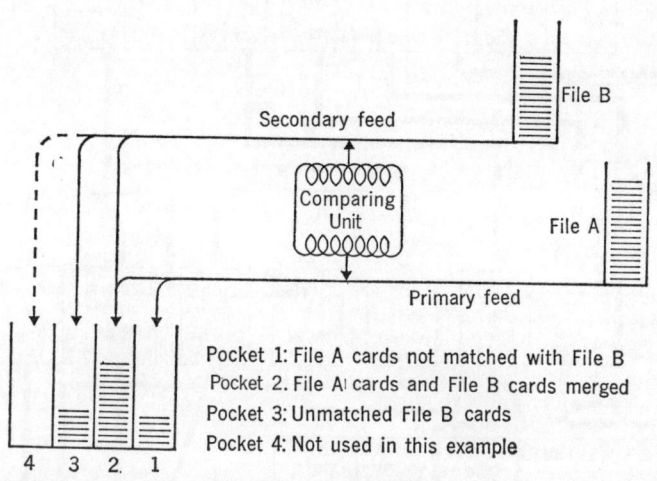

Pocket 1: File A cards not matched with File B
Pocket 2. File A cards and File B cards merged
Pocket 3: Unmatched File B cards
Pocket 4: Not used in this example

70. Merging operation using a collator

Reports providing a line of print for every card are usually referred to as *lists,* whereas reports only providing several levels of totals derived by summarising details from individual cards are known as *totals only* tabulations. Continuing with our sales analysis example, we assume that the cards have been previously sorted into the sequence of product code within salesman no. within area no. A report is needed to show the total value of sales achieved by each sales area and each salesman, giving a breakdown of individual salesman by product code.

Counters are used to accumulate details from the cards, and three levels of totalling are needed: these can be effected by use of a comparing control feature provided on the tabulator. This device is capable of interrupting card feeding whenever a change is detected in fields that have been nominated as control data. In this case control fields are allocated as follows:

Field	Columns	Control Level
Sales area	11 to 12	Major
Salesman number	13 to 18	Intermediate
Product code	19 to 21	Minor

Thus, for example, if the comparing control feature detects that two

201

successive cards have different data in columns 11 to 12, the tabulator performs the activities programmed to occur at a major control change. Figures 71 and 72 illustrate this operation.

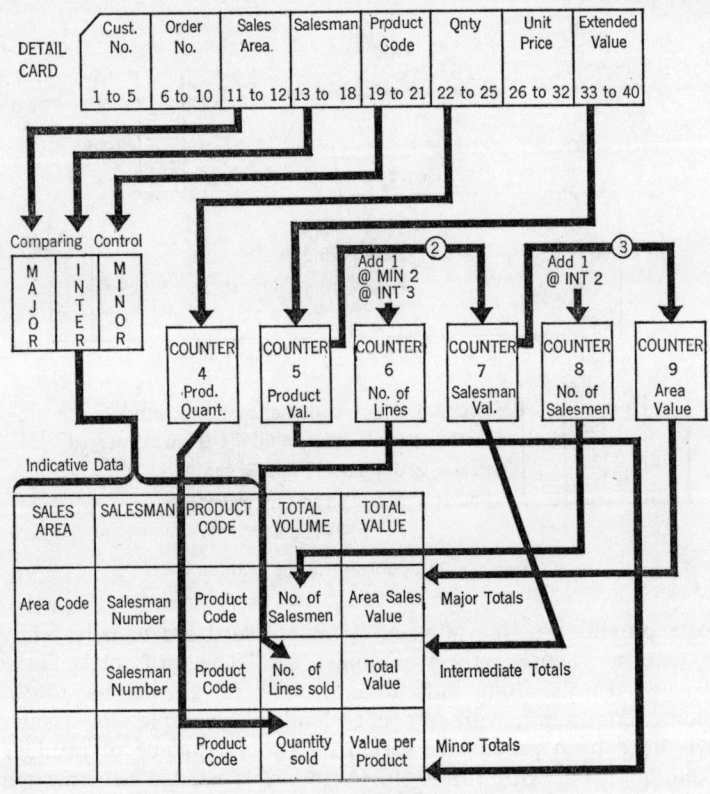

Notes

① Indicative data is listed from cards as follows: (a) First card of each Major group Area No. Salesman No. Product No.. (b) First card of Inter group Salesman No. Product No. (c) First card of Minor group Product No. ② At Minor control change contents of C5 are added to C7. Totals from C4 and C5 are printed ③At Intermediate control C7 is added to C9 Totals from C6 and C7 are printed ④At Major control totals from C8 and C9 are printed⑤ See Figure 72 for further basic functions at control stages

71. Tabulator operation

Speed of operation

Using continuous stationery a tabulator can produce reports or documents at listing speeds of about 100 lines per minute. For totals only applications the card feeding rate can be as high as 200 cards per minute. It is also possible to connect a reproducer to a tabulator so that summary cards can be produced at control stages. A summary card punch can operate at about 100 cards per minute.

CONTROL STAGE	TOTAL PRINTING	ADDING	ZEROISE	LINE SPACING
Minor 2	From C4 C5	1 to C6 C5 to C7		
Minor 1			C4 C5	✓
Inter 3	From C4 C5	1 to C6 C5 to C7		
Inter 2	From C6 C7	1 to C8 C7 to C9		✓
Inter 1			C4, C5 C6, C7	✓
Major 4	From C4 C5	1 to C6 C5 to C7		
Major 3	From C6 C7	1 to C8 C7 to C9		✓
Major 2	From C8 C9			✓
Major 1			C4, C5, C6 C7, C8, C9	✓

72. Control functions (as required in fig. 71)

PART 4
Design of Computer Procedures

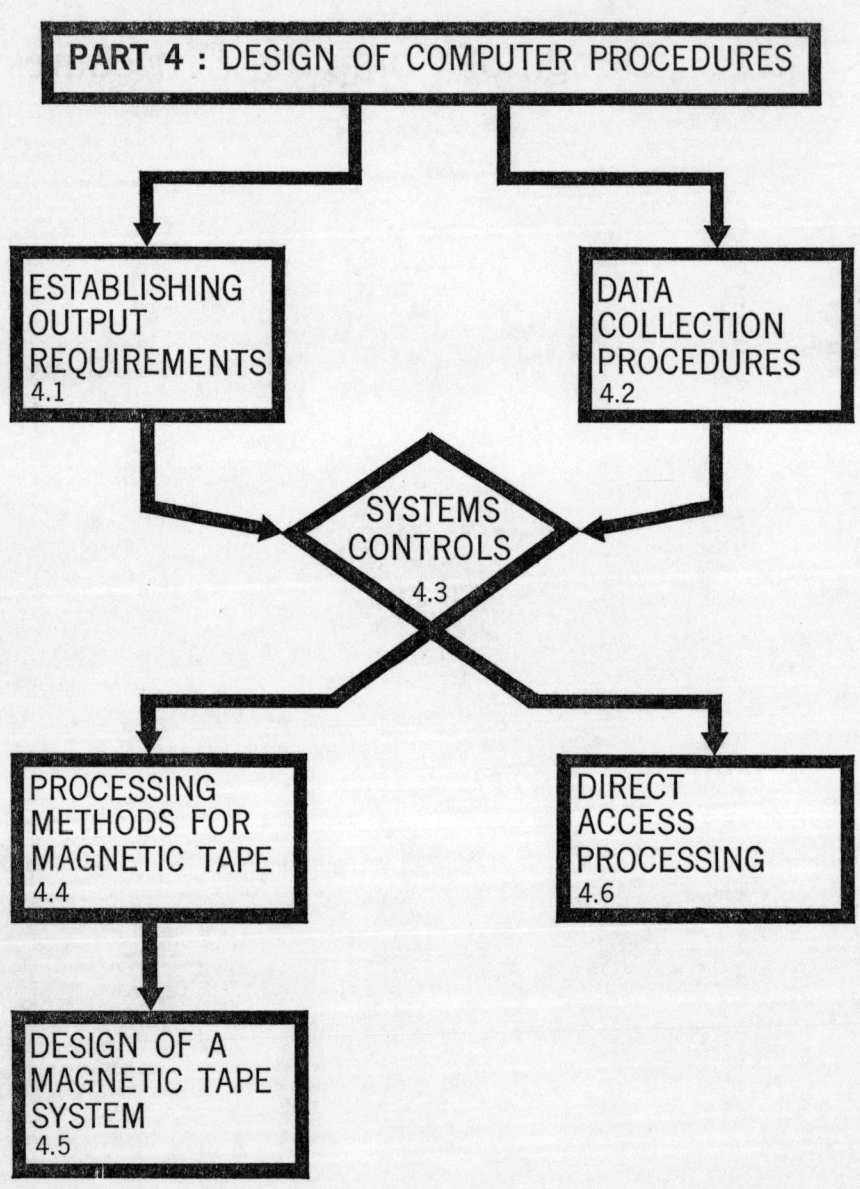

PART 4 : DESIGN OF COMPUTER PROCEDURES

ESTABLISHING
OUTPUT
REQUIREMENTS
4.1

DATA
COLLECTION
PROCEDURES
4.2

SYSTEMS
CONTROLS
4.3

PROCESSING
METHODS FOR
MAGNETIC TAPE
4.4

DIRECT
ACCESS
PROCESSING
4.6

DESIGN OF A
MAGNETIC TAPE
SYSTEM
4.5

INTRODUCTION

There is never a precise moment when a systems analyst can be said to start designing a new system. Even at the stage of receiving an assignment brief, which he himself may have helped to write, he may well have started to formulate ideas about how the new system might operate. If the project is one that calls for a lengthy initial investigation, it is probable that some of those ideas will be committed to paper in working files long before the investigation is completed. As the investigation is brought to a close, the systems analyst will gradually start the design phase in earnest.

In Part 4 of this book the design phase is considered as a separate stage in which certain basic tasks are conducted, including:

1. Specification of output requirements.
2. Identification of data elements required in the system.
3. Specification of the data collection requirements and design of data collection procedures.
4. Design of the file structure and record formats.
5. Design and specification of the controls needed to check the accuracy of the system when it is in routine operation.

This part opens with some general guidance on how a systems analyst should approach the design phase; it is concluded with a number of sections which describe some of the main principles to be observed in establishing a file structure for magnetic tape or disc based data files.

4.1 ESTABLISHING THE OUTPUT REQUIREMENTS

4.1.1 The importance of a clear objective

It is essential that the output requirements of a system are firmly established. A system has little value in its own right; it is the information it produces that justifies its existence. The output has to be designed to give the people using it control over the functions for which they are responsible, and it is necessary to consider the content, volume, format and timing of the information to be supplied.

The output formats required from a system are sometimes specified at a very early stage in the systems investigation. Sometimes line managers express their initial system requirements by setting down outline printer formats that they expect to receive from the system. Any requirement of this nature must always be thoroughly reviewed at the beginning of the design phase.

As the project continues, all those concerned with its development will probably improve their understanding of the original problem, and they will be aiming a little higher than originally. Where design sessions are conducted as project teams involving the users and systems staff, a much wider grasp of the nature of the system and the associated organisation structures may emerge.

Timing of data supply
The individual data elements in output reports are the means by which information is conveyed to users. By establishing what these elements are the systems analyst is beginning to specify the data that must be held or generated within the system. This is straightforward, but information is only of value to the user if it is supplied at the correct time to help the user take necessary action to achieve his particular goals. Take as an example a stock controller in a warehouse—it is little use a system telling him that demands cannot be satisfied because certain stock items have not been ordered. It is true that the system has supplied information, but in this case the information has been delivered too late resulting in loss of customer satisfaction and potential loss of sales. The system should indicate items that are running low so that orders can be placed to replenish stock before it is used. This implies that the system needs to forecast future demands, needs to know whether any orders have been placed, needs to record the current stock levels, and know how long it takes to procure more stock of particular items.

Thus by a study of the problem of a particular user, it is possible to identify which data he needs and to know when he needs it. In fact we specify the response criteria of the system and in so doing establish the information that must be collected, when it must be collected, and how it is to be processed.

Volume and sequence of data
It is important to keep the user's objectives in view the whole time; one aspect of this concerns the volume and sequence of data required by him. Consider once more our stock controller. When he is engaged in expediting existing orders on suppliers, it is not necessary to list all the orders that have been placed; first of all he will want to see only those that are overdue, and perhaps only a subset of these (i.e. those where stock is at or below a determined safety level) and therefore the aim is to cut down the volume of records that the user needs to examine.

Furthermore, the stock controller will need to contact or write to the supplier, and will need to see all overdue transactions arranged in supplier sequence. Taking this a step further the systems analyst may consider designing the system so that it produces documents which can be despatched directly to urge the suppliers concerned, and which also highlights critical items so that the stock controller can telephone the supplier to apply a more personal pressure.

All these principles are common sense, particularly when put into the context of a simple everyday function, but it is often quite difficult to appreciate the problems of a user and to express them in terms of such selective criteria. One important way of developing an appreciation of the problems facing a user department is for the analyst to go into that department and work as a member of it during the investigation stage. This allows an accurate observation of day to day problems which may otherwise not have been fully noticed by the analyst.

Approval of output design
As the output formats are finalised they should be prepared as samples on layout sheets with supporting narrative wherever necessary. The users should be asked to check these thoroughly and should be questioned fairly searchingly by the analyst to ascertain whether they understand and accept documents. It is also a good practice to ask users to sign acceptance sheets for each format.

Finally, it is always necessary to check the formats against the aims set down in the original assignment brief. If there has been a departure from the brief then it must be agreed with those responsible for authorising the systems project. Otherwise there may be a danger of unwittingly extending the bounds of a project, or of developing it with a bias towards particular managers who have used their personality to influence the analyst or the project team.

4.1.2 Definition of output formats

As indicated in the previous section it is important to verify the user's output requirements in detail, before designing the final file specifications and programs. An apparently minor change to an output format might have a considerable affect on a report program or a print file. There should be a formal approach to defining and documenting output requirements so that users can become familiar with the methods of the systems departments, and so that systems staff are aware of all factors affecting a particular report (see also section 2.1.6—project teams).

Print definitions should include a *sample format* drawn up on a print layout chart as shown in figure 74 and this should be accompanied by *narrative* explaining any conditions that cannot be represented on the chart—e.g. notes on the accuracy of particular fields of the chart.

Dummy reports
Charts of this nature are very useful for indicating to a user the layout and general appearance of the output he will be receiving. However, it is not unusual to find that when the report is eventually generated on a line printer that the user reconsiders the format and asks for changes. For example, he may ask for line spacing to be changed so that particular fields or totals stand out. One way of avoiding such changes of mind is to create a dummy report using a general purpose program to generate printer formats from data supplied to create any particular report sample. A program for this purpose should be written to accept heading lines, detail

209

PRINT LAYOUT CHART

PROGRAM/SYSTEM *PAYROLL*

LAYOUT NAME *VARIABLE DEDUCTIONS A TABULATION*

Design by V. WELLS

PERIOD **

PAY CODE

PERSONAL NUMBER

INS. CODE

NO. OF STMPS

SEL. EMP. TAX

STAFF PAYROLL

VARIABLE DEDUCTIONS A

NAT. INS. EMPLOYER

NAT. INS. EMPLOYEE

PERSONAL ACCOUNT

PERIOD

OT DEDU

X X X   XXXXXX   XXX   XX   XXXX   XXXX   XXXX   XXXX   XXXXXX   XXXX

          XX   XX   XX   XX   XX   XX XXX   XXXX

NOTE

PRINT POSITION 26=R OR ▽

PRINT POSITION 86=C OR ▽

74. Output layout chart

210

lines and totals prepared as parameters on paper tape or punched cards. Once a general purpose program has been written for this task it could be used over and over again to create actual printer formats for obtaining approval of users.

Print schedules
Both the print layout charts and dummy printer formats are useful in gaining user approval, and in generally specifying the content of a report, but they are not so useful for the systems analyst or programmer who wishes at a later stage to write a print program. For this purpose it is recommended that a simple schedule should be produced to accompany output formats as shown in figure 75. This can be developed as print files are specified and is then incorporated in the analyst's working files to emerge as part of the final systems specification.

Field limitations
To complete the task of output definition the systems analyst has to consider the maximum size of all fields and he therefore has to calculate the maximum values that may arise for particular data elements. Also he must consider whether values may be positive or negative and allow for the printing of sign indicators.

It is worth spending time to make sure that all field signs are accurately assessed, particularly quantity or value fields. It is not a good idea to rely purely on the advice of user departments in this matter, because people unfamiliar with the stringency required will generally talk in terms of average values and be unaware that the system has to take care of all exceptional conditions that may arise.

The analyst can usually find sources of information to assist in making these assessments; the number of staff on the payroll, the turnover of an organisation, the budget of an operating division, the number of products marketed, the maximum hours an employee can work including all over-time, etc.—factors such as these can be ascertained from company records. Remember that every system must allow for an expansion of the activities that it supports.

4.1.3 Approval of output formats

Approval of output has to take place at two main levels within user organisations:

1. the divisional management or policy-setting level
2. the line management or operating level.

It is important to ensure that approval is secured at both these levels before attempting detailed design work for files or programs.

In practice it is usual to expect the approval at policy-making level to be made during an initial investigation or during feasibility studies (section

HEADING LINE

Data Element	Size in Characters	Printer Format	Print Positions	File Address	File Format	Sign	Remarks
Salesman No.	6	N	6 to 11	1.0 to 2.1	Character		
County	12	A/N	26 to 37	2.2 to 2.3	Generated from 2nd Location Code		
Week No.	2	N	44 to 46	3.0 to 3.1	Character		
Date	8	N	50 to 57		} Generated by Print		
Page No.	3	N	74 to 76		} Program		

N = NUMERIC
A/N = ALPHANUMERIC

DETAIL LINES

Data Element	Size in Characters	Printer Format	Print Positions	File Address	File Format	Sign	Remarks
Product Code	7	A/N	8 to 14	4.0 to 5.2	Character		
Volume This Week	4	N	19 to 22	6.0	1 word binary		
Volume Year to Date	6	N	27 to 32	7.0	1 word binary		
Volume Target Year to Date	6	N	37 to 42	8.0	1 word binary		
Target % Achieved	3	N	47 to 49		Generated by Print Program		
Value This Week	5	N					
Volume Year to Date	7						

75. Output format: data elements

2.1.1). The aim at this stage is not to specify every factor affecting the individual data elements, but rather to identify the data elements needed, and their relevance in relation to the overall aim of the assignment.

The detailed specification of output formats usually takes place in project teams, involving the operating management and the systems staff. Usually a systems analyst will present layout proposals in print layout charts or as dummy print formats for consideration by the users (see 4.1.2). After critical examination by those needing to use the formats, the final versions are drawn up and can be formally approved by signatures of the systems analyst, the systems manager, and the line management.

Approval at this initial stage may be obtained at formal meetings in which the systems analyst presents the result of the project team's investigations to show how the aims of the assignment can be accomplished. It is better to conduct these presentations finally in lengthy sessions at which there is opportunity for systems staff, supervisory management, and the policy-making management to cross-question one another. The systems analyst must be satisfied that there is a clear understanding of the suitability of these general output proposals at the beginning. If the development of the system is spread over a long period then it is important to review these requirements, particularly when new managers are brought into the organisation, so that the aims coincide with the current plans of the management.

Supporting data
Approval should be obtained not solely on the basis of the format but should also include consideration of the accuracy of the data and the volume of the output. Most output reports are designed to draw attention to conditions warranting action by management or clerical staff; in order that an affective response can be made the users will need to know how many such decisions they will be expected to make during a day, or week, etc. The frequency with which the system produces results is also important; it is often better to have a short list every day than a long list relating to events that occurred, say, in the preceding month.

Appearance of output
The appearance of printed output is an important factor in the efficiency of its eventual use. Attention given to small details such as format, line spacing and general presentation can create confidence by users and gain enthusiasm for future developments.

Multiple copies
Where reports are of a general statistical nature and have a wide circulation the analyst may do well to consider methods of reproducing the required number of copies. Stencils or hectographic masters are relatively cheap and effective to use for smaller quantities (say 20 to 50) and can be produced on line printers. Probably the best method is to produce a photo-

graphic master by simply printing with a clear black type onto white stationery; this can then be used as a master for lithographic or xerographic work; photoreduction can then be employed to create reports of a handy size.

4.1.4 Choice of output medium

Many output requirements can be seen as simple lists or tabulations produced on a line printer each day, week or month to provide specific information. They are produced on continuous stationery, perhaps pre-printed, and possibly in several copies. However, there are many situations where output documents can be employed to trigger off events which give rise to transactions that must re-enter the system.

Turn-around documents

Some examples occur where OCR, MICR, or mark coded documents are used (section 3.2.10). Computer printers can print marks on continuous stationery documents, and can sometimes be modified to print characters to the type specifications of some optical reading devices. Where this is the case the computer can generate turn-around documents bearing identifying keys which are issued to circulate as part of some external procedure, as shown in figure 76. They are marked up with entries incurred in the external procedure, and are returned to be read by an OCR reader directly into the processor once more.

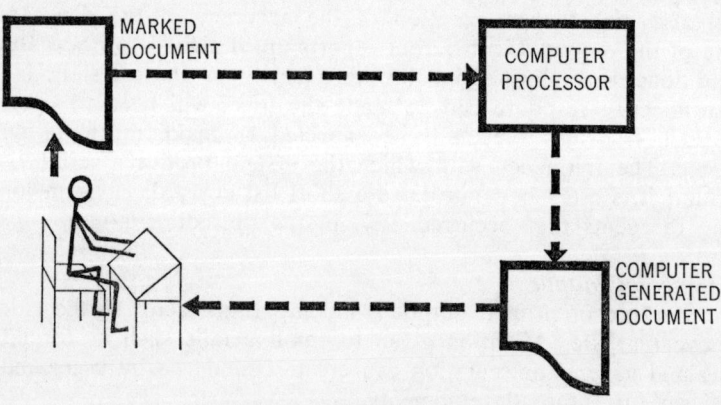

76. Turn-around documents

Turn-around documents can be employed to avoid delays, costs, and errors that would be incurred in conventional data preparation methods. The turn-round principle is applicable to situations where a large volume of fairly standard transactions arise, e.g. in gas and electricity billing, census applications, and in retail stock ordering situations.

Magnetic tape output
Sometimes a system may be used to generate its principle output in a form that can be accepted by another system, in which case one has to consider the most suitable output medium. Punched cards or paper tape may be useful if the volumes are relatively small, but for higher volumes a magnetic tape file may be more appopriate. The choice is usually governed by a consideration of the extent to which a particular input or output peripheral may effect the efficiency of a computer system; some guidance on this aspect is given later in sections 4.4 to 4.7.

On-line output
Sometimes the demands of a particular problem may involve giving the user the ability to obtain answers to situations at very short notice; and here the analyst has to consider the use of some on-line peripherals to enable the user to interrogate files and immediately receive information displayed in character form for his consideration. Visual display units (section 3.2.8) or teleprinters can be used to present information in this way and they can also be used as a means of entering data directly into files held on-line in direct access storage devices.

Visual display units can also be used to display graphs or drawings of data stored in memory; some models incorporate light pens to enable modifications to be made to pictorial data and instructions can then be given to store the data so that the drawing can be subsequently displayed in its amended form. The systems analyst has to keep abreast of developments in the creation and utilisation of output media, and must always seek to give the user facilities that meet his needs effectively and at an appropriate cost (see section 7.2). Probably the most important single factor is to meet the response requirements of the user to make sure that he has the opportunity to take effective action at the right time.

4.1.5 Principals of exception reporting

We have already stressed the importance of producing output reports which satisfy the functional needs of users, and that the analyst must guard against the tendency to produce long tabulations which people do not have time to read. In general it is best to consider a system not solely as a means of recording facts but as a tool to enable management to set standards of operation, to maintain performance against such standards, and to report on conditions which fall outside the nominated standards so that action can be taken. This implies that the system should be flexible to enable line management to set criteria which govern the nature and volume of the output according to the current conditions prevailing in their particular area of responsibility.

For example, say that a maintenance manager has to control installations over a wide area. His concern, we will assume, is to ensure that each installation provides a minimum standard of performance in terms of operational hours in a given period, and that this standard should be achieved within specified cost factors for direct materials, and man hours in different categories. Initially we might assume that this system has been established not only to monitor the performance of installations but to find out facts about the costs incurred in the maintenance function so that standards can be reviewed to assist in establishing better budgets for the whole operation. In the first six months of operation it may be that the management are concerned to see details relating only to those installations that consistently fall below standards of performance, or that incur exceptionally high maintenance costs. In this situation an exception report will be generated to show if the machines at a particular installation required excessive maintenance. (There would be no point in highlighting a bad situation one week if it was due to some abnormal condition.) If there was widespread evidence of installations falling below standard then the criteria for reporting might initially be set to ensure that only the worst cases were reported for a time. There would probably be a need to generate data in greater detail for these installations, perhaps analysing breakdowns by equipment units, by nature of faults, showing number and durations of breakdowns, and components replaced, etc. Perhaps man hours would be reported for these sites by categories, e.g. preventative maintenance, breakdown repairs, routine testing, standby maintenance, field modifications and so on.

It can be appreciated that to list all this data for every installation might make the output unmanageable as well as using large quantities of computer stationery, and valuable printer time on the computer. As management action caused improvements in performance—perhaps by better training, more routine preventative maintenance, or improved equipment design—then reporting criteria could be altered to reflect new standards. Later new equipment may be installed and for a time detailed analyses might be required separately for these units.

It is perfectly reasonable for a manager to expect the analyst to design procedures which enable this sort of flexibility to be available in reporting. It is important for them to study together the potential characteristics of the reporting system in great detail during the feasibility study for the project. Only in this way can the analyst know what data he will need to collect and how he will need to process it and arrange it within the file structure.

Flexibility—program techniques
There are a number of techniques that can be adopted in programming to permit this sort of flexibility. One method is to examine each record of the file with an interrogating program that will create a print-file of those records that exhibit the desired characteristics, a further run being used to list the records. In other situations it might be better to create an exceptions file during a run to update the main file; some methods

are suggested later in section 4.4. Whatever technique is chosen it is best to consider the use of parameterised routines, so that reporting criteria can be changed easily, and so that major program amendments are not needed when new management requirements arise. This significant technique is further described in section 3.5.5—data management software.

4.1.6 Error reporting conditions

Requirement for error reporting
Error reporting is needed where data in a computer system is found to be invalid according to particular rules written into the system during system design. These rules would normally be framed to act as a check on possible error situations arising in procedures external to the computer. Let us take as an example the updating of a file where master records are changed by means of transaction records having the same identification key. A transaction record entering the system may have a key which cannot be matched against a master record; the implication here is that the transaction key has been incorrectly coded or that the master file has not been maintained correctly.

In this situation the file updating program may not be able to decide which of the two conditions is applicable, and in any event it will not be able to process the record in the usual way. Details of the transaction will have to be printed out to enable data control staff to trace the error, perhaps by reference back to original documents. Sometimes rules can be framed so that errors can be categorised into types, and error reports arising from a run can be sequenced to enable data control staff to deal with all errors of a particular type at the same time. Certainly the system designer should specify programs to diagnose the particular error conditions as far as possible, and should attempt to provide as much information as is pertinent to the error conditions.

Acceptance of error
Some error conditions may be of such significance that the run cannot proceed, but in practice such a condition is rare. It is generally necessary to reject error transactions and list them so that action can be taken to correct the conditions at the next updating cycle. This will depend on the nature of the system and will be related to uses to be made on the output.

In other situations the error in a transaction may be such that it does not invalidate its principal use in a system, but it does affect some subsidiary use in which some degree of error can be tolerated. In this case the transaction need not be rejected but perhaps an error list would still be produced to focus attention on the condition.

Guide lines for error reporting
Further discussions of validation and error reporting principles are given

217

in sections 4.4.2 and 4.4.6. The main things to remember can perhaps be summarised as follows:

1. Try to foresee all potential error conditions and incorporate checks to trap them.
2. All error reports should be designed with care to ensure that data control staff are able to deal with them with sufficient accuracy and speed to maintain the desired quality in the system.
3. Where error reports turn out in practice to be excessively large, then seek out better methods of data collection and control.
4. Do not report trivial conditions if no action is required—keep error reports concise and accurate.

It is advisable to monitor all systems regularly (even though they may have been operational for some time) to ensure that error reports are still relevant and within tolerable bounds agreed between the systems staff and user department (see also section 7.4—system review).

4.1.7 Appraisal of output effectiveness

In the final analysis the output generated by a system is only of value if it is being used regularly to monitor and control operations and to assist in formulating methods for improved performance in the organisation concerned. At each stage of system development it is advisable to review the effectiveness of the output to see what impact it has on the departments using it. Some points to check include:

1. Does the output satisfy a functional need within the organisation?
2. Is it arranged in the best sequence for each user?
3. Is the volume too much for the user to deal with; can the information be effectively summarised and still satisfy the user?
4. Is it in a suitable format for the user?
5. Can the content sequence and format be readily altered to suit varying or changing needs?
6. Is the output produced in time to enable effective action to be taken; e.g. what is the time lag between file updating and output generation?
7. Is the user organised to make use of the output?

These principles should be set before members of project teams designing and approving output formats, so that a determined effort is made to design relevant and effective reports. Where the implications of a report entail reorganisation of existing departments, or of the functions of staff external to the computer procedures, make sure that the department managers concerned agree to the proposals before wider publicity is given. Try to gain the confidence of line managers and give them every opportunity to carry out the necessary consultations with their staff in dealing with staff on organisational matters. Above all make sure that users are ready to deal with reports as the system goes operational, and that there are agreed standards of action for exception reports and error conditions.

4.2 ANALYSING DATA COLLECTION REQUIREMENTS

4.2.1 The impact of output requirements on data collection

Classification of data elements
Once the output reports have been specified, the systems analyst should compile a list of all the data elements that are needed in the file system to meet the output requirements. In so doing he can classify data elements under the following headings:

1. Input data — data elements that must be collected on a routine basis to provide information for updating files or for direct printing on an output report. For example, a quantity field in a customer's order record.

2. Key data — data elements maintained on main files within the system may serve to identify a record or to provide a description for the record, e.g. an account code or a name and address.

3. Generated data — data elements generated by the processing of elements in input transactions or file data by various arithmetic or logical processes. For example, a total field appearing either on a main file or in an output report; e.g. Stock Quantity in an inventory file.

Reporting frequency
The analyst will now be able to identify the data elements he must collect on a routine basis to maintain files. The next important step is to consider the frequency of the report and the uses to be made of it. If a report is required once a week it may be possible to batch all input transactions and apply them to the files just before the production of the report. However, it may be advisable to schedule the take on of data so that invalid transactions can be identified, corrected, and re-applied to the file before reporting takes place.

If the application is one that requires accurate up-to-date output reports on a minute to minute basis (e.g. an airline seat reservation system) then transactions affecting main files must be collected and applied to them as soon as they arise.

As a general rule we can say that the data to be collected and the frequency of its collection and preparation are governed by the content, accuracy, and frequency of the output reports.

219

Transaction sources

The data elements forming the input data are then further classified to identify the likely sources of transactions. To some extent this stage of classification is straightforward; the nature of the data itself indicates its source. If we consider a stock file in an inventory management system the following relationships might present themselves: data elements concerned with materials ordered from external suppliers will appear on buying orders, details about materials entering stock are on goods inwards documentation, and information about prices of products may come from the commercial department.

Data to be collected can be related to transaction events, procedures, and documents. Details of existing procedures may already have been noted during the initial investigation, and these can be studied to improve and amend them so that data is derived to meet the requirements of the new system. Where the input requirements are not covered by existing methods then new procedures must be developed.

4.2.2   Developing data collection procedures

Data characteristics

When the data elements that must enter a system have been identified it will be necessary to study the attributes of each data element, so that file specifications and input formats are designed to cater for the characteristics of the data. For example, fields on input media or main files must allow sufficient positions to cater for the maximum size that a particular data element may reach. Value and quantity fields may sometimes be difficult to assess, but the systems analyst must study existing procedures and live documents to check the maximum and minimum values that arise. When in doubt it is best to allow a generous margin for error.

Choice of input medium

Having classified data elements and grouped them according to transaction types the analyst must next consider means of getting the data into the computer system. Some medium is required to bridge the gap between the letters, numerals, and symbols that human beings use and the internal representation of the computer. The analyst must therefore consider the many types of input media that are available.

The first step is to consider whether or not transactions can be collected as a by-product of some external operation, so that the delays and costs of conventional punched card or paper tape preparation can be avoided. In retail systems, for example, the day to day transactions can often be captured by automatic paper tape punches attached to cash registers.

Mark scanning techniques may perhaps be used to advantage; here transactions are recorded by marks made with a pen or pencil in predetermined positions of specially designed documents. These documents can be read by mark scanning devices to input data directly into the computer. Such documents are particularly useful where large volumes of transactions occur, and where the content and format of each transaction falls within a standard framework. A sample of such a document is shown in figure 77. MICR or OCR coded documents may also be good candidates for such applications.

ORDER FORM	Group	6 3 2 1	6 3 2 1	6 3 2 1	Sheet	Issue No		1
	Branch	6 3 2 1	6 3 2 1	6 3 2 1	2 1	6 3 2 1		2
		Tens	Week Units	Day	Hour			3
		6 3 2 1	6 3 2 1	6 3 2 1	6 3 2 1			4

Department				Ordered	Supplied	
Code No.	Description	Pack	Price			
				10 9 8 7 6 5 4 3 2 1	10 9 8 7 6 5 4 3 2 1 0	5
				10 9 8 7 6 5 4 3 2 1	10 9 8 7 6 5 4 3 2 1 0	6
				10 9 8 7 6 5 4 3 2 1	10 9 8 7 6 5 4 3 2 1 0	7
				10 9 8 7 6 5 4 3 2 1	10 9 8 7 6 5 4 3 2 1 0	8
				10 9 8 7 6 5 4 3 2 1	10 9 8 7 6 5 4 3 2 1 0	9
				10 9 8 7 6 5 4 3 2 1	10 9 8 7 6 5 4 3 2 1 0	10
				10 9 8 7 6 5 4 3 2 1	10 9 8 7 6 5 4 3 2 1 0	11
				10 9 8 7 6 5 4 3 2 1	10 9 8 7 6 5 4 3 2 1 0	12
				10 9 8 7 6 5 4 3 2 1	10 9 8 7 6 5 4 3 2 1 0	13
				10 9 8 7 6 5 4 3 2 1	10 9 8 7 6 5 4 3 2 1 0	14
				10 9 8 7 6 5 4 3 2 1	10 9 8 7 6 5 4 3 2 1 0	15
				10 9 8 7 6 5 4 3 2 1	10 9 8 7 6 5 4 3 2 1 0	16
				10 9 8 7 6 5 4 3 2 1	10 9 8 7 6 5 4 3 2 1 0	17
				10 9 8 7 6 5 4 3 2 1	10 9 8 7 6 5 4 3 2 1 0	18
				10 9 8 7 6 5 4 3 2 1	10 9 8 7 6 5 4 3 2 1 0	19
				10 9 8 7 6 5 4 3 2 1	10 9 8 7 6 5 4 3 2 1 0	20
				10 9 8 7 6 5 4 3 2 1	10 9 8 7 6 5 4 3 2 1 0	21
				10 9 8 7 6 5 4 3 2 1	10 9 8 7 6 5 4 3 2 1 0	22

77. Direct input document for mark scanning

Volumes

It is important to ascertain the volume of each transaction type and to estimate the potential growth rate for the years ahead. The procedures developed to deal with data collection must be capable of extension to meet an increased volume. This applies to the external clerical procedures that deal with source documents as well as to the data preparation routines and the input transcription programs.

No matter what input medium is used it is essential to see that there is sufficient capacity to meet future growth or sudden peaks of activity. The development of the data collection routines should be undertaken in project teams in which representatives of the line management, data control and data preparation staff play a prominent role. It is important to assign responsibility for the maintenance of data files to various user departments and to ensure that the full implications of this are understood by them from the beginning. The systems analyst will often need to encourage line managers to take particular care to develop data control facilities within their departments: a computer system is only as good as the data that is input to it.

4.2.3 Source documents and data preparation

Source documents
Source documents are usually initiated during an external clerical procedure. The source documents may initially be completed by staff in line departments, and a copy of each document will be sent to a data preparation centre for punching and verifying into cards or paper tape (see section 3.9). The information can then be fed directly to a computer.

Design of source documents
Source documents designed for punched card or paper tape data preparation have to serve the purposes of the people operating the particular procedure (e.g. salesmen, stores clerks, van drivers and so on) as well as provide the punch operators with an easy punching document.

For data preparation operations the following points should be considered:

1. As many input records as practical must be contained on a single page, in order to minimise page turning.
2. Fields requiring to be punched must be easily identifiable and should be presented in the sequence of punching.
3. Where punched cards are used column numbers on documents should be shown.
4. Boxes on documents to contain data fields are desirable, and should be large enough to enable characters to be entered.
5. Punching documents should if possible be designed so that entries are made in black on a white background; i.e. avoiding colours that tire the eyes.
6. The number of manual entries required should be limited by the provision of pre-coded fields wherever possible.
7. Carbon copies should not be used for punching purposes; if they are, bottom copies (where entries are likely to be smudged or faint) should be avoided at all costs.

From the user's point of view it is important to see that documents contain the data elements required in performing their particular function. The form layout must be suitable for use in the appropriate location and should be of an appropriate size; e.g. to fit into working files, or onto a clip board, etc.

If there is a need for a number of carbon copies to be created, make sure that users have an adequate surface to complete the forms on, and that a suitable pen is available to make the entries, otherwise the bottom copies may be too faint.

222

Source documents should if possible be tested in a pilot scheme in an actual working environment (see section 5.4.2).

Responsibility for control
Obviously if a document is not well adapted to the procedure that initiates it, then there is a strong likelihood that it will not be dealt with accurately and efficiently in the line departments concerned. It is a good objective to design systems procedures so that the user departments are responsible for controlling the data that they themselves most depend on. This will ensure a full awareness of the need to control all data with the utmost stringency.

In the early operational stages of a new system it is not unusual for user departments to be uneasy about giving up their existing procedures, and the analyst sometimes has to be quite forceful in persuading the users to support the new system fully at its critical opening period.

Routing of source documents
It is for this reason that when an organisation is introducing its first computer there is also a tendency to avoid changing the organisation and its associated procedures until the computer is running and paying its way. This may be sound common sense from one viewpoint, but there are dangers in this approach, which gives rise to the type of data collection

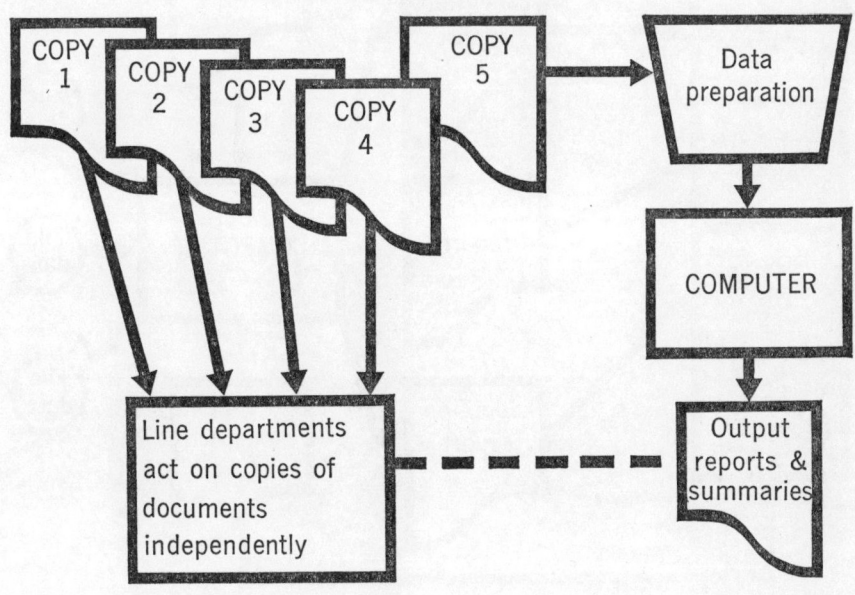

78. Data collection. All copies are acted upon before computer file is updated. (Compare fig. 79.)

223

procedure shown in figure 78. Here we can see a source document which has four copies used independently in different parts of the organisation, one copy being sent to data preparation to create input transactions for the computer files. The multiple parts of the original document are still circulated and may be amended by clerical action at different stages. Different people in the organisation are using these documents independently to perform their function and are able to do this by associating their copy with, perhaps, their own working files to decide on actions to be taken.

Figure 79 on the other hand shows the better approach; only one source document is initiated. Data from this document enters the computer procedures and is validated, and then a number of reports are generated for distribution to the user departments, having been associated with any other file data, as required by the individual departments. Thus the power of the computer is utilised to simplify channels of communication, and to correlate information to achieve the functional aims of each line department. It will be noticed that this method emphasises the interdependence of the line departments and the computer systems they use, and this underlines their responsibility for correct maintenance of data.

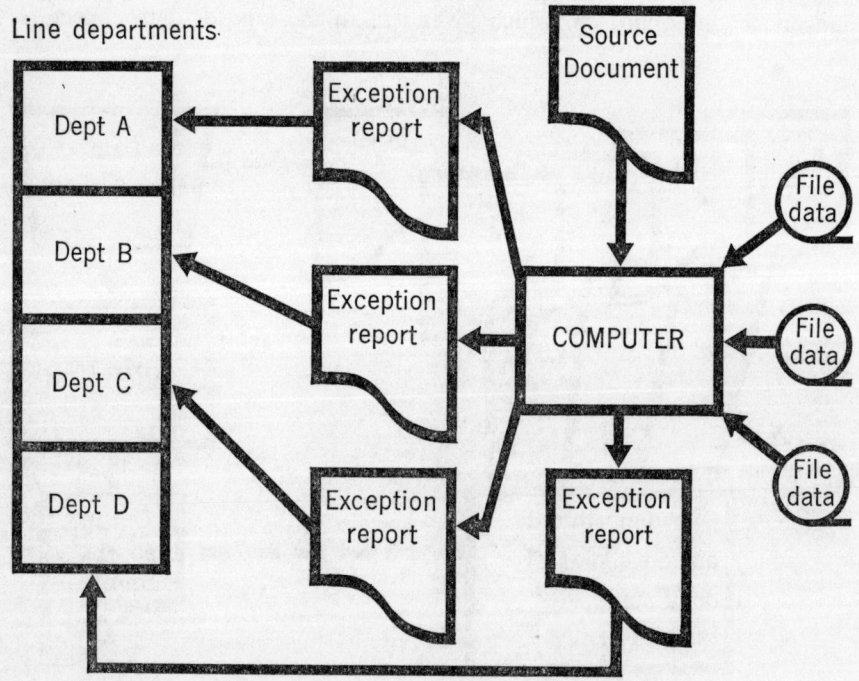

79. Data collection: computer file is updated and generates appropriate forms. (Compare fig. 78)

Batch totals

To control the preparation of input it is customary to divide the source documents into batches. This enables work schedules for data preparation to be more easily controlled, and facilitates the subsequent manual selection of documents should this be necessary later when dealing with invalid records rejected by the computer. For each batch of documents it is usual to assign a batch number and to calculate the number of cards (or record entries) that will result from the punching operation. Sometimes special batch totals are taken for specific fields in the source documents; e.g. a quantity or value field may be accumulated.

These batch totals are used to check that documents on cards are not mislaid during the data preparation procedure, and act also as an extra check that the fields concerned have been correctly punched. These totals are maintained and checked through all stages of data collection up to and including the final input programs. Indeed, control totals can also be maintained throughout all stages of processing as is shown in more detail in section 4.3.

Basically the steps in preparation of cards are as follows, and it is usually necessary to check batch totals at the end of each phase.

1. Collect source documents, batch them, and create batch totals using an adding machine if necessary. Enter details of each batch into log book.
2. Punch cards, check card count.
3. Verify cards, check card count, pass error cards back for repunching.
4. When the punching and verifying process is completed tabulate batches on a balancing tabulator to control totals.
5. Where totals disagree examine the appropriate batches to locate errors.
6. Present cards to the input transcription program, which should generate a list showing batch number, card count, and batch control totals.
7. Check for missing batches and check card counts and control totals.

Batch sizes

The size of batches is an important factor; if a batch is excessively large, data preparation staff may feel that the job is never ending, and as a result accuracy may suffer or speed drop below acceptable limits. If a batch is too small on the other hand there are too many checking operations and unnecessary delays are incurred.

For punched cards a tray of 2,000 cards is probably a reasonable size for a batch. With punched cards, error cards can be extracted and replaced without too much difficulty at the various stages of preparation, but with paper tape the problem is more acute since the whole tape may have to be repunched if errors are found in the middle of a batch. For this reason there may be a leaning towards smaller batch sizes for paper tape.

225

H

Invalid data

It is important to observe a distinction between *errors* in data, and *invalid* data. The whole object of data preparation is to check that all input records have been accurately transcribed from the original source documents. After a batch has successfully completed the verification process we can say that the batch is correct, but it may still contain invalid records.

Valid records are those which conform to logical rules about the format and content of the constituent fields comprising the record; e.g. account numbers may have to be within the range 052 to 250, therefore codes less than 052 or greater than 250 are invalid.

Invalid records are usually detected by input programs and have to be displayed in error reports for data control staff to take action. Separate control totals are usually accumulated for both valid and invalid transactions during input programs so that a complete check on control data can be made.

4.2.4 Summary of data collection principles

1. Ascertain what data has to be collected, and when; make sure that data collection frequency is suitable for the output response required from the system.
2. Data preparation is at best a necessary evil, try to develop automatic data collection procedures wherever possible.
3. Evaluate the likely failure or error rates associated with the particular input medium chosen, and ensure that the data control facilities are adequate to cope with them.
4. Make sure that all source documents satisfy the needs of both users and data preparation staff.
5. Make sure that all data collection systems are tried as a pilot operation in the actual working environment before going live.
6. For all live systems, monitor the operation of data collection and validation procedures and check that the output generated is within agreed tolerances of accuracy and response time.

4.3 SYSTEMS CONTROLS

4.3.1 Standard hardware and software controls

Hardware controls

Computers are designed to provide automatic checking facilities, to safeguard users against machine malfunctions. For example, each time data is transferred to or from a central processor and one of its peripheral units the internal circuitry checks to see that characters have not been lost or multilated (see section 3.1.1—parity checks). If a transfer is incorrectly executed the processor may halt to enable an operator to restart the program concerned, or it may repeat the operation automatically—e.g. by re-reading a block from magnetic tape or provide a signal to the user program enabling alternative action to be taken.

File processing software controls
Manufacturers of computer systems usually provide housekeeping software (section 3.5.1) to take care of the standard operations affecting backing storage devices such as magnetic tape or disc files. These software modules usually incorporate sections that check file labels to ensure that correct generations of a particular file are in use during all processing operations.

For example, each file on magnetic tape has to have a header label which will state:

1. name of the file
2. date file was written
3. generation number
4. purge date
5. reel number

The name of the file is checked every time that the file is opened for use in a program, to see that the correct file is being used. The generation number is also checked to ascertain that the correct version of that particular file is being referenced. If the file occupies more than one reel of tape it will also be necessary to check the reel number.

The purge date is the date after which the file may be scratched or overwritten. It is generated when the file is initially created by the addition of a specified retention period to the current date. In this way the contents of the file can be preserved, because the file processing software will prevent it being opened as an output file until the purge date is exceeded. Other safeguards provided by file handling software include: block counts to provide a check that all data blocks on a file have been read, and on direct access devices file definitions that reserve areas on the storage device for particular files, and thus make sure they are not overwritten.

Most lengthy file processing operations should also be protected by using *dump and restart* facilities incorporated in housekeeping software systems (see section 3.5.8). These features enable the contents of the computer's internal memory to be dumped from time to time to a backing store, so that, should the program be interrupted, it will be possible to restart it from an intermediate point (see section 4.4.9).

Write permit or inhibit rings
Another way of safeguarding the contents of magnetic tape files is to attach to the reels themselves devices which prevent the reels from being opened as output tapes. In some computer systems a 'write inhibit' ring

227

is such a device: when it is fitted, a reel can be opened only as an input file. Other systems use an opposite convention and a 'write permit' ring has to be present to allow the file to be overwritten.

Program lockout

Within multiprogramming computers it is necessary for programs operating concurrently to be prevented from violating storage areas required by one another. A hardware device, or an executive program, is used to limit the areas available to each program, and to signal an illegal condition should any program attempt to address an area of memory outside its own nominated bounds.

Monitoring live operations

The features described above are examples of the many standard facilities provided by manufacturers if hardware and software. The systems analyst should be familiar with the principles and rules concerned with the operation of these facilities. Particularly he should ensure that optional facilities are not overlooked once a system has passed over to full live running. For example, failure to use generation numbers in addressing file labels can result in a wrong version of a particular file being used in an updating run.

Check digits

One of the major problems in any coding system is to ensure that a code is *valid*. It is easy for checks to be made at the data preparation stage to ensure that what is written on a document, for example, is punched correctly on cards or paper tape. It is also straightforward to check that a code falls within a prescribed value range when it enters the computer system. But these checks do not guarantee that the code has been correctly written out in the first place, or that a badly written or printed code value has not been misinterpreted at the data preparation stage. The technique of check digits enables the validity of a code to be checked by examining the code value itself. The types of error which can be checked by means of a check digit system can be summarised as follows.

Transcription errors are those caused by misreading badly written or printed digits, e.g. confusing 2 and 7. A single transcription error involves misinterpreting one digit in a code; where several consecutive digits are misinterpreted the error is a multiple transcription error.

Transposition errors are errors caused by interchanging two digits in a code, e.g. writing 13762 as 17362.

Shift errors are caused by shifting the digits of a code to the left or right, deleting or adding zeros as the case may be, e.g. 01376 instead of 13762.

Check digit techniques have been devised to provide the automatic rejection of up to 100 per cent of errors caused by one or more of these types of error. Many different types of check digit generation formula are available. In essence, a check digit is a digit calculated by a formula applied to the digits in a code. The formula is devised so that if the digits in the code are transposed in any of the ways mentioned, a different check digit will be generated. The check digit is added to the code when it is allocated, e.g. added to an employee number or customer number when issued. When a number is input to a computer system, the check digit is recalculated and compared with the digit as entered; if these do not match, an error will have been made in entering the code.

Modulo 11 check

As an example the calculation of a 'modulo 11' check digit for the code 124683, using a weighting system, is given. This system involves the following steps:

1. Write out the code digits, isolating the check digit:

   12468 check digit 3

2. Allocate weight to each digit (the weights are chosen by a statistical method designed to give maximum error detection).

code	12468
weights	12536

3. Mulitiply each digit by its weight

code	1	2	4	6	8
weights	1	2	5	3	6
code × weights	1	4	20	18	48

4. Add up the results of multiplying code and weight

   $$1 + 4 + 20 + 18 + 48 = 91$$

5. Divide the sum obtained by 11 and find the remainder

   $$\frac{91}{11} = 8, \text{ remainder } 3.$$

The value of the remainder is the calculated check digit.

If the reader performs a similar calculation on the code 12486, i.e. 12468 with the last 2 digits transposed, he will find that the resulting check digit will not equal 3, thus indicating that an error has occurred.

Check digits are particularly valuable in preventing errors in an updating system, where transactions and main file records are matched on a coded key value. If a coding error occurs undetected on a transaction file, movements may well be posted to the wrong main file record, and it may well prove impossible to prevent this form of error by any internal system or logical check other than by using check digits.

4.3.2 Control accounts

In every computer system control totals must be maintained throughout all processing operations to ensure that transactions have been processed and have been related to their appropriate records. Every file must therefore have one or more control accounts which are maintained during any processing run to give an overall picture of the actions that have taken place during the run.

In the simplest case there may be one basic record type on a main file, and at the beginning of a processing run the file may contain, say, 10,000 records. A transaction file used to update this main file may contain 2,500 records of which 1,500 are additions, 500 are deletions, and another 500 are amendment records altering certain fields of main file records. At the end of the file updating run the following control figures could be listed:

RECORD COUNTS:	*Main File b/f*	*Transaction File*	*Main File c/f*
	10,000	1500 additions	11,000
		500 deletions	
		500 amendments	

These figures would not in themselves be sufficient to check the accuracy of the run, but would provide a general indication that all records have been processed.

To be more certain that actions had taken place correctly it would be necessary to maintain totals of the fields altered by the amendment records. For each of these fields the control figures should show the previous balance on the file, the amendment figures, the remaining balance on the file.

Thus, if every record in the previous example had contained a value field, control figures might have been generated as follows:

	£
Input main file	19,901,028
Amendments—debits	2,781,027
credits	5,010,602
Output main file	22,130,603

The figures used for such control accounts must be variable for each record, but if there was not a suitable quantity or value field it would be just as effective to maintain counts of some other field containing numeric data, e.g. salesman number, or stock account code.

If the file is one in which a number of different record types are maintained the control figures could be generated to reflect a number of control accounts—for example, in a stock file stock quantities and values could be maintained for a number of independent stock accounts. In this way, during the checking for error conditions attention can be focussed on the errors in particular stock accounts.

It is not difficult to program a computer to maintain such accounts throughout all processing runs, but often considerable effort has to be expended outside the computer environment to ensure that relevant figures

230

are developed for transaction data, as it is collected and prepared for input. It is essential that departments supplying data to a file system should have the opportunity to check that their data collection procedures are working satisfactorily and that their data is being correctly prepared and processed by the data processing department.

Control stages

The procedures for conducting such checking operations will vary according to the means of data collection employed. Where there is some automatic data collection process, e.g. collecting punched paper tape as a by-product of some operation, then facilities should be incorporated to permit control accounts to be captured at the same time. For more conventional data preparation procedures the following stages of control can be identified.

1. Coding and collection of source documents
2. Batching of documents and creation of control totals
3. Punching (i.e. cards or paper tape)
4. Verification
5. Input to computer runs
6. Checking by program at each processing stage
7. Checking of totals or documents at output stage

In this situation user departments would be expected to submit batched documents with control slips containing relevant control figures. The control figures themselves could be developed using a simple adding machine, and would be used to check each subsequent operation. The data processing department should be expected to maintain control accounts in a control ledger and to check and verify these with the users after each operational run. During data preparation special balancing tabulators can be used to create check totals of control fields on punched cards.

Coupled with adequate validation programs on input to the system, these methods should provide very tight control over data processing operations. The important thing is to create control accounts so that they can readily prove the accuracy of files: they should therefore be easy to interpret and figures generated by the computer should be brief and to the point.

Maintenance of control procedures

In designing control systems the analyst should take into account conditions that will prevail when the system is live and he has moved onto another project. He should ensure that control procedures are written into operations guides (section 6.7) produced for use by line departments and the data processing operations staff. The programme of education conducted to launch each new system should place particular emphasis on control procedures, so that error conditions are identified and dealt with as soon as they arise.

231

4.3.3   File control and file security

The files within a computer system are usually stored on magnetic tapes or magnetic discs; these files become extremely valuable assets to the organisation concerned and they have to be safeguarded to prevent loss or damage to the file media. Magnetic tapes and discs can be damaged if not handled correctly, and they must be handled by experienced people and maintained generally in an air-conditioned environment.

Some organisations would find it extremely difficult to remain in operation should their major files be damaged, and they therefore keep several copies of their current files and usually ensure that some copies are held in fireproof cabinets.

In section 4.3.1 we discussed some of the common hardware and software facilities used to safeguard files during file processing operations. It is also essential to consider the means whereby physical control can be extended over tapes or exchangeable discs. All magnetic media of this sort should be kept in a library close to the computer room and data files should be issued to computer operations along with the basic running documentation, just before the job is run. In a well run installation it should not be necessary to have tapes or disc packs lying about the computer room.

Processing operations on tape files are usually conducted so as to preserve at least three generations of each file; the current version, and two other versions resulting from the preceding two updating runs. This technique is referred to as the *Grandfather, Father, Son* principle. The relevant transaction files have also to be preserved so that they can be used to recreate the current version of the file in the event of irretrievable damage to the current file. The relationship of the three generations is shown in figure 80.

Disc files can be preserved by dumping the contents onto magnetic tape at frequent intervals to create back-up files. The relevant transactions have again to be retained to enable the current position to be re-established in the event of loss of file data.

The librarian who has the responsibility for issuing all file media, will generally maintain manual records for all tape reels to enable him to ascertain the contents of any particular reel. It is the librarian's job to ensure that the correct files are issued with each program suite. The librarian has also to release old files (e.g. Great-Grandfather files) so that the tapes can be overwritten as new files or as work tapes.

The records maintained for these purposes must be accurate and permit access to relevant information within a few minutes. The records must be arranged such that a tape containing the current version of a particular file can be identified and selected for issue. In the same way it must be possible to identify the contents of any particular reel, to ascertain whether it can be issued for use as an output file or a work tape.

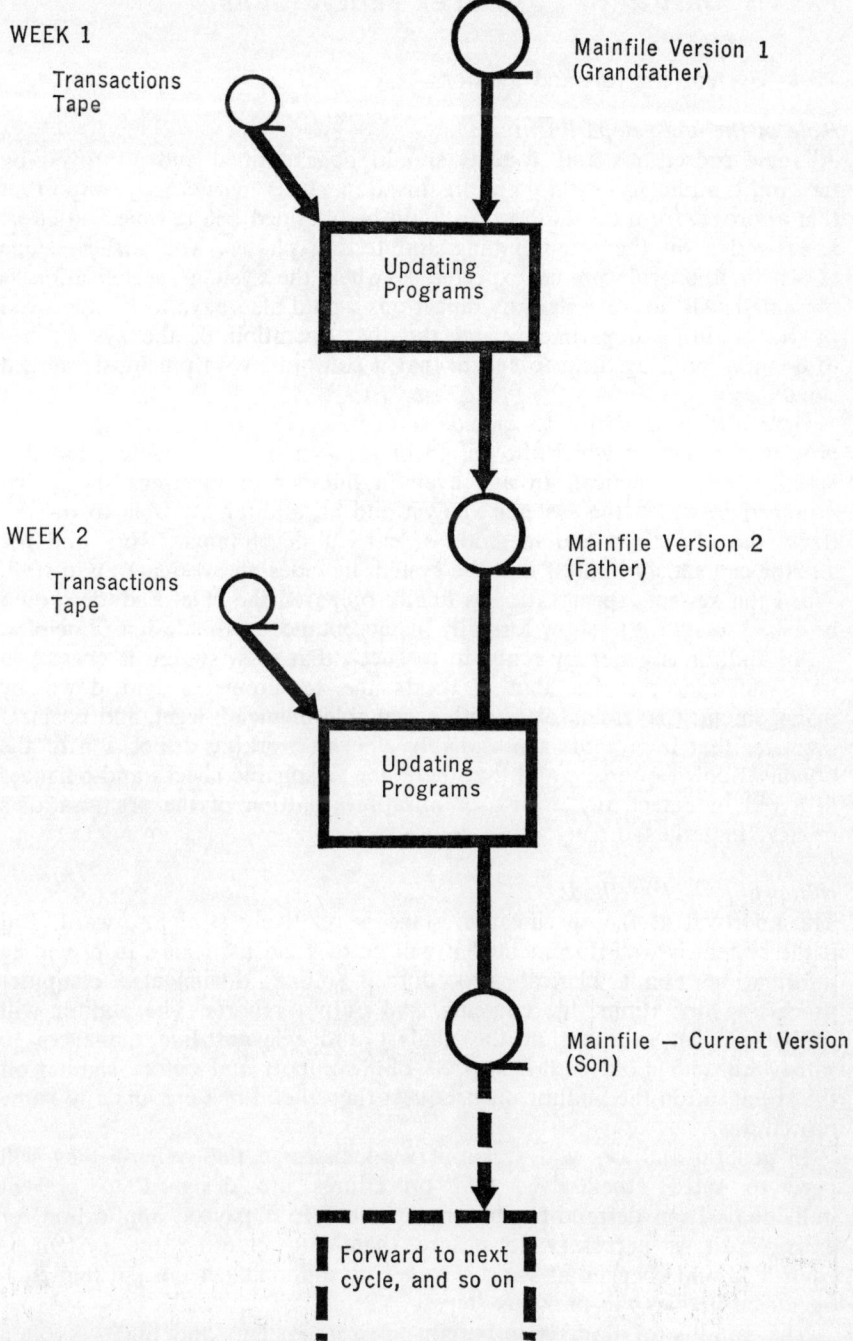

WEEK 1

Transactions
Tape

Mainfile Version 1
(Grandfather)

Updating
Programs

WEEK 2

Transactions
Tape

Mainfile Version 2
(Father)

Updating
Programs

Mainfile — Current Version
(Son)

Forward to next
cycle, and so on

80. File generations: grandfather, father and son

233

H*

4.3.4 System controls and auditors

Role of the audit department
All new procedures and systems should be examined and approved by the audit authority within an organisation. It is particularly important that approval from this authority should be obtained before time and effort is expended on the programming and testing phases. The earliest stage at which approval can be expected is when the systems specification is presented. All future system modifications would also have to be approved by the auditing department, and the live operation of the system has to be monitored by them to ensure that it conforms with previously agreed standards.

Generally it is useful to include a member of the audit team in any project meetings in which the design of the system is considered, but this is not always practical. In any event a number of meetings should be arranged in which the systems analyst and an auditor are able to discuss freely the objectives and methods of current development work and the auditor can satisfy himself that the system includes an adequate audit trail. When the systems specification is finally prepared the chief auditor should be asked to sign it to show formally his acceptance of the design principles.

An auditor is generally required to check that each system is correct in its basic concepts – i.e. that it meets the requirements laid down by management; that it conforms with acceptable financial, legal, and business practice; that it presents a reasonably efficient working disposition of the organisation's resources; and that there are reasonable checks and balances to avoid or detect any misuse or misappropriattion of the organisation's money, materials or any other resources.

Initial approval methods
The approval at the specification stage is relatively straightforward, and if the system is well documented it will be of great assistance in providing information about clerical procedures, source documents, computer processing operations, file contents, and output reports. The auditor will probably wish to question the analyst and relevant line managers to satisfy himself about certain aspects of the report and before signing off the specification the auditor may request that alterations are made to some procedures.

In general auditors will seek out weaknesses in the system – they will need to satisfy themselves that procedures are designed to prevent individuals from defrauding the organisation. In a payroll application for example, it is necessary to ensure that data entering the system is controlled and checked at various stages by different persons, so that each operation verifies the previous steps.

The validity of data is important in any system, and auditors often need to examine in detail the way in which transactions arise, in order to

234

satisfy themselves that illogical or inconsistent data is identified and corrected.

The processing methods specified within the programs have also to be checked to ensure that the desired accuracy is maintained. These checks are repeated when the program is written and tested, and if necessary the audit department should have programing staff who are able to conduct acceptance tests on programs. The audit department should be allowed to test programs using their own test data and where this facility is available the data should include examples of all conditions of the program.

The acceptance of programs should once more be signified by the signing of a formal acceptance document. From then on the audit department should be allowed to keep a copy of the operational program; and when any program amendments are made a new version of the program should be supplied along with documented details of the amendment for testing.

Verifying control procedures

The systems analyst should document all control procedures used to check the accuracy of a running system. This document should take the form of a procedure manual for use by computer operations staff and data control staff in the line departments concerned. The audit department will need to check and approve these procedures before the system goes live, and will have to be satisfied that all persons concerned are familiar with their responsibilities in this matter.

Verification of a live system

Hitherto we have considered how an internal audit department should ensure that a system under development is the right one. Another important duty of such a department is to check that the systems in ordinary day to day use are being operated accurately and correctly. The following paragraphs describe some techniques which may be useful for this sort of verification.

If the checks and tests previously described have been carried out satisfactorily, the internal auditors should be satisfied with the accuracy of the system. However, they will want to conduct tests to see if the correct program is still being used, and there are several possible ways in which this can be done:

1. The standard test pack originally prepared for the acceptance tests should be run, using the program that is in everyday use, and the results obtained are then checked against standard results held in the program documentation file. The results should clearly be identical. Where the system has undergone amendment, it is of importance that the standard test pack and standard test results should also be kept up to date.

2. The version of the program in everyday use can be compared against

a version of the program held by the internal auditors. This can be done by printing out the two versions in full and comparing them visually. The internal auditors must clearly be able to trust the print-out program absolutely, and it would be wise for them either to write such a program themselves or to use a standard software program. Since the visual comparison of two versions of a large program can be very laborious, it may be preferable to use a computer program to make the comparison. Such a program should print-out any differences between the two versions, or print an *all clear* report if they are found to be identical. Such a comparison program should also be written by the internal auditors or selected by them from standard software. These two techniques have the advantage that they check the entire program, whereas running the standard test pack, even if the same results are obtained, may not disclose unauthorised alterations to the program.

3. Maintaining a complete copy of the computer's console log—wherever this is available—provides a very detailed means of ascertaining what has happened at each run.

Conclusion

The vital point is that the internal auditors or accountants should involve themselves in the development of systems from the earliest possible stage. Thereafter, they are in a good position to maintain adequate control over the operation of the systems. The key to both duties is the use of a methodical procedure such as that described.

4.4 PROCESSING METHODS FOR MAGNETIC TAPE

4.4.1 Characteristic structure of magnetic tape systems

In section 3.3 the general principles of magnetic tape storage have been described; in this present section we show what impact these principles have on systems using magnetic tape as their main storage medium.

Serial medium

Magnetic tape is a serial medium. That is to say, to read any specified record from a magnetic tape file it is necessary to search through the file from the beginning until a record of the specified key is located. When the required record has been read into main memory it can be amended as required by the program, but to preserve the amended version of the record it is necessary to write it to another reel of magnetic tape where it will be stored until required once more.

It is obviously uneconomic to search all records on a reel of magnetic tape simply to locate a small group with the same key value: however, if the reel is sorted into the sequence of the required key, once all records with the required value have been found, no further records need be examined. If they were in random sequence, they would all need to be examined. Therefore it is necessary to maintain records in serial files in the sequence of some particular key; for example a personnel records file may be sorted to the sequence of staff number. To amend records on a master file or to add or delete records, it is necessary to apply transaction records to the master file. These transactions can be collected over a period of time (a day, a week, or a month) and can all be applied to the master file at the same time, thus creating a new master file. The transaction records are sorted into the same sequence as the master file, so all the required processing is carried out while the files are passed once from beginning to end (see figure 81).

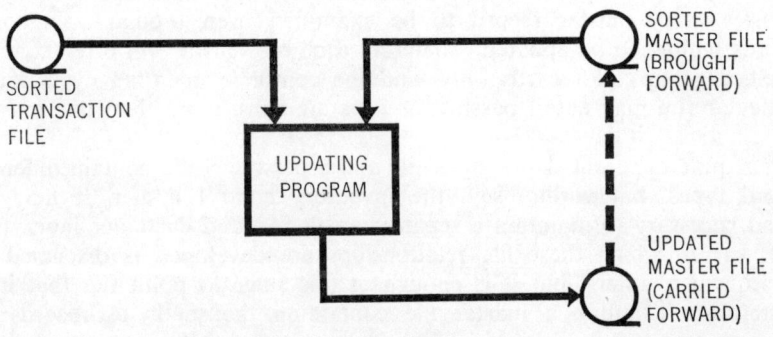

81. A file updating run

A master file is re-written each time an updating run takes place and previous copies of the master file are preserved along with their relevant transaction files to safeguard against tape damage or error conditions in a particular updating operation.

It is usual to maintain three generations of the files (as discussed in section 4.3.3) and each new generation of a master file is carried forward to the next updating cycle.

Batch processing
Such an approach, with records being batched and then applied to a master file, is known as batch processing. It is a convenient mode of operation in many commercial and financial systems, but has the inherent disadvantage that the information in the master file is only an accurate record up to and including the transactions posted in the latest updating run.

It is usual to establish a regular schedule for updating master files and

237

the frequency of the updating cycle is governed by the nature, volume, and frequency of the output reports required from the system. For example, a file maintained to report on financial expenditure by departments in an organisation may require only monthly updating and reporting, whereas an inventory file in a production control system may require daily updating with daily reporting of critical stock conditions.

The effect of batch processing systems is to force a routine pattern of events on those departments feeding data into the files or receiving data from them. A rhythm is imposed on some line departments and on data control and data preparation centres, and to some extent this rhythm helps to create a discipline for the system as a whole.

File relationships

The capacity of a reel of magnetic tape is in the region of three million characters. Files can be spread over two, three or more reels of tape, but they become unwieldy to use and consume much machine time if allowed to become too large. It must be remembered that each updating cycle requires every master record to be examined even though only some records require to be updated. Therefore each master file will often contain specified groups of records only, and the complete updating cycle for a particular run may entail passing, as separate runs, more than one master file.

It is quite apparent that a stock file and a personnel file contain different record types, but within, say, the inventory control system it may be found necessary to maintain a separate stock file, and customer index file. The way in which these file relationships are developed is discussed in subsequent chapters, but it is enough at this stage to point out that it is wasteful to examine a master file containing thousands of records of different categories, when it is desired to access small numbers of these categories.

Basic run structures

Magnetic tape decks operate at very high speeds when compared with, for example, card readers. Therefore it is usual to attempt to take all transaction records into a system or sub-system in one run so that the main updating runs utilise magnetic tape only. The nature of the data may not always permit this but it is generally inefficient to slow down main updating runs by restricting the passing of master files to the speed of slower peripheral units.

In any event transaction records must usually be sequenced before being applied to master files and this implies a sorting operation before the main updating runs can be undertaken. Figure 82 illustrates this point; run 1 transcribes transactions from cards and creates a transaction file on magnetic tape, run 2 sorts the transactions to sequence before updating commences.

It is assumed in this example that the updating run (run 3) gives rise to a

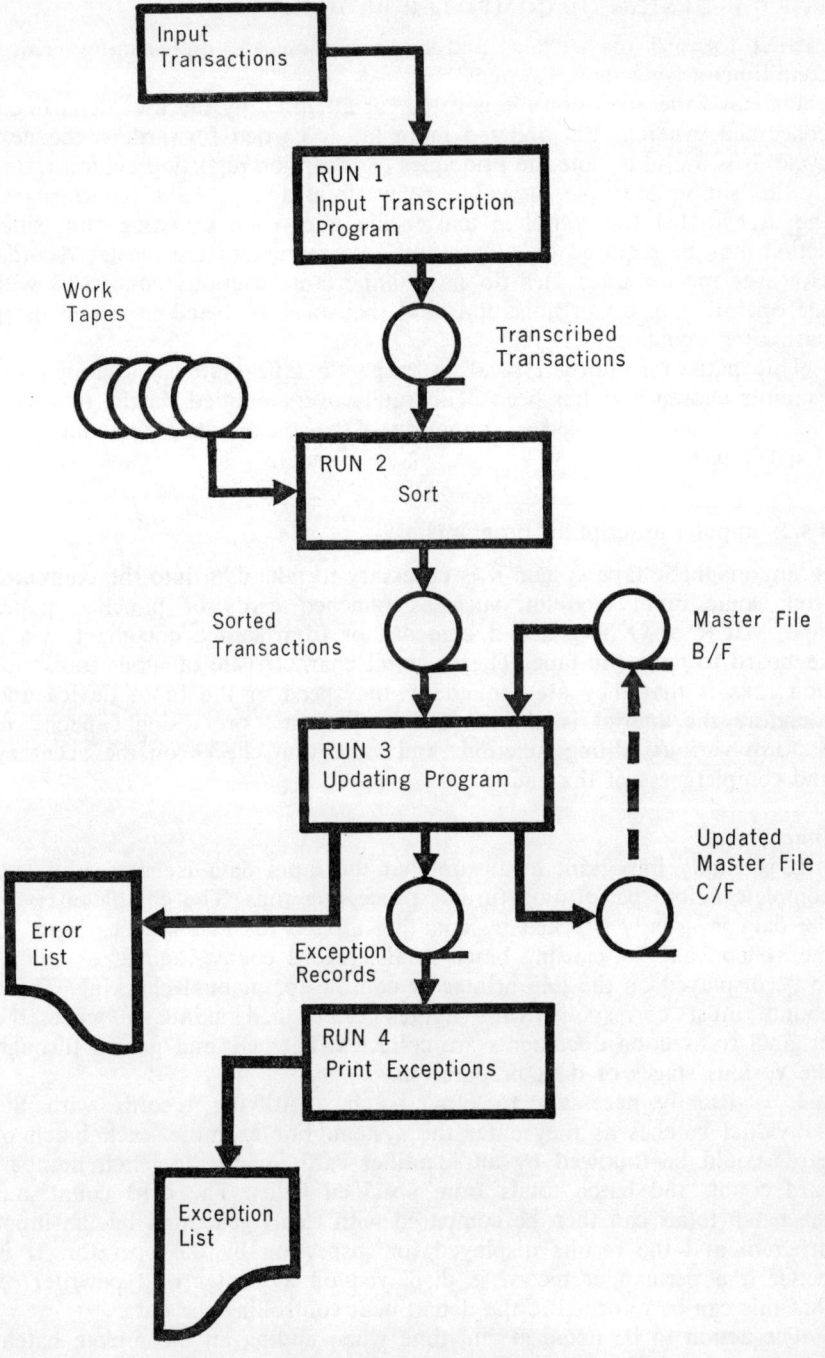

82. An updating procedure

239

carried forward master file, and an exception file representing critical conditions arising in the system.

On run 4 the exception file is listed for attention by the user department concerned, whereas the updated main file is carried forward to the next cycle. It is useful to note the principles of exception reporting demonstrated by this simple example. It would be wasteful and probably confusing to the user to list the complete master file after each updating run, since action may be required on only a small percentage of the master records. However most master files do have important functions concerned with the operation of an organisation, and they may be listed in whole or in part after updating.

This pattern is fairly typical in magnetic tape systems, although the example chosen here has been deliberately oversimplified. In the following chapters a more detailed examination of basic file types and run types is developed.

4.4.2 Input transcription programs

In any magnetic tape system it is necessary to take data into the computer from some input medium, such as punched cards or punched paper tape, MICR or OCR coded documents, or from source documents via a keyboard to magnetic tape. The essential characteristic of input transcription runs is that they are limited by the speed of the input device and therefore the analyst is able to utilise the spare processing capacity to perform various editing functions and validation checks on the accuracy and completeness of the data.

Batch totals

It is vitally important to ensure that the input data is both valid and complete before performing further processing runs. The completeness of the data is usually checked by counting all records and batches input to the system, and by causing batch totals, record counts, and error counts to be displayed on the line printer or console for manual checking. These counts must correspond with figures established manually when the original transaction documents are collected, batched, and passed through the various stages of data preparation.

It is usually necessary to input batch identifying records with the individual batches as they enter the system. For example, each batch of cards could be followed by an identifier card giving the batch number, card count, and batch totals from specified fields. The card count and the batch totals can then be compared with totals generated by the input program and the results displayed for inspection by the operator. It is better if a permanent record is displayed on a printer or typewriter so that this can be returned to the department controlling the data.

The action to be taken at run time when finding an incomplete batch, or when a batch is missing, depends on the nature of the data. Generally it is better if the card reader can be kept running at high speed, one batch after another being entered until all the input data has been dealt with. The cards and the batch control list are then returned to the department concerned to enable omissions or error records to be re-inserted later. Such corrections can perhaps be held over to the next processing cycle, or they may have to be written to tape and merged with the earlier records before the completion of the current cycle.

The following information can be usefully displayed for each batch:

> Batch number
> Expected number of records
> Expected batch total
> Number of records read
> Batch total for rejected error records
> Batch total for accepted records

The rejected records mentioned above would be those which are not written to tape because they fail validation checks—more will be said about these checks later in this chapter. The invalid records would generally be listed above the batch totals with the error conditions highlighted in some way.

Check sums

Apart from these batch counts there is often a need to accumulate certain fields from the input records in order to create totals for use in check sums at later stages in the system. Such totals would normally be made from specified quantity or value fields and may relate to all or only certain of the record types input. Sometimes these check totals can be made to correspond with the batch totals accumulated to check the completeness of the input records.

Another type of check sum may be created to check against some predetermined factor entered into the system as a parameter at the beginning of a run or at the beginning of some time period. For example, the input run to a labour analysis system may check that all employees in a plant have returned a time sheet; a simple record count being generated to compare with the previously stated number of staff in that category on the payroll.

Validation

It is preferable to incorporate as many validation checks as practical into the input run, so that erroneous records can be rejected before they can enter the system. Sometimes validation checks are incorporated as a separate run which immediately follows the input routine, but generally this is wasteful. It can be a justifiable approach where input records have been created on magnetic tape by a separate off-line transcription unit, which perhaps has few facilities for validation checking. Again there could possibly be a case for minimising the processing required on an input run

241

in order to ensure that the input device is maintained running at full speed; but most modern processors have very high memory speeds and provided the input routines are efficiently designed can accommodate all the validation processing required. Some validation checks can be performed only if the transactions are first sorted to sequence, and such a validation run is therefore used after both input transcription and sorting have been completed.

Validation checks are designed to ensure that the data received into the system is correctly recorded. This may entail examining individual fields within the records to see that they conform to specified rules. Also checks may be made on the logical relationships between fields within records.

The simplest check is to ensure that fields designated as numeric contain numeric characters only, alphabetic fields alphabetic characters only and that certain columns are blank, or that certain fields do not exceed a given number of character positions. Sometimes it can be stated as a rule that a field may exhibit certain values or ranges of values only.

Generally speaking on punched cards numeric values should be right justified within the bounds of the allowed card columns; as a validation check therefore the program can reject records which contain value fields with blank columns to the right of a number.

An illustration of a logical validation is afforded in the following example. A plant hire company may keep a file of all items of equipment that they have on their inventory. Transactions entering the system represent changes in the status and or location of the machines concerned. Any transaction record entering the file updating system to notify the installation of the equipment may be expected to contain a hiring date, a hiring charge, customer number, and customer address in predetermined fields of the input record: whereas such records might be expected to contain blank columns in a field reserved for outright sale price.

Certain fields of input records may contain codes that can be checked for legitimacy on input. For example transactions entering a sales analysis system may be expected to contain in a field a product code within a predetermined range. Where the range is a continuous set of integers, say from 01 to 99 a simple comparison against the upper and lower bounds of the range can be made. On the other hand if there are gaps in the range perhaps a table look-up technique can be used to validate individual codes (see section 3.4.7).

Transaction types entering a system are identified usually by transaction codes in pre-arranged input fields, these codes must always be validated to eliminate ambiguous movements.

The construction of validation checks requires considerable thought by the analyst to ensure that only correct data is admitted. A thorough understanding of the data and the relationships of one data element to another is required. The analyst must remember that the computer programs can only cope with conditions that are specified and he must anticipate the nature of the errors that can arise in clerical, data collection, and data preparation procedures.

Input editing

Once records have been accepted as valid they are generally edited by program into the format required on the magnetic tape files. This process may include the conversion of relevant numeric values into binary form, i.e. all values to be subject to arithmetic within the system.

The re-arrangement of the fields within a record is often necessary to reconcile the differing formats required for data preparation and the format requirements of later processing runs.

Often it is necessary to translate certain character codes; for example spaces may need to be converted to zeros where they occur in fields that are to be used as keys in sorting operations.

Some input records may be expected to give rise to more than one record when stored within the system, for example, the spread card shown previously in figure 62 where a separate record will be created for each item ordered. Another example may occur in a financial accounting system where a particular input record may represent a pair of debit and credit transactions; the record can thus be duplicated in the input editing process, appropriate record type designations being added to each.

The adding of record type codes is quite common during input runs, and sometimes other data fields can be added such as *transaction date* or *transaction serial number*. Another form of input editing includes the indexing of certain codes against *look-up tables*, to select further descriptive or identifying data that cannot be readily or economically captured during data collection. For example, in an order processing system products can be identified by unique numeric codes used for input, which can be indexed via a table to select related product descriptions and prices etc. within the system.

It would probably be uneconomic to tackle this task during input editing if the number of entries in the table were large, and one might expect therefore to perform such indexing later against a series of master records on a separate magnetic tape index.

During file creation it is sometimes necessary to take on records from cards or paper tape using character codes not normally handled by the computer concerned. Tables can usually be used to examine each input character and to translate to the appropriate coding used within the computer.

Input processing

The process chart for an input transcription run is shown in figure 83. Typically we have a slow peripheral such as a card reader or paper tape reader linked via the processor to a magnetic tape deck, records being read from the slow speed peripheral to the processor where they are

243

validated, edited and output to magnetic tape. Such a run should be designed to operate at or near to the maximum speed of the slow peripheral.

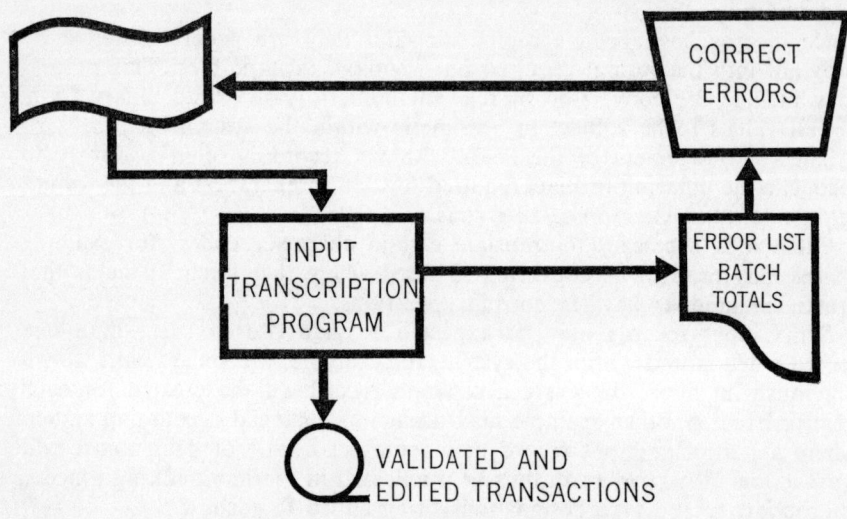

83. Input transcription with error list

Error records occurring in the run should be relatively few in number, but details of these and of batch control totals will require to be output on a line printer, or perhaps be transferred to a paper tape punch to be listed as a subsequent operation.

If error records have to be re-input before further processing operations with the transactions are performed, the input run can perhaps be designed to include an optional facility permitting a magnetic tape file containing the valid transactions to be copied to an output file, then allowing the corrected errors to be re-inserted at the back of the transaction file. A process chart showing this operation is given in figure 84.

4.4.3   Sorting and merging

Once transactions have been accepted into a system they must usually be arranged into sequence by some key field, preparatory to further validation and editing or before an updating run against a main file. Standard routines are generally provided by computer manufacturers to enable various sorting and merging functions to be performed (see sections 3.5.2 and 3.5.3). Systems analysts will find that these standard routines are convenient and efficient to use, but the analysts must guard against the tendency to use sorts without giving careful thought to the overall economics of the system. A sort run usually does nothing more than arrange items into sequence as a preparation for further processing, and in so doing occupies a considerable portion of the hardware perhaps for a long duration. The analyst should attempt to reduce sorting to a minimum by achieving a practical file structure.

Most sorts will enable a file to be arranged into either ascending or descending key sequence, and usually several keys of different significance can be specified for the same run. Often it is possible to arrange a transaction file into a sequence that will satisfy the requirements of several subsequent master file updating runs.

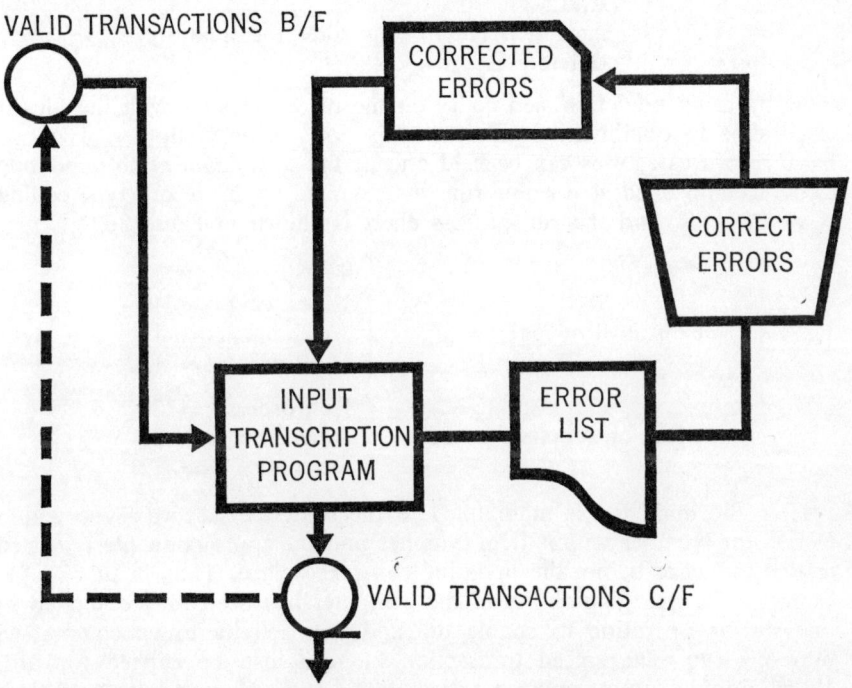

84. Re-entry of corrected transactions

The speed of a sorting operation depends on the number and size of the keys, the number of records in the file, perhaps whether the records are already in a useful sequence, the number of tape decks to be used, the number of input/output channels available to serve the decks, the speed of the tape decks, and the efficiency of the sort routine itself.

Record type coding
The problem of *what to sort* and *when to sort* may be quite complex. In a simple routine as shown in figure 82 the question seems fairly straight-

245

forward, but in practice it may not be quite so simple. Let us assume that a distribution organisation has to maintain records of items owed to its consumers, and also of items owed by suppliers to itself. These could easily be maintained as separate files and separate transaction records entering the system on different runs, can be sorted and run against their relevant master file as indicated in figure 85. In this case assume that each record on the master file contains the Item No. and Document No. as the basic keys, and that one of the data fields on each record indicates the quantity on order. Each transaction record may result in one of three basic actions on the relevant master file.

1. Add a record
2. Amend a record (e.g. alter the quantity)
3. Delete a record

Now by using predetermined codes on the records held within the files it is possible to obtain more efficient utilisation of the computer. The two basic record categories can be held on one file thus enabling the updating to be accomplished as a single run. An example of the record type coding is given below and the run process chart is shown in figure 86.

Main file

Consumer owing = 10

Order on supplier = 20

Transaction file

New owings   = 10
Amendments = 11
Deletions   = 12

New orders   = 20
Amendments = 21
Deletions   = 22

Here the main file is maintained in the sequence Record Type within Document Number within Item Number and the transaction file is sorted to this sequence before the updating run takes place. Thus in this simple example the arranging ability of the computer has been fully exploited in one sorting operation to enable the updating activity to be compressed into one run. The sorted transaction file can also be carried forward, already in the correct sequence, to maintain a stock file as a later operation.

The use of such transaction coding can ease the structure of a system considerably and can help to minimise sorting operations. In some systems, transaction coding is vital to the logic of the updating process: in the example described above it is obvious that additions to the master file must occur before amendments or deletions bearing the same keys.

The nature of the data obviously dictates the sequence of files, but the most important aspect of system design for magnetic tape computers is to achieve a file structure that avoids unnecessary runs and minimises operator activity. Each run has to be set up by loading the required transaction files, work tapes and main files, and this consumes time otherwise spent productively.

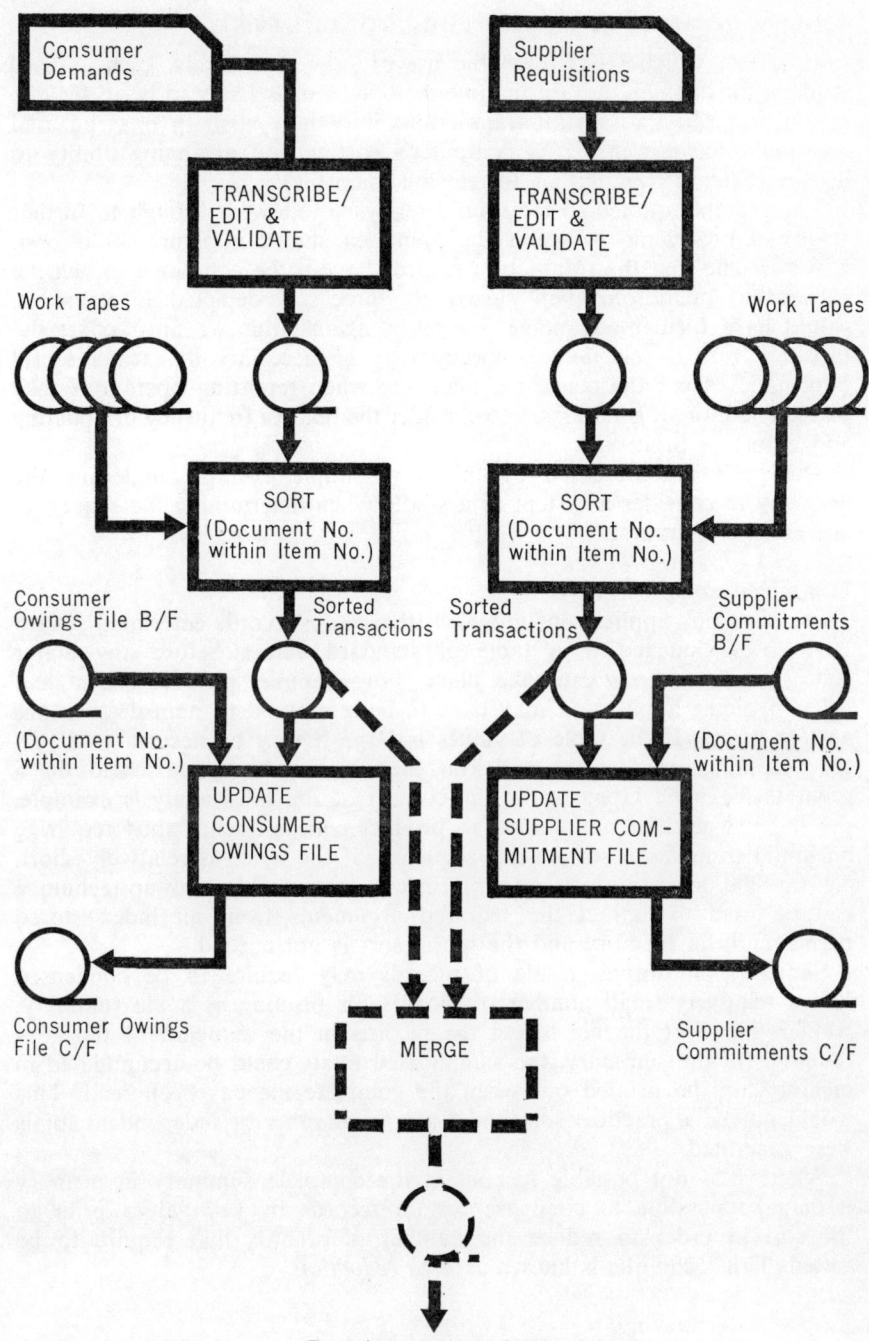

Consumer Demands

Supplier Requisitions

TRANSCRIBE/ EDIT & VALIDATE

TRANSCRIBE/ EDIT & VALIDATE

Work Tapes

Work Tapes

SORT (Document No. within Item No.)

SORT (Document No. within Item No.)

Consumer Owings File B/F

Sorted Transactions

Sorted Transactions

Supplier Commitments B/F

(Document No. within Item No.)

(Document No. within Item No.)

UPDATE CONSUMER OWINGS FILE

UPDATE SUPPLIER COM- MITMENT FILE

Consumer Owings File C/F

MERGE

Supplier Commitments C/F

To subsequent routines

85. Maintaining files with independent updating

247

A system which relies upon the use of slow peripherals, such as card readers, for full time use during intermediate runs is likely to be inefficient. It is usually best to get all transactions into the system in one run, and rely upon the power of the computer's sorting and arranging ability to deliver required transactions to relevant main files.

The example quoted so far should really be followed through to further stages of processing to see if the approach shown in figure 86 is best. If we assume that the transaction records have to be combined to achieve subsequent operations then clearly the processes depicted in figure 86 would have further advantage. However, against this we must offset the fact that further selection or sorting may be necessary to extract useful information from the combined main file when reporting operations take place. Therefore it is necessary to consider the relative frequency of updating and reporting processes.

The problems presented by this very simple example underline the necessity to consider a system as a whole when determining file sequences and sorting operations.

Minimising sort operations
There are many applications in which transaction records entering a system have to be indexed to a table of standard values before any major processing operations can take place. For example, product codes in a sales invoicing application may have to be referenced to item descriptions and/or prices. If the table of values is large it may be necessary to sort the incoming transactions and run them against an index held on a separate magnetic tape; e.g. in the context of the sales analysis example, the index would be maintained in product code sequence thus requiring an initial transaction sort to this sequence. If the index is relatively short, e.g. say 200 or so product codes are in use, then a table look-up technique can be used to extract the required arguments from an index stored permanently in memory and the initial sort is not needed.

Similarly on output, a file of records may require to be condensed into a relatively small number of records for printing as a file summary. Rather than sort the file to get the records in the sequence of the keys required for the summary, the summarised totals could be accumulated in memory and be printed out when the complete file has been read. This would not be a practical solution if a large number of independent totals were generated.

Where it is not possible to contain the complete summary in memory it may be possible to condense certain records by key values prior to the sort in order to reduce the number of records that require to be sorted. This technique is known as *data reduction*.

Editing while sorting
Many of the standard sort generators provided by computer manufacturers allow the user to perform his own processing operations on data during

248

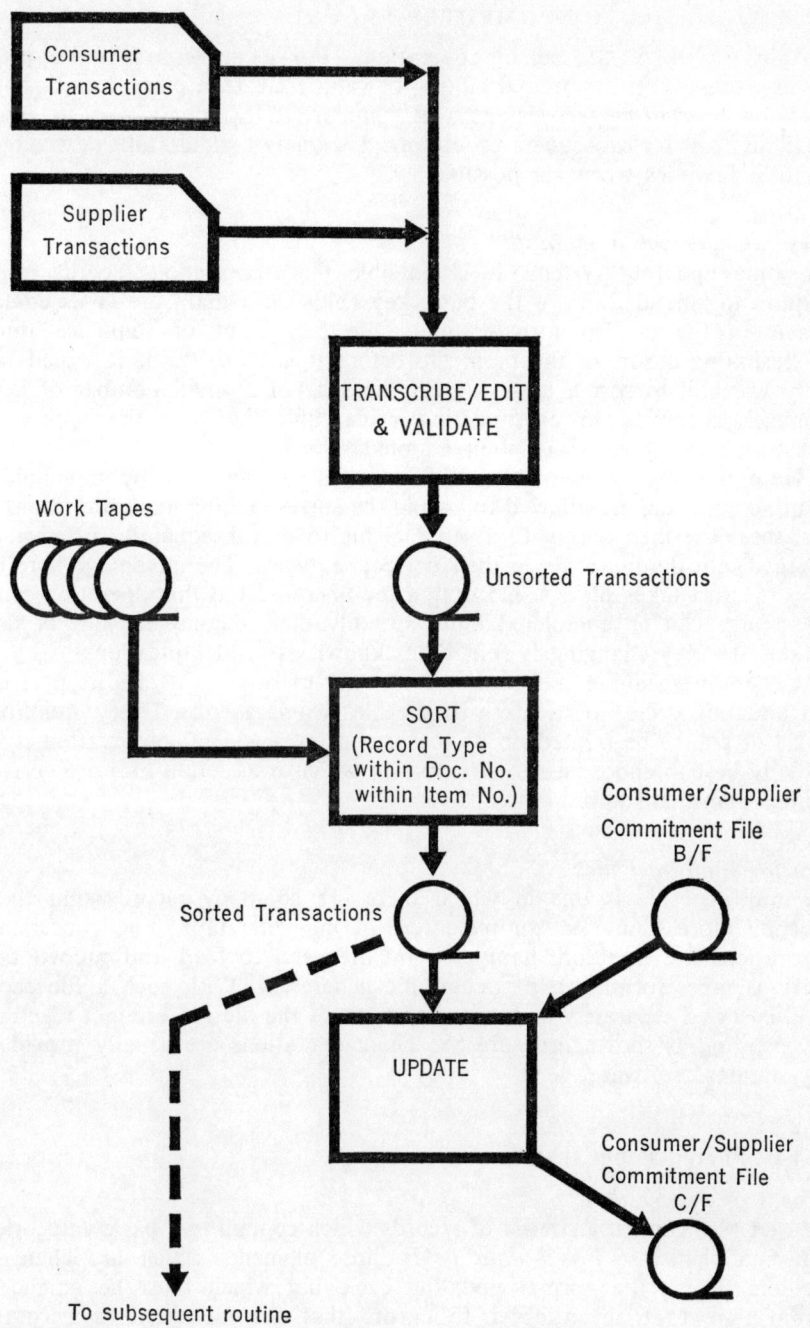

86. Maintaining different record types on one file

certain phases of the sorting operations. For example, on initial pre-stringing runs conversion and editing of values can take place, and during merging operations records may be summarised by key values to consolidate items for subsequent processing. The analyst should take advantage of these facilities whenever possible.

Key changes on a main file

In some updating systems it is possible that transaction records may require to amend some of the basic key fields on a main file. This could result in the carried forward main file being out of sequence thus necessitating a sort of the main file before it is used again. It would be very wasteful to sort a large file for the sake of a small number of key changes occurring in an updating cycle, and therefore the approach illustrated in figure 87 is often adopted.

Here the key change transactions (probably identified by a suitable transaction code) are allowed to amend the corresponding main file records, but these are then output to a separate file to be subsequently sorted and merged into the main file in their correct sequence. The advantage here is that the sort takes place using only a few records, and this operation plus the merge can be completed more speedily than a complete sort of the whole file. Key changing is sometimes known as re-identification.

Generally speaking it is wasteful of time to sort main files as part of an updating cycle—particularly if the file is a large one. Every situation must of course be treated on its merits, but as a general observation it is usually best to choose a run structure in which transaction files are sorted rather than main files.

Sorting multi-reel files

A multi-reel file is one in which there are so many records that they occupy more than one complete reel of magnetic tape. File processing operations are therefore hampered by the need to load and unload the various tapes forming a particular file generation. With such a file each reel is sorted separately and sorted sections of the file are brought together by merging as shown in figure 88. These operations are usually provided by standard software.

4.4.4 File updating methods

Types of master data

Master files generally consist of records which contain two basic categories of data elements: *Fixed data fields* those elements which are changed seldom during the normal updating cycle but which must be amended from time to time in order to ensure that the file contains accurate descriptive, historical, or identifying data. *Variable data fields* those fields which represent conditions or values describing current conditions in the

250

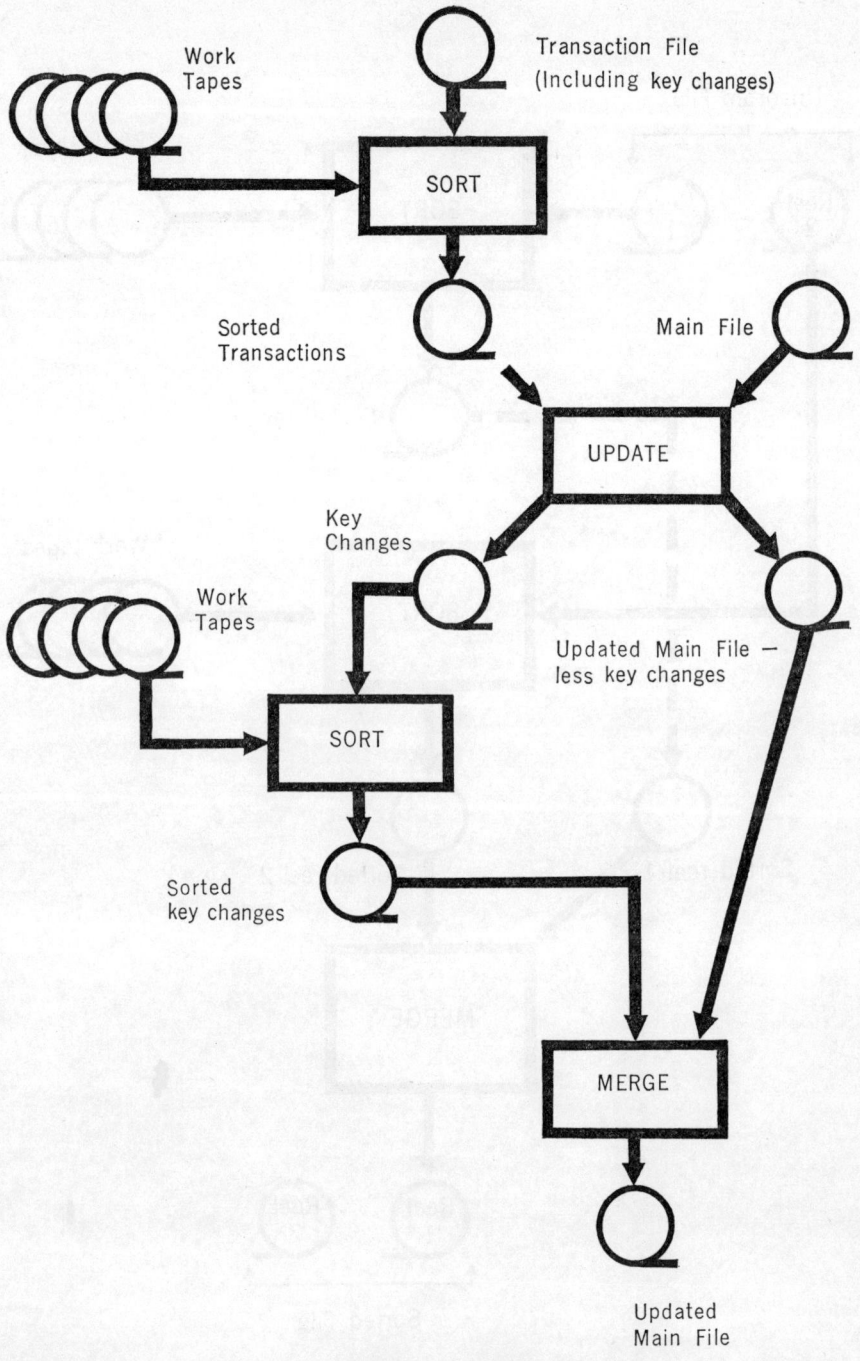

87. Dealing with key changes

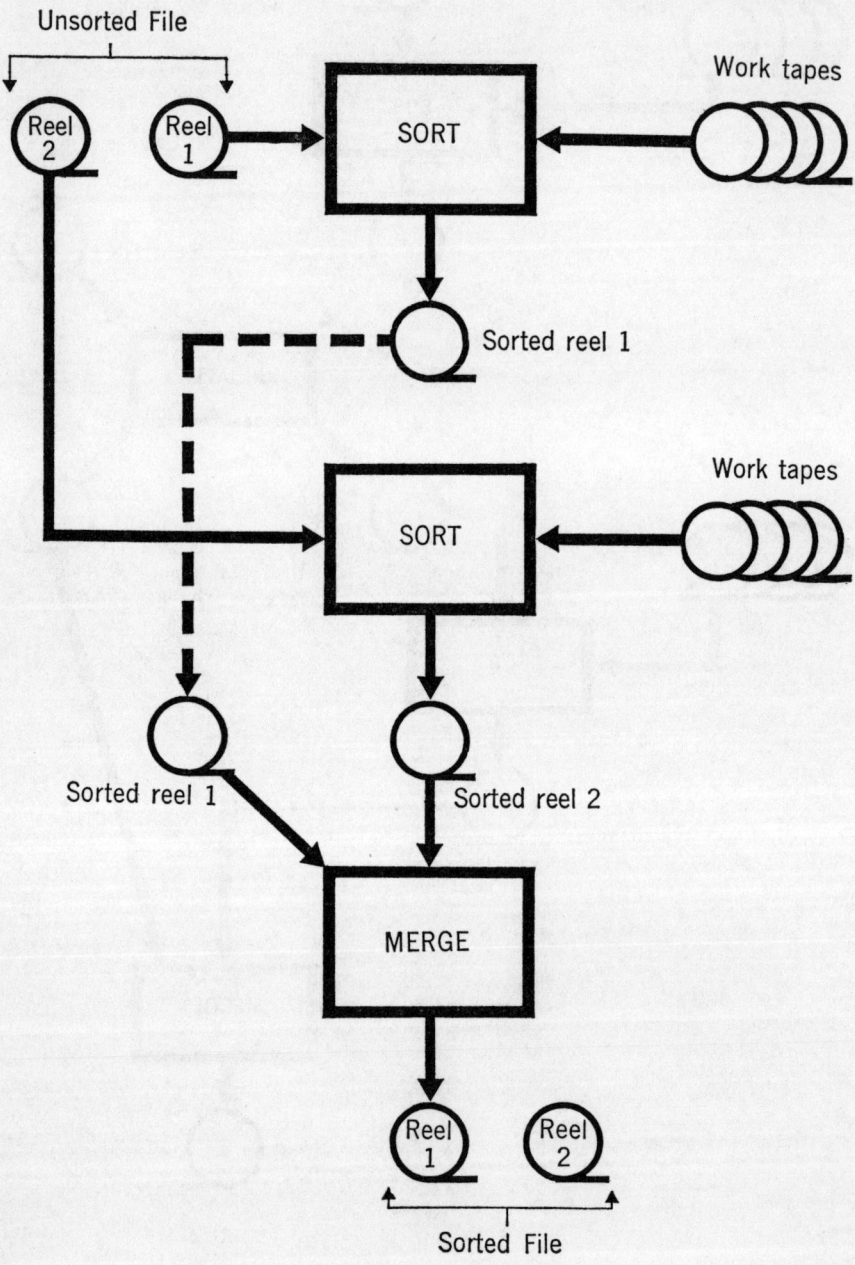

88. Sorting multi-reel files

physical system under consideration, and which can be amended by transaction records during each updating cycle.

To illustrate this principle refer to figure 89 which shows the data elements for a stock file in a simple stock control routine. The fields designated as fixed data may be altered from time to time—e.g., a part-number may be superseded, a new cost may be established, descriptions

STOCK MASTER FILE — Sequence : Part No. within Depot No.				
DATA ELEMENTS	Fixed or Variable	FORMAT	CHARACTERS	WORDS
Part No.	Fixed	BCD	12	} 4
Depot No.	..	BCD	2	
Description	..	BCD	20	5
Cost	..	binary	—	1
Purchase Tax	..	..	—	1
Total Stock	Variable	..	—	1
Total Shortages	..	..	—	1
Total Orders on Suppliers	..	..	—	1
Demand this Month	..	..	—	1
Average Monthly Demand	..	..	—	1
Total Value of Stock	..	..	—	2
Economic Re-order Quantity	..	..	—	1
				19 words

BCD = BINARY CODED DECIMAL

SORTED TRANSACTION FILE — Sequence : Type within Part No. within Depot No.				
DATA ELEMENTS	Fixed or Variable	FORMAT	CHARACTERS	WORDS
Part No.	Fixed	BCD	12	} 4
Depot No.	Variable	BCD	2	
Transaction Type	..	BCD	2	
Customer Order No./Doc No.	..	BCD	6	2
Customer Code/Supplier Code	..	BCD	6	2
Quantity	..	binary	—	1
Date	..	binary	—	1
				10 words

89. Example of record structure

may be inaccurate and require amendment. On the other hand many of those fields designated as variable may be expected to change regularly as materials are issued or received into stock perhaps several times a day. Some of the variable fields on this record could on the other hand be re-calculated monthly only—e.g. the Average Monthly Demand.

Distinctions between updating and maintenance

Sometimes the term *file maintenance* is used exclusively to refer to the amendment of fixed fields in a file—such a run might include adding new records or deleting old ones. The term *file updating* is often being reserved to describe the routine processing operations associated with the so called variable fields. In fact these terms are used in this book because they help to draw attention to the different problems and techniques associated with using computer files, but it should be noted that in practice the distinctions are not always apparent. The file maintenance and updating operations are often combined in the same run, and certainly the terms are often used in either context. In any event the basic file handling techniques are the same.

The nature of basic updating runs

So far it has been described that transaction records enter the computer system from some input media, and must be transcribed to magnetic tape, be validated and possibly edited, and then sorted to the required sequence for running against a master file.

One of the most important aspects of design is to choose a suitable sequence for the master file bearing in mind that this will have an effect on the preparatory runs described above, as well as an immediate impact on the updating and reporting functions. It has already been stressed in the previous chapter that sorting and merging should be kept to a minimum. The analyst must consider the number of preparatory sorts required prior to updating, and any subsequent arranging runs, to deliver error prints, exception reports and file prints in the correct sequence to users.

The keys chosen for sequencing a master file should not be too susceptible to change otherwise file maintenance problems are exaggerated. For example, a file might contain records of equipment on hire to various customers—it would be best to maintain the file by equipment serial number rather than customer number since the latter will change fairly frequently thus requiring frequent sorting operations.

In the previous chapter the use of transaction coding was shown to have advantages when performing preparatory sorts; such codes also simplify the logic of updating runs since they enable transactions of various types to be presented to their master records in an appropriate sequence for the updating process. For example, additions to a master file should be processed before any change records requiring access to the addition records. Again there is the possibility that certain transaction types may require a large subroutine to be called down from backing store into main memory. It is more efficient to minimise this operation by grouping the transaction types requiring this subroutine wherever possible.

A process chart representing the updating of the stock file shown in figure 89 is given in figure 90. Notice that the master file is held in the sequence Part No. > Depot No. (Part No. within Depot No.) and that the corresponding transaction file is sorted to Transaction Code > Part Number > Depot No.

Notice that this updating run gives rise to three output tapes, the updated master file, an error file containing for example unmatched transactions, and a changes journal containing transactions which have been successfully applied to the file. The changes journal is probably required for use in a further stage of the system, for example in producing demand statistics, financial ledger entries, or auditors reports.

The error report in this example is listed in depot sequence for the attention of data control staff at the depots. These errors must be investigated so that the transactions can be corrected and resubmitted on the next processing cycle, perhaps highlighting that some maintenance transactions require to be applied to the main file.

Processing operations in a simple updating run

The processing required during an updating run is relatively simple. Essentially the run will start by reading in a block of records from the master file and a block of records from the transaction file. The records in each block must be examined one by one and their keys are compared by the updating program in accordance with the rules shown in the flowchart in figure 91.

If at any stage the keys of the current transaction record exceed the keys of the current master record (see B in flowchart) then the master record is placed in an output area ready to be written to the carried forward master file, and the next master record from the input area is retrieved and compared with the transaction record. If the current transaction record has the same keys as the current master then the appropriate updating routine is entered to amend the master according to the type of transaction encountered. Any further transactions relevant to the master record will then be retrieved and applied one by one to the master record; as each transaction is actioned it is transferred to an output area ready to be written to the changes journal. The master record is retained throughout this process and is continually updated until a key of higher significance is encountered on the transaction file and the updated current master is transferred for output and the next master record is selected.

Transactions appearing on the transaction file and not matched on the master file may be errors (e.g. erroneous keys have been punched) or may be additions. On reference back to figure 90 it can be seen that additions on the transaction file were distinguished by the code 01. Thus such transactions would be written to the carried forward master file, but other unmatched transaction types would be written to the error file.

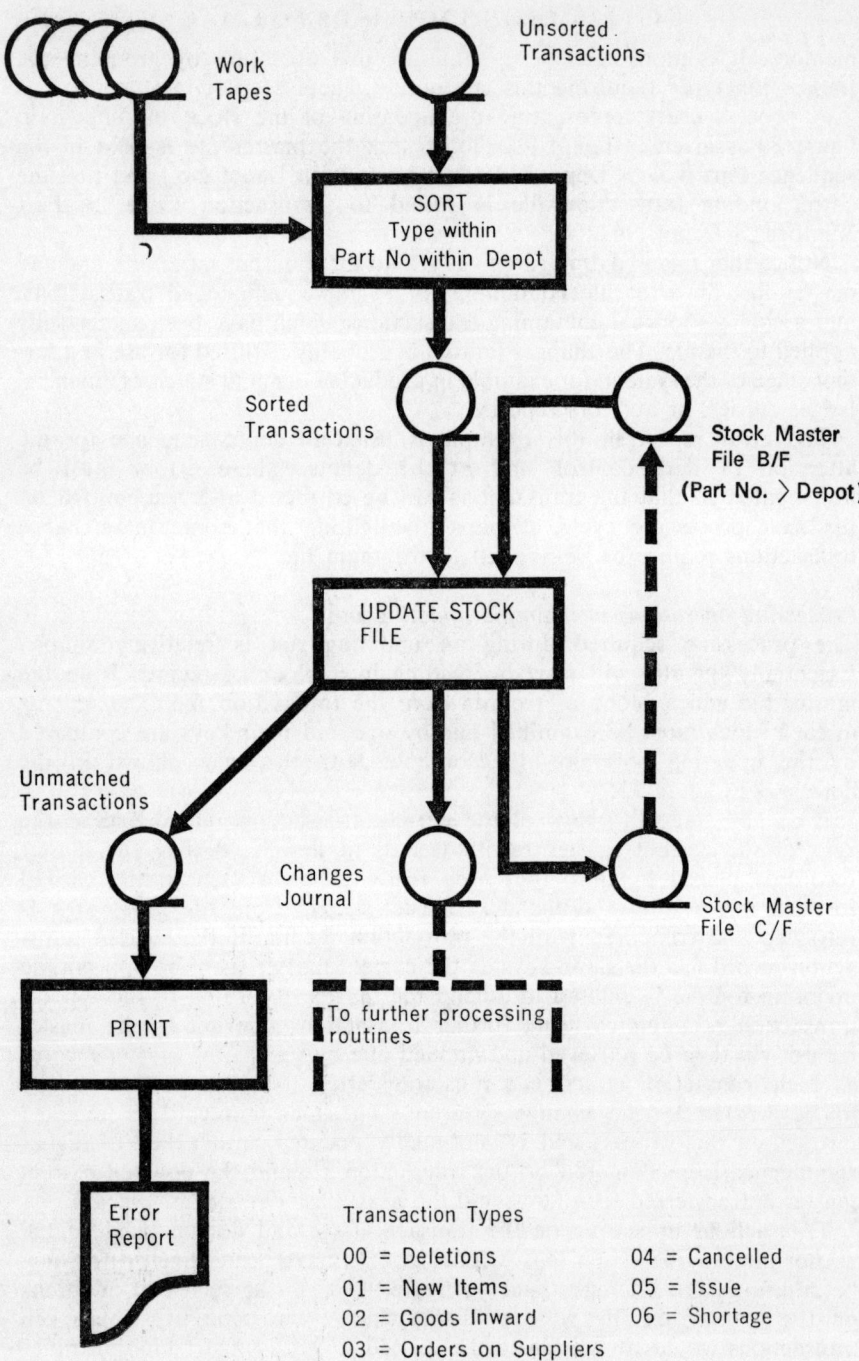

Work Tapes

Unsorted Transactions

SORT
Type within
Part No within Depot

Sorted Transactions

Stock Master File B/F
(Part No. > Depot)

UPDATE STOCK FILE

Unmatched Transactions

Changes Journal

Stock Master File C/F

PRINT

To further processing routines

Error Report

Transaction Types

00 = Deletions   04 = Cancelled
01 = New Items   05 = Issue
02 = Goods Inward   06 = Shortage
03 = Orders on Suppliers

90. Updating with errors file and changes journal

Note 1 : When this routine is entered the first transaction and first master have been read and are available for comparison.

Note 2 : For simplicity end of file conditions are ignored.

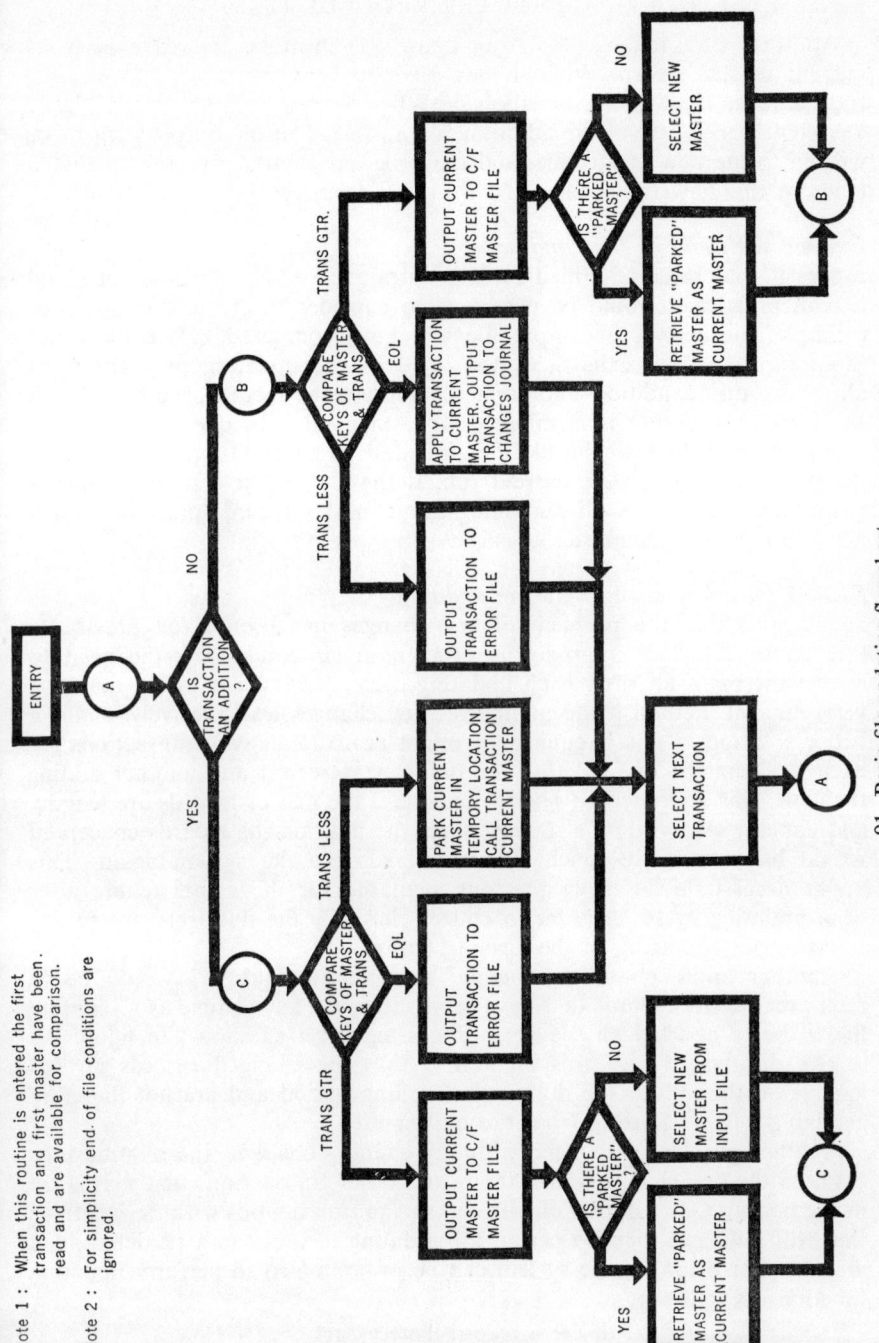

91. Basic file updating flowchart

257

Addition transactions (see A in figure 91) must be treated exactly as current master records, in that they have to be retained so that ensuing transactions, if any, can be applied. When a new transaction of higher key significance arrives the addition is transferred to the output area to be written to the new master file, and the program continues to seek a match between this new transaction and a master from the input master file.

Dealing with end of file conditions

Figure 91 has been simplified to demonstrate the basic principles involved and in practice it would be necessary to consider further conditions. For example, the end of one input file will be encountered before the other input file has been exhausted; the logic of the updating program must allow for this condition and treat all subsequent records accordingly. If the input master file is terminated first, then all subsequent transactions should be additions to the file, or amendments to additions coming from the transaction file in the current run. If the end of the transaction file is encountered first, then all remaining input master records must be copied across to the output master.

Dealing with key changes during updating

An approach to the problem of key changes has been given previously (see figure 87). This approach is often used since it avoids the need to re-sort the main file after each updating cycle. However it would not be a very efficient method if the number of key changes was relatively small; a better solution in this circumstance might be to input two transactions for each key change, one deleting the old master record and another adding the new. This approach however is tedious if the master records are lengthy and contain several data elements; since the data on the records concerned would have to be re-punched and entered into the system again. This might necessitate having a printout available for the complete file after each updating cycle, in order to collect clerically the data relevant to any key changes occurring in the ensuing period.

Another approach to the method outlined in figure 87 would be to retain the master records that have been subject to key change as a separate file to be re-inserted on the next processing cycle as shown in figure 92.

The disadvantage of this method is that the changed records do not appear on the master file during the ensuing period and are not therefore available when reporting from the master file.

Another problem associated with key changes concerns the relationships between the key change transactions, and other transactions that may arise in the period. One solution entails coding the transactions with designations that will sequence them to permit all updating to occur in a predetermined manner. For example, the system can be programmed to perform updating functions as follows:

1. delete a record if necessary
2. add a new record if necessary

3. apply amendments to the record
4. change keys where required.

This merely provides rules for the updating program but doesn't help to control events that arise outside the computer procedures. Unless the data collecting procedures are able to ensure that all amendment transactions are given keys corresponding to the key to be used in the particular updating cycle, the problem still exists. Perhaps it may be suitable to simply allow the offending records to be mismatched, and to appear on an error list to be inserted correctly on the next cycle.

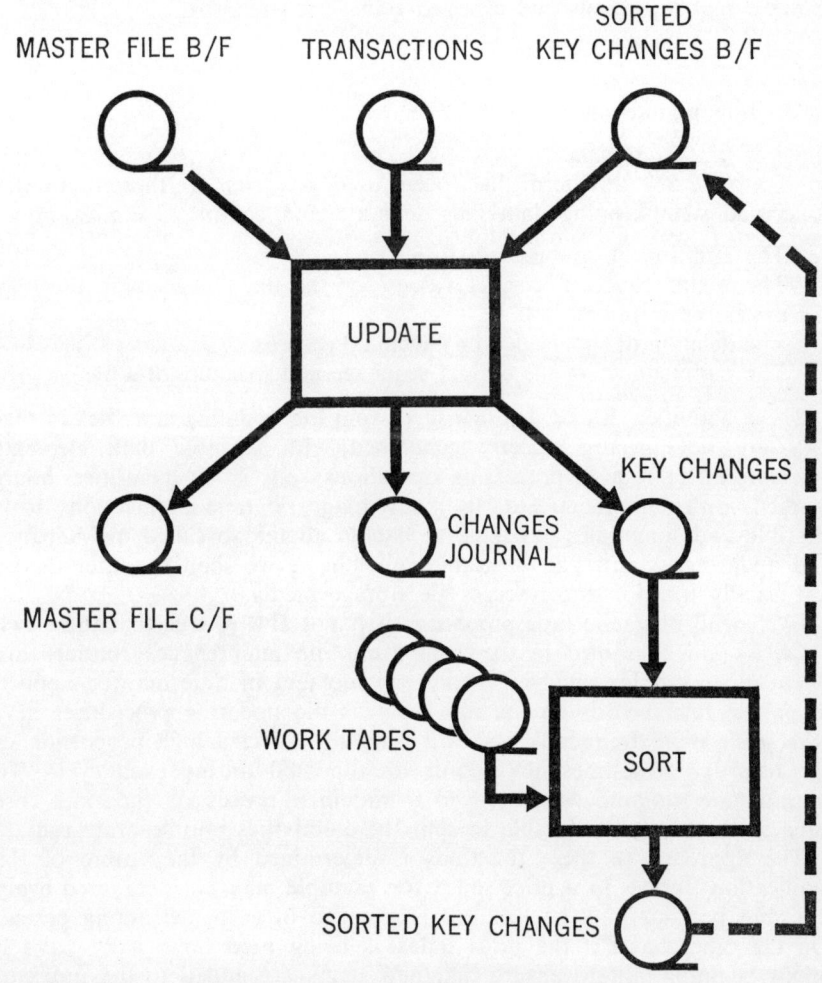

92. Updating with key changes carried forward

There is no doubt that magnetic tape systems are relatively clumsy when dealing with key changes and if such transactions form an important aspect of the system further complexities may need to be included.

Probably the best solution involves keeping a *ghost* record on the master file for all superseded master records; ghost records will contain an appropriate code and the keys of the new master only.

Thus if any transactions quoting superseded keys are entered into the system, their keys will be changed on being matched with the ghost item and they will be output to the key changes file. This file may then be sorted and used to update the main file once more as a secondary updating run. At any moment in time the current key changes files may hold both changed master records and changed transaction records.

4.4.5 File maintenance

What is file maintenance?
In section 3.5.4 this term has been used to refer to those activities concerned with keeping data files accurate and complete. For example:

1. The addition of new records to a file.
2. The maintenance of correct values for the fixed fields of a file (e.g. prices, descriptions).
3. The deletion of superseded or unwanted records.
4. The maintenance of the correct sequence and structure of a file.

These activities can be distinguished from file updating activities in that they are not usually directly concerned with variable data elements encountered in routine processing operations—e.g. stock quantities, hours worked, order numbers, etc. In many magnetic tape applications it is possible and sometimes desirable to handle all the so called maintenance and updating activities in the same run, but as we shall see later this is not usually true for direct access file storage media.

Even with magnetic tape processing it is not always true that these two activities are combined in the same run. File maintenance routines are often quite complex and yet at any one moment in time may be applied to only a few records on the file, whereas file updating procedures may take place more frequently and will generally affect a high proportion of the records—sometimes all records are updated during each cycle. To simplify the program structure and to minimise processing time and core capacity it may be preferable to split these activities into separate runs.

The approach to these functions is determined by the nature of the application—prices in a price index for example may not be altered every day, but perhaps may be received at the end of each accounting period. On the other hand if the price index is being used on a daily basis it might be important to ensure that new items are added to the index at the beginning of each day.

User participation

The real problem with magnetic tape systems is not how to maintain the file from a programming standpoint, but how to ensure that user departments keep the files notified of conditions arising outside the system. This is particularly necessary for files used as indexes to other data files or transactions.

The important aim is to ensure that user departments have responsibility for maintaining files in respect of new items or changes to existing items. Generally it is preferable that these amendments arise directly as a by-product of work done routinely in the user departments concerned. Where such changes are likely to affect the results of routine data processing, facilities must be available to permit users to amend files before routine updating transactions are applied. These may be done during an updating run by dealing with maintenance transactions before the updating transactions—i.e. by suitable transaction coding to give precedence to file maintenance amendments. We have already seen (4.4.4) that additions, deletions and key changes can be effected during file updating.

Techniques of file maintenance

It is important to ensure that additions of master records are recognised, before transactions are received relating to them. Otherwise unnecessary error reports will be generated for the unmatched transactions. This essentially entails picking up the additions from the department originating them. One way of doing this is to make the originating department dependent upon the computer system for carrying out their function. As an example, consider a personnel records system; assume that a separate record is maintained for each employee showing his personal statistics and details about his salary, leave entitlement, hours of work, department, industrial classification, etc. This central file can be the basis for the payroll and pension fund systems, as well as providing general management information for manpower analysis. Such a file could contain a great deal of data for each employee, but only some of this data can be considered essential for legal or administrative purposes. One way of ensuring that records are created for new employers is to make the computer system generate the formal offer made to each new member of the staff. In this way the system can be inhibited from issuing such an offer until certain standard data elements have been supplied to it—e.g. name and address, basic salary, leave entitlement, etc.—and other conditions of employment. On receiving this data the system can automatically create a record for the new employee if necessary generating a staff number. Subsequent changes in the status or working conditions of staff members can be handled in a similar manner, all such notifications being routed via the computer files.

Maintenance of values for fixed fields

By their nature fixed fields are not changed daily by routine transactions,

261

but remain more or less permanent features of a record, although changes may take place from time to time. In most records there may be several fields falling within this category and a routine has to be established for dealing with them. In a personnel records system for example the 'hours worked this month to date' may be updated every day or at least once per week, but the employee's name will change only rarely—perhaps never in the case of male employees. It would not be unusual in such a file to allow fields concerned with the generation of the payroll to be changed before the payroll e.g. on a weekly basis, while other data elements, less immediately important, are maintained once per month. There could be a large number of the so called fixed fields in each record and a separate maintenance run might therefore be more economic.

The maintenance of such fields is usually straightforward; sometimes a field has to be deleted, or a new value has to be substituted. Separate transactions records may be needed for groups of data elements on a main file record; where a field is to be deleted the corresponding transaction field can be punched with all zeros; where a new value is to be inserted it can be input in the appropriate transaction field; but if a field on the main file is to remain unchanged then, should a corresponding transaction record otherwise occur for the group of fields concerned, the particular field will be blank (all spaces).

Sometimes certain fixed fields are directly dependent upon other fields —e.g. in a personnel records system the leave entitlement may be directly derived from *pay scale* and *years of service*. Thus a change of one field implies an automatic change of another. Changes of this nature should be generated by letting the computer program do the work: the user department which maintains the file should not be expected to codify more fields than are absolutely necessary.

Nevertheless it is as well to allow for the possibility of all fields being independently amended by direct input. A person's *date of birth* will obviously be a fixed item, but it is as well to permit amendments to this field since coding errors can create incorrect values.

Deletion of records
The choices open to the analyst depend upon the nature of the system—in some cases redundant records can be permitted on the file without affecting the results produced, but in other cases they cannot. For example, in a hire purchase accounting file, it is essential to ensure that records are not active after final payments on contracts have been made. Even this does not imply necessarily that the record need be deleted, since the status of a particular field of the record may indicate that the account has been completed. Sometimes inactive records can be automatically written to history files so that access is still possible should some historical analysis be required.

Obviously the main reason to strip inactive records from main files is to improve the economics of file updating and reporting. Where there is some formal event that designates a record as redundant it is as well to strip it from the file during routine updating; if on the other hand such a condition can be distinguished only by an analysis of its updating history then, perhaps, appropriate routines should be incorporated as a separate maintenance function at appropriate intervals.

Maintenance of correct sequence and structure
In magnetic tape file processing the sequence of a file has to be strictly maintained according to the keys of the records in the file. This is guaranteed by the basic file updating techniques previously described. The main problems concern key changes and some solutions to this have been discussed in section 4.4.4.

However there are other problems affecting file structure against which the analyst must be safeguarded. One of these concerns the duplication of records—i.e. the existence of two records with the same keys on the same master file. Such a situation can cause confusion during file updating and it is as well to search a file for this condition as part of the standard file maintenance routines. Master files newly created from punched card files are often prone to this condition, but thereafter it is possible to prevent duplicate master records being created during routine updating.

Sometimes data elements have to appear in more than one master record; where this condition applies the analyst must be sure to update the relevant files to the same level of modification. This is particularly important where the elements thus superseded are used as keys. For example, in a manufacturing and distribution organisation *item codes* may be used as keys for several files. If one item code is superseded by another, all files using the item should be amended.

Need for maintenance
If the general principles given above are not applied the results produced by a system can often be inaccurate or at least be inconsistent. The economics of file processing may be impaired by the production of needless error reports or through time-consuming reading and writing of redundant data. If the maintenance routines can be accommodated during routine updating without loss of efficiency or without unnecessary processing complexities then they should be so incorporated. Otherwise a routine analysis and/or amendment of all files, should be undertaken at predetermined intervals.

4.4.6 Error reporting

Error conditions
In previous discussions concerning data validation (4.4.2) we described some of the ways in which data has to be checked when it enters a computer procedure. This type of checking is generally aimed at checking the accuracy of data elements within an individual record and serves only as an initial

check upon the item. During later stages of processing other error situations may occur relating to the relationships between the transactions and other records stored within the system. For example, in an updating run a transaction may fail to match a corresponding main file record and it must then be assumed that either the main file has not been adequately maintained, or the transaction has entered the system with the keys incorrectly coded. In either event it is necessary to display the record for the data control staff, so that it can be examined and re-entered into the system on the next cycle.

Some error situations may be more complex, being governed by some logical condition existing between a transaction and its corresponding main file record. In this situation both the main file record and transaction must be displayed for correction.

To illustrate these conditions assume that a file is maintained showing orders placed upon suppliers for commodities. Every time an order is placed, a record is posted to an Order File on magnetic tape showing the commodity code, order number, quantity ordered, supplier code, date placed, expected delivery date. When consignments are received from suppliers, a copy of the goods received documentation is used to create computer transactions quoting, commodity code, order number, quantity received, supplier code, date received, and suppliers' dispatch numbers. The goods inwards transactions have to be matched against the original orders so that satisfied orders are deleted from this Order File during an updating run; the keys are commodity code and order number. Now consider the error conditions that could arise when processing, say, just the goods inwards transactions:

1. The input validation could reject a goods inwards record—e.g. because the commodity code was invalid, or the order number was outside the range of numbers assigned.
2. A goods inwards transaction could fail to match against an original order on the Order File.
3. The transaction could match but give rise to a spurious condition, e.g. the quantity received could exceed the original quantity ordered.

Now these errors can be difficult to trace, and will certainly cost time and money in rectifying. Take case 2 above—has the goods inwards transaction been coded with the wrong keys, or was the original order transaction lost or even written to the file originally with incorrect keys? Again in case 3—has the goods inwards transaction been coded with the wrong quantity field or was the original order record given the wrong quantity? has the supplier genuinely made an error, or are we again troubled with wrong keys? And so on.

Obviously it is not possible in this situation for the computer system to correct the condition—the choice of errors is too wide, and the potential sources of error in the external procedures may be numerous. Clearly with efficient punching and verifying routines many transcription errors can be eliminated during data preparation, but no system is foolproof and clerical errors and illegibility of external documentation will cause such conditions to arise.

Dealing with errors

The simple example above demonstrates that error conditions in most commercial systems have to be reviewed by persons in the line departments responsible (e.g. the buying office, or goods inwards department). The systems analyst must however ensure that all error conditions are displayed with sufficient detail to ease the job of correction. The format and content of the error report are vitally important, and the computer should be utilised to draw attention to the nature of the error whenever possible.

Error reports can be made directly onto a line printer during the updating run as shown previously in figure 82. Such a report would then be in the sequence of the keys used in the updating run, which might be useful if, say, one department were handling all error types. However, it might be advantageous to output errors onto paper tape or magnetic tape so that a sort run may be employed to sequence errors by type before printing them. The automatic generation of error codes helps to identify and arrange error records.

To assist the data collection and preparation staff it is generally useful to carry a batch number as part of each transaction record, so that this can be output as part of any error record to speed up the selection and checking against original documents.

It is important to encourage staff concerned with data control to keep records of error conditions and to analyse their cause and frequency. Where persistent errors occur despite training and consultation with the originators, the systems analyst should attempt to improve the data collection techniques employed. For example, to use turn around documentation employing, say, OCR or MICR techniques. It is essential for the analyst to monitor the error reporting functions of any system that he may have installed, even though the system may have been running productively for many months.

Re-entry of corrections

Corrections must always be re-entered into the system, with due regard to the activity previously incurred in processing the original error transaction. For example, where systems are integrated so that basic transactions form input to several systems they may pass successfully through several updating phases before causing an error condition at a later phase. The method of re-entering the transaction must avoid duplication of the transaction at earlier stages. Figure 93 indicates such an example.

Errors and control totals

The subject of control totals has been dealt with in sections 4.3.1 to 4.3.4;

265

I*

it is worth stressing at this stage that all control totalling routines must be able to take cognisance of error conditions occurring in a system. With magnetic tape systems it is generally required to accommodate record counts and totals of selected value fields from records so that control totals can be established during file processing. All error transactions must generally be accumulated as a separate category so that a reconciliation of records input and output can be achieved for each run.

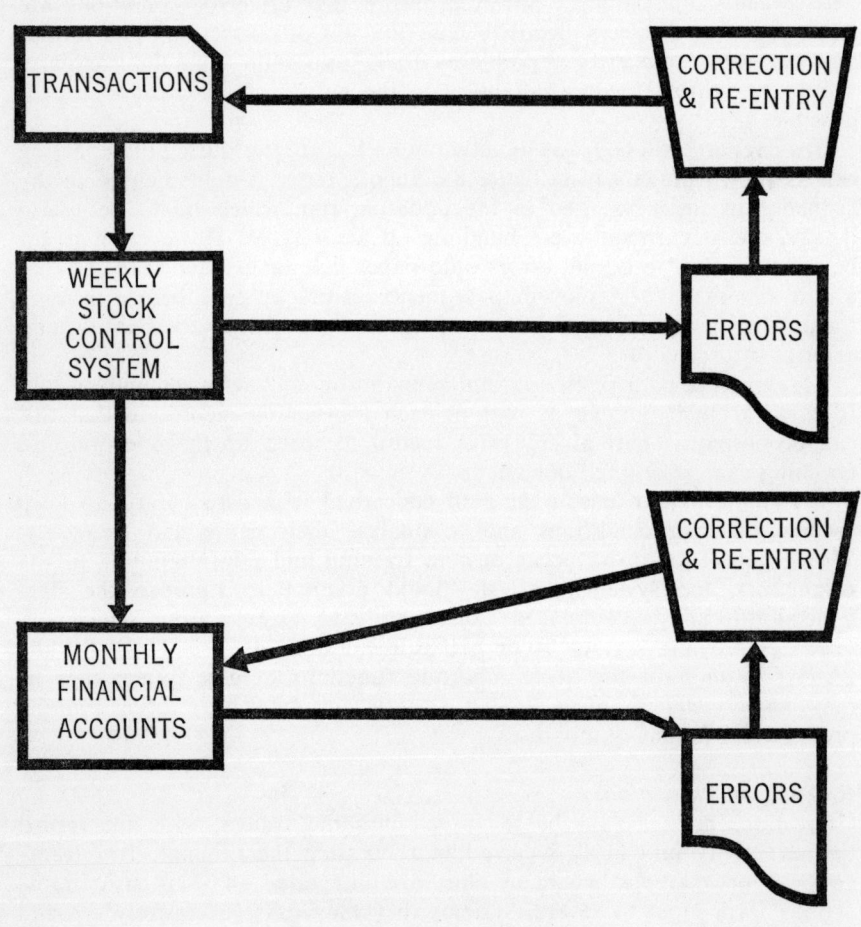

93. Re-entry avoiding duplication

Maintenance of an error file
The treatment of errors varies according to the nature of the system. In some cases it is sufficient simply to produce an error list and rely upon

the data control staff to expedite corrections. However, in certain cases it is best to retain error transactions on magnetic tape until they can be positively removed from the file. Figure 94 shows a method of retaining an error file.

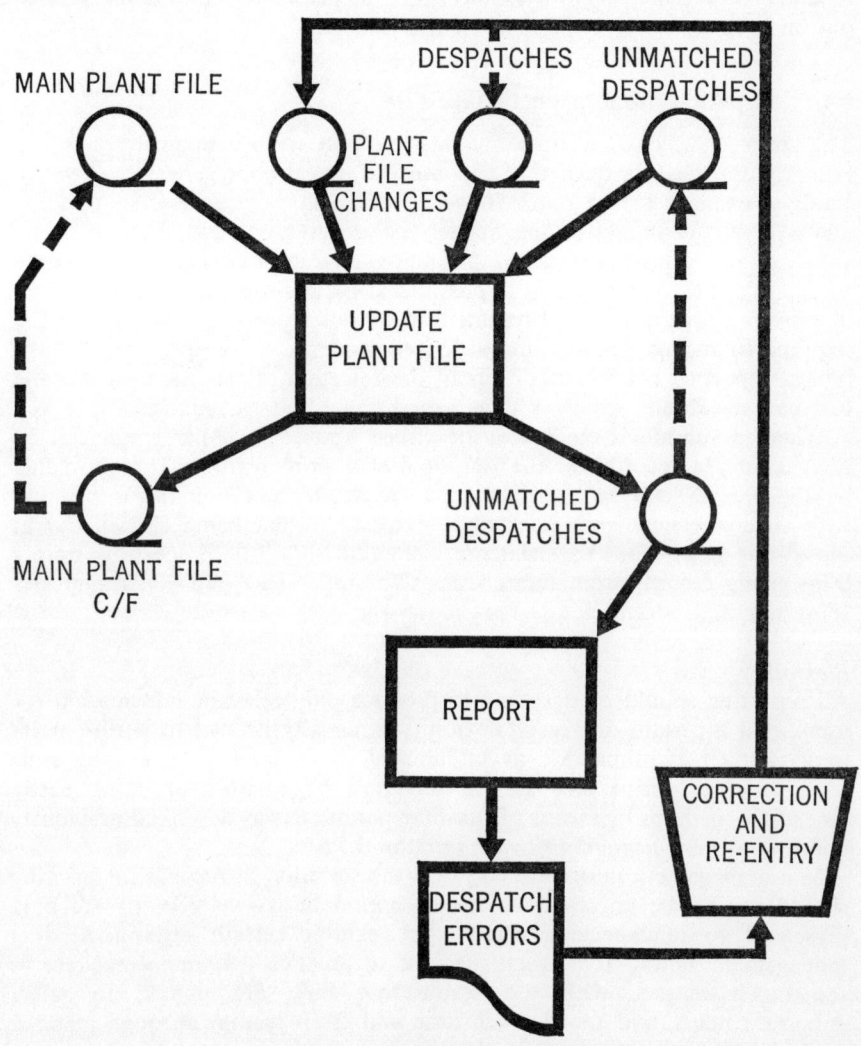

94. Updating with circulating error file

The example is for a plant hire company; there is a master file which contains individual records for every item of capital equipment in the company's inventory. The errors are transactions representing machines

267

despatched on loan to customers, but which are not matched during an attempt to update the master file. Since the despatch transactions represent revenue to the company concerned, it is important to retain the error records until positive action can be taken to update the master plant file or until the despatch record is cancelled. The errors are continually printed out on each updating cycle if no action is taken.

4.4.7 Reporting from magnetic tape files

The simplest way to generate an output report from a magnetic tape file is to list the records from the file onto the line printer generating control totals at changes in key data. However most master files tend to be large and a complete listing of the master file is rarely needed. Usually some subset of the records is required or a summary report is needed of the data held on the file. Sometimes it is possible to maintain a file in the sequence in which records are to be presented on output reports, but often a sort is required to arrange records into a desired sequence before printing. Thus a typical reporting routine might entail the selection of records from the file to create a sub-file, which is then sorted and listed as required.

Where a sub-file is created as described above, the editing run can be used to create records on the output file in print format. That is, fields on the records can be arranged into the sequence required for printing with all conversion from internal to external format being carried out at the editing stage. Thus the print run will entail very little processing, simply transferring records from input areas to output areas and generating any sub-totals and control totals required.

Flexibility

All reporting should be designed to produce only relevant information for users, and a certain degree of flexibility is usually needed to permit users to vary their requirements as circumstances demand. The editing runs used to create print files should therefore be capable of being easily amended—perhaps by means of run-time parameters as described previously under *data management software* (section 3.5.5).

In a management information system the selection of records for printing should usually be governed on the exception basis—i.e. records are only presented to management where they exhibit certain criteria needing management action. It is often the case in practice that managers receive too much output, with the result that they are unable to select important items, and spend much time and effort wading through massive reports. The decision rules for selecting records for printing must be clearly established during the design of the system even where it is known that run time variations are permitted.

It is best to anticipate all variants of the report, so that run-time variations can be catered for. A thorough discussion and understanding of the reporting capabilities is necessary—so that both computer staff and user departments are fully aware of the type of analyses available from particular computer files.

Economy of print out

Printing on a line printer is always a comparatively slow operation and the analyst should ensure that the output formats are designed to minimise the number of lines to be printed. Of course such economy should not be achieved by simply cramming information into every line thus rendering the report unreadable. One method of minimising printer time is to produce two output documents simultaneously side by side on the printer; e.g. pay advice slips along with a payroll analysis, or invoices along with shipping documentation.

Line printers often require a lot of attention during operational running, and the analyst can minimise this unproductive time by ensuring that stationery can be set up easily. First, it is best to use standard paper sizes wherever possible and to standardise on the widths of margins at the edge of the stationery. If possible design forms to use standard paper-tape control loops for controlling vertical spacing and throwing of paper.

It is a good idea to insist that all printing programs incorporate a standard alignment routine whenever preprinted stationery is required. For example, the program could on entry be arranged to print a block of test characters and then stop allowing the operator to position the stationery. The program is then re-activated and the test block printed once more; this process continues until the test characters appear in the right position, whereupon the program is entered at another point to commence printing the file data.

Most print programs require headings to be printed at the top of each page: it is often best to enable such headings to be loaded into the program as parameters at run-time, thus permitting variations to be made from one report to the next. Page counts are usually maintained and permit a page number to be printed at the top of each sheet of the report.

It is important to ensure that restart facilities are incorporated in any lengthy reporting programs, so that it is not necessary to restart from the beginning of the run in the event of line printer failure. Restarts can be made by simply re-opening the input file and reading without printing until some condition nominated at run-time is reached.

Such conditions might include a particular block number, a record number, a specified key, or a page number (see section 3.5.8, Dumps and Restarts). During the run in to the specified condition all control totals and control breaks would have to be recognised and maintained.

4.4.8 Multiprogramming with magnetic tape files

In the earlier description of the principles of multiprogramming (section 3.1.6) it was shown that multiprogramming presents opportunities to utilise the power of a central processor and its peripherals more extensively. In a

single programming machine the processor is often momentarily idle awaiting the termination of some activity concerned with a peripheral unit,
but in a multiprogramming system these intervals can be used to perform
processing in another program or to initiate peripheral operations for
another program.

To obtain maximum operational benefit from these characteristics, the
systems analyst should design his computer procedures to ensure that
particular programs do not utilise hardware resources uneconomically.
For example if a processor has a main memory of 32,000 words and
8 tape decks, there will be little opportunity to secure these advantages if
a single program occupies say 28,000 words, requires 7 tape decks, a paper
tape reader, and a line printer. Certain standards have to be observed in
designing computer runs to suit the hardware available.

Peripheral-limited and processor-limited runs
Programs which utilise slow speed peripherals such as card readers, paper
tape readers, and line printers are generally peripheral-limited: that is,
the total time required for the run is governed by the physical speed of
the device concerned. During such runs the processor will constantly stop
processing data while it awaits the completion of a transfer to or from
the device concerned. If the volumes of input/output data are reasonably
high a great deal of processing time remains unused.

A processor-limited run is one in which the amount of processing
required for each record is such that delays are experienced in issuing
peripheral instructions, because processing for the current record is not
completed in time. Severely processor-limited runs are not generally
advisable where large volumes of input and output data are concerned,
since it becomes impracticable to drive the peripheral units at their
maximum speed. This may be true for programs using either fast or slow
peripherals, but generally applies to fast peripherals such as magnetic
tape. The best policy therefore is to aim at making major processing
runs marginally peripheral-limited so that the magnetic tape units can be
driven at their maximum speed. Such programs are known as balanced runs,
but in the following discussions we will classify this category as processor-
limited for the sake of simplicity.

Mixing run types
If a multiprogramming machine permits two basic programs to be
operated concurrently, it is best operationally if a peripheral-limited run
and a processor-limited run are chosen to run together. The peripheral
limited run will be given higher priority so that the executive program can
continue to let it operate until it is delayed awaiting data from one of
its peripheral units. At this point control is transferred to the processor-
limited program. Clearly if the processor-limited program were given
highest priority there would be few opportunities to pass control to the
other program.

4.4 PROCESSING METHODS FOR MAGNETIC TAPE

Generally magnetic tape to magnetic tape programs will be processor-limited; and card to magnetic tape, or magnetic tape to printer programs will be peripheral-limited. Thus a good mix can be obtained by running tape to tape programs, in conjunction with input or output transcription runs. Figure 95 depicts three programs operating in the same system in this way.

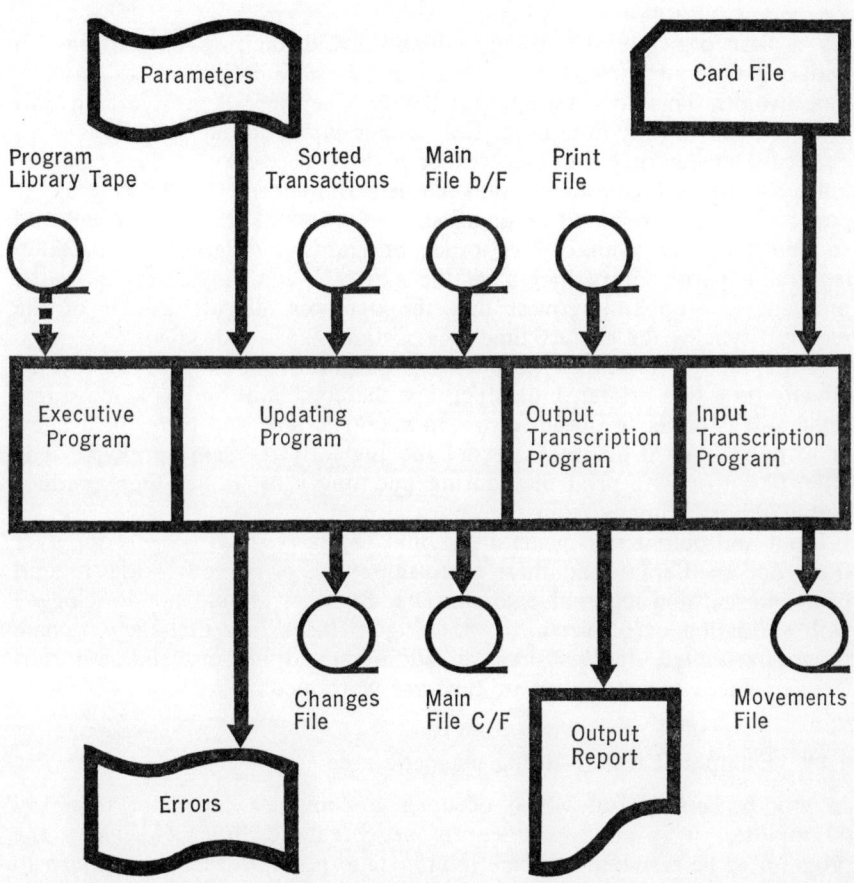

95. Core and peripheral allocation for a multiprogramming mix

Scheduling runs

Runs designed on this pattern provide standard arrangements for the utilisation of peripheral units, thus easing the task of scheduling. If programs are allowed to use a wide variety of peripheral types it is very much more difficult to obtain a good mix, because different programs may be requiring the same peripheral units. To some extent these principles

271

can be relaxed where the particular configuration has a wide range of peripherals available—e.g. twelve tape decks, two paper tape readers, two card readers, two line printers, and a paper tape punch. Nevertheless careful consideration should be given to such operating situations when designing runs. Where an operating system is in use such scheduling takes place automatically (see section 3.6.5).

Peripheral utilisation

As a first principle the analyst should avoid limiting any editing or updating program through unnecessary use of slow peripherals as input or output units. This does not apply if the slow peripherals are required only spasmodically during the run for low volume input/output operations.

A card reader or paper tape reader may be necessary at the beginning only; e.g. to load parameters into a run at initial entry to a program. If possible such units should be de-allocated as soon as they have been used so that they are available for other programs. Similarly if a program requires a particular peripheral at the end of a run only, then allow the program to stop and request that the operator allocate a unit of the required type at the correct time (see sections 3.6.3 and 3.6.8).

Error reports produced in updating programs should not be printed directly on a line printer, unless perhaps the configuration has a spare line printer. It is better to output errors to magnetic tape and print them later, or to paper tape if volumes are not too high. By the same principle it is often best to create print files during updating runs, rather than produce output reports directly from the run.

Input and output runs generally should be reserved for transferring data from one medium to another, performing necessary conversion to and from internal and external data formats. But they should not be clogged with validation or conversion processing to the extent that they become processor limited. In this situation additional editing or validation runs from magnetic tape to magnetic tape are warranted.

4.4.9 Dump and restarts using magnetic tape

In any processing run which occupies a computer for more than say 40 minutes, it is good practice to incorporate facilities to enable the program to be restarted at some intermediate point—rather than return to the beginning of the run should a failure arise. A dump routine should be entered, possibly every 20 minutes, to write away the internal state of the memory to a magnetic tape dump file. A restart routine can subsequently be used to restart the program from an appropriate dump point as required.

Dump and restart routines are usually provided by computer manu-facturers as part of the basic software for any magnetic tape house-keeping system (see section 3.5.8). A dump routine can usually be incorporated in a program when it is initially compiled, and the systems analyst must take care, when he requires this facility, to specify how the routine is to be called. A simple method is to arrange for the dump routine to be entered after a specified number of magnetic tape blocks have been read. It is also necessary to consider any operating conditions that may occur at dump points, so that operators can have the opportunity to reset slow speed peripherals accordingly when restarting. For example, it is difficult to reset a paper tape reader to read a particular character in a block of input data, therefore it is far better to coincide dump points with an end of block condition. Similarly, where printing is concerned it is advisable to coincide dump points with the head of form condition on the line printer.

The positioning of magnetic tape files at dump points is not so critical since the restart routine can open each file and reposition the tapes so that relevant block transfers are imminent, although end of reel is an efficient dump point. On restarting, the internal condition of memory for the program is re-established from the dump file.

Dump routines
When a dump routine is entered it may write the following information to the dump file:

1. A dump file label—which identifies the file and specifies also the particular dump number so that a restart can be made at the appropriate dump point.
2. A description of the magnetic tape files currently allocated to the program, and details of other peripheral units allocated to the program.
3. The contents of the internal memory allocated to the program, written away as blocks of data to the dump file.
4. An end of dump label including a count of the blocks in the dump file so that the restart routine can check that all data has been read.

As well as the dump file some information will usually be printed on the console to enable parameters to be prepared for the restart routine; alternatively the restart parameters may be punched directly on to paper tape or punched cards. The main purposes of the restart parameters are to provide information for the restart routine about the dump file itself, and to provide details for enabling operators to reset the peripheral units for restarting.

Restart routines
A restart routine is usually kept permanently on the program library tape; it is then called into memory when required and activated by loading restart parameters resulting from a particular dump point. The restart routine then searches for and opens all magnetic tape files for the program concerned and repositions the tapes. The information relating to other peripherals is acted upon, and finally the memory conditions of the

original program are re-established by reading from the dump file. The console operator is notified by the restart routine that the restart has been completed, and the original program is then activated to recommence operations from the last dump point.

Use of dumps and restarts

Most runs involving slow speed peripherals (e.g. card reading or line printing) should always be carefully considered for restarts. A card reader failure at the end of a long input transcription run can be particularly wasteful and frustrating. The same can be said for a reporting program: a failure of the line printer midway through a long printing program would necessitate restarting the run from the beginning of the print file.

Input and output runs can often be restarted in a simple manner without the necessity of dumping the current state of the program. For example, a card transcription run could be programmed to start in two basic modes—reading from cards, and reading directly from a magnetic tape input file. Thus if an input run is interrupted the operator can be requested to close the output file and stop the run. When the run is (subsequently) restarted the previous output file is used as input, and is read copying records across to a new output file. When the end of file label on the input reel is reached, the program reverts to card reading mode to transcribe further data to the output reel. The program can be written to generate all necessary control totals during both modes of operation.

Similarly a reporting program could be written to allow a print file to open and read without printing, until a specified condition were detected during a restart. For example, until a specified record is reached, identified either by a unique key or perhaps by its position in the file or until a particular block is reached, or perhaps a specified page of the report. When operating in restart mode the print program must be programmed to maintain relevant control totals, control breaks, page numbering, and line counts.

It is unnecessary to incorporate dump and restart facilities in short updating or editing runs but any lengthy run will require the full facilities described previously. Computer time will otherwise be wasted, perhaps by quite trivial incidents in the machine room, stationery will be wasted if output printing is involved and production schedules may be disrupted.

4.5 THE DESIGN OF A MAGNETIC TAPE SYSTEM

4.5.1 Magnetic tape files—structure and relationships

General aims

The systems analyst's basic objective in designing computer procedures is to minimise the overall processing time and the amount of hardware required in producing the specified results to a desired accuracy and frequency. There are several areas in which the analyst may search to achieve economy in his computer procedures:

1. Minimise the amount of operator activity required in setting up running and taking down a job.
2. Ensure that runs are structured to facilitate easy scheduling particularly where multiprogramming facilities are available.
3. Design programs to have a balance between input/output and processing operations so that peripheral units can be driven at their maximum speed.

Scheduling for multiprogramming

Aims of scheduling are not easily reconciled with one another, and may also clash with other principles apparently just as important.

First consider the problem of minimising operator activity. The easiest method is to minimise the number of runs involved by compressing processing activities into as few runs as possible, e.g. by updating two main files in one run from the same transcription file. The general effect of this is to increase the number of peripherals required during the run thus demanding more complex logic in the program, in turn implying more storage in internal memory, and perhaps, making it more difficult to establish a balance between input/output and processing activities.

In a multiprogramming environment it may follow that such a program becomes hard to schedule, because the number and variety of peripherals required and the high storage capacity demanded may make it incompatible with other work.

Immediately we see that the analyst must become involved with areas outside the orbit of his current system, and begin to evaluate the effect that his work will have in computer operations. If the job is to be processed in a multiprogramming system then clearly he must structure runs bearing in mind the principles laid down previously (see 4.4.8). It is better for instance, to allow input and output transcription programs to be peripheral limited rather than attempt to absorb all spare processing capacity with editing and/or validation functions.

Generally speaking, it is advisable to divide processing into distinct functions to be performed as separate runs, thus taking advantage of multiprogramming facilities in the manner previously described. In so doing it will be found that a suite of programs can be more readily maintained in the light of changing management requirements. For instance it is far better to achieve editing of input data in a particular run rather than spread editing functions throughout the suite of programs. Obviously one should not go to extremes, say, isolating input validation and editing as separate runs when usually they can be accommodated together in a preparatory run.

Following this line, the basic run structure of all systems developed within a particular organisation will be standardised. For example, all input editing will be achieved at a particular phase preparatory to any updating of main files thus permitting standardised data formats at various stages in the systems.

Run scheduling in a multiprogramming situation must usually be a tactical exercise; it is rare for an operations manager to be able to make rigid schedules in which certain jobs are always run together. Thus it is doubly important to observe general standards for run types so that the computer hardware can be used optimally. (But see also section 3.6).

Input/output balance

It is unnecessary to ensure that all runs are balanced in respect of input/output and processor time—in a multiprogramming situation it is generally best to balance all magnetic tape to magnetic tape runs, and to allow runs involving slow input/output devices to be peripheral limited.

Where tape to tape runs are clearly tape limited, it may be possible to render them more efficient by introducing further processing from adjacent runs. This policy should be avoided where it destroys the basic functional relationship between runs in the suite, but otherwise can help to even out an unbalanced suite of programs. For example, editing while sorting or merging can be used to create balanced runs and to obviate the need for separate editing runs in some cases.

Frequency of file processing

Magnetic tape files are not suitable for answering random inquiries— unless the inquirer is prepared to wait until it is convenient to load the file concerned, and have a program read through the records searching for the records fulfilling the specified criteria. The computer operations manager may not, in any case, consider this an economical use of machine time: it depends on the nature and benefits of the inquiry.

The tendency therefore, is to produce systems which are updated, say, once per week (or once per month) prior to reports being produced for management. This is particularly true of financial control or administrative systems as compared with say operational systems such as inventory control. It creates the situation where a heavy processing load is incurred at the end of the period. The analyst should therefore attempt to structure his system to minimise this peak. For example, all input transcription, editing and validation may be achieved at off-peak times, thus enabling errors to be corrected and re-inserted before the reporting phase is reached. It may even be worth continuing with sorting and some updating functions depending upon the nature of the work.

4.5.2 Record design for magnetic tape files

General influences on record design
We have already discussed the basic influences on file design of magnetic tape systems—the prime consideration is to minimise the time required to read or write data to and from peripheral units and to minimise internal processing time. Contributing functions in achieving these objectives are the application of sensible design principles in the internal formats of tape records. Record design is influenced by the nature of input and output data for the system and by the general hardware characteristics of the equipment to be used.

The output requirements dictate fairly rigidly the data to be collected on a routine basis, and the data to be maintained permanently within the files. Bearing in mind the input and output media available and the nature of the processing involved the analyst is influenced in selecting the code structure for particular data elements, their inter-relationships and perhaps their arrangement within records.

The code structure is conditioned not only by the immediate information requirements of the system, but also by future requirements that can be foreseen to stem from the same transactions or main files.

The hardware factors affecting record design include the methods of addressing data in memory, the type of storage available and the economic factors concerned with storing and handling raw data or files. The repertoire of characters available within the internal code of the processor is also important and the range of logical functions available within the computer's order-code.

All the considerations involved with these functions generally amount to straightforward economic arguments to minimise the overall computer time and storage capacity required for the application. They involve the reconciliation of conflicting requirements to minimise both input/output time and processor time.

Minimising input/output time
The time required to read or write a tape file depends on the size of individual blocks. Block size is related to the size of individual records and the simplest way to minimise record size is to store data elements as concisely as possible. For example, when using a machine which operates on operands in pure binary form, it is best to convert quantities and values into binary form as soon as they enter the system. Binary data occupies fewer tape characters than data held in character form format, thus efficiency is obtained throughout the system if such data can

be carried in binary form through several processing runs to be finally converted to character form when producing output reports.

Record size can also be reduced by packing data elements as consecutive characters within the record format on tape, but if the computer has a word-oriented internal addressing system the savings required in input/output time may be more than offset by the processing time required to unpack data elements for processing in internal store. The savings to be made are related to file size and the complexity of processing required per record. Obviously where a magnetic tape file contains a large number of records it is important to keep the records as small as possible; e.g. if a file has 100,000 records a saving of, say, four characters per record would provide an appreciable overall saving in input/output time. The saving would be particularly worthwhile if the file had to be read/written daily or more frequently. In the final analysis it is necessary to consider the implications of processing the record format within internal store.

The method of addressing data within memory has a considerable importance for the design of magnetic tape record formats. Some computers can address data as single characters whereas others must address complete words of say 4 or more characters. Some machines can handle pure binary operands whereas others cannot. Many of the latest processor models can operate in several modes and therefore allow a variety of internal formats to be used, but in every case it will be found that the basic design of any hardware system tends to favour efficiency of internal operation in a particular mode.

If the efficiency of a program is limited by the volume of input/output data to be handled, then the analyst will probably choose to store data within the magnetic tape record as concisely as possible. If on the other hand the program is processor limited the data on the tape format can be more loosely arranged to permit operands to be selected for processing directly from input areas without editing.

In short if a run tends to be input/output limited it is best to go for packing on the tape file, because the increase in processing time will not increase overall running time. On the other hand a processor limited program will be further limited by packing and it is better to organise the tape file to minimise the input editing functions.

Hardware/software implications
In character oriented addressing systems the user is normally allowed to use variable length operands within the overall record format—the beginning and end of each data element being marked by special *item separator* codes recognised automatically by the hardware system. Obviously the analyst should expect to exploit this characteristic to the fullest advantage. However this characteristic will not permit the analyst to vary the overall structure of his record format unless he gives consideration to the methods described later in this section under the heading *Fixed length and variable length records.*

278

Data elements can be stored in a concise fashion if full advantage is taken of both the character repertoire and the range of instructions permitted for the computer concerned. For example, descriptive or indicative data is not usually subject to arithmetic operations but it is operated upon logically. Thus alphabetic characters or symbols can be used within the system for such data elements allowing them perhaps to be stored more efficiently. If a computer permits bit patterns to be easily manipulated special codes consisting of perhaps only a few bits can be generated internally to represent required codes or conditions.

The sequence of items within records is not particularly significant except that a certain arrangement may simplify programming and thus possibly reduce processing time. Certain general purpose programs (sort routines, etc.) may expect to find some data elements in particular locations of the record, or there may be advantages to be gained by observing particular formats. For example, if there are several keys for a particular file, processing can be simplified by arranging for the keys to appear in appropriate sequence as consecutive characters on the record—thus enabling the keys to be considered as a single operand during comparing operations.

Any data elements that are to be operated upon by standard subroutines should be designed to observe the rules of the subroutines, otherwise extra processing time will be incurred in editing to appropriate formats. Such rules will apply particularly to variables containing decimal points or sign indicators.

Fixed length and variable length records
In some files the nature of the data will be such that not all fields will be present in all records of the file. The systems analyst therefore has to consider whether he will retain a fixed record length, with redundant fields on some records, or a variable record length with each record containing just the number of characters or words necessary to hold the relevant data elements. Examples of record format are given in figures 96 (fixed length) and 97 (variable length).

Variable-length records reduce the amount of magnetic tape storage required and therefore minimise the input/output time, but they present the following processing problems:

1. The selection of complete input records from an input area requires more complex logic.
2. Fields within records must be located and identified before processing can commence.

The first of these problems can usually be overcome by simple means depending upon the general hardware/software philosophy of the computer. For example, in systems employing a fixed word-length the first word of every record may contain a word-count for that record, whereas systems which allow variable field sizes generally allow for an end-of-record marker to be stored as the last character in the record. However additional processing time is incurred in searching and acting upon these conditions.

Modules	1	2	3	4	5
Record 1	A	B	░	D	E
Record 2	A	B	C	░	░
Record 3	A	B	C	D	E
Record 4	A	B	░	D	E
Record 5	A	░	C	D	E

96. Fixed length record format

Modules	1	2	3	4
Record 1	A	B	D	E
Record 2	A	B	C	
Record 3	A	B	C	D	E
Record 4	A	B	D	E
Record 5	A	░	C	D	E

97. Variable length record format

The second problem above may present more difficulties and the solution to some extent depends upon the nature of the data. The usual method is to consider each record to consist of certain basic modules: permanent modules which contain all those data elements found in every record (e.g. keys), and non-permanent modules which contain those data elements found in certain record categories only. Items which occur in a high percentage of the records may be placed in the permanent module even though they are not used in every case. All modules contain a pre-determined subset of the fields in a record, and within each module a fixed framework is maintained. Thus the program logic must first ascertain which modules are present in each record. Usually modules contain a code which not only provides identification but serves also to sequence the modules present within each record.

The fixed-length format is simpler for the programmer to handle, and requires fewer instructions to be performed at run time when processing records. The variable-length format occupies less space when stored on magnetic tape and therefore implies less input/output time when processing the file. These factors have to be considered one against another when determining the strategy for a particular file. However, the variable length format also implies extra complexities for the programmer, occupies more processor time, and requires more storage elements for holding program instructions. The analyst also needs to establish whether software available will operate on variable length records.

The analyst must first consider the general nature of the program. If the program is likely to be input/output limited it is worth investing in variable-length formats since the overall running time for the job can be reduced by minimising input/output operations. On the other hand if the program is likely to be processor limited there will be no advantage in attempting to minimise input/output time.

If the variations in record size are in any case likely to be small again it is not worth introducing extra programming complexities. The same argument would apply if the number of records to be compacted proved to be a small percentage of the overall number to be handled, or if the file itself was of small volume.

Variable record size also presents difficulties when grouping records into blocks for writing/reading to magnetic tape, requiring variable block sizes to be adopted (see 4.5.3). However all these complexities may be well worth facing if the economics of a run can be improved, particularly if large master files employed in daily operations are involved.

4.5.3 Block organisation for magnetic tape files

General influences on file organisation

In the previous section we have discussed the basic mechanics of record design. In this section we shall extend this discussion by considering how they can be organised into *blocks*. A block is a logical unit of data read from or written to a tape; it may consist of a single record or of several records, it may be either of fixed or variable length, and the records within the block may be of fixed or variable length.

It is customary for manufacturers to fix an upper limit to the block size permitted for a particular computer. Obviously it would not be practical to read a whole reel of tape into main memory, in one read operation, because even if the memory were large enough to hold all this data there might not be space to store a program segment let alone contain an output buffer of equivalent size. The limitation usually depends in practice upon the software used; e.g. the manufacturer's housekeeping software will usually specify an upper limit for the block size expressed in words or characters. There may also be a minimum block size.

Usually blocks on tape are physically delineated by an inter-block gap which is created at the beginning and end of a block as it is written to the tape. Thus when a read instruction is given the tape starts from an inter-block gap and reads characters from the tape into memory until a further inter-block gap is detected. The input area allowed in main memory must therefore be large enough to accommodate the largest block size expected for the particular file.

If a particular file uses fixed length records then clearly the block size can be set as a fixed number of fixed-length records. However, if variable-length records are used the block size can be allowed to vary within the upper limit set for the computer's software as a whole, or within some particular limit established for the program itself. Establishing the block size (or the upper limit for block size) involves thinking about some of the following points:

1. The space available in main memory to accommodate the input output buffers and program.
2. Obtaining a good input/output balance for programs using the file.
3. The minimisation of the amount of magnetic tape required to store the file; and consequently minimising the time required to process the file.
4. Consideration of the need to conform to requirements of standard software to be used when processing the file.

In general these arguments reduce to a discussion of the general effect that a particular policy has on the efficiency of computer running.

For any given file, the larger the block size the fewer the inter-block gaps. These gaps occupy space on the tape and require the tape to be stopped and started while passing the file. Thus where there is an upper limit upon block size it may be best for the system designer to use a block size that equals, or most nearly equals, this maximum. Against this it must be remembered that the efficiency of a run may be impaired by using a very large block size. Therefore the designer must explore points 1 and 2 above very carefully. These two points must be considered for all programs that use a particular file, and it may be that a conflict will arise between differing requirements of programs. The system designer must then seek the format that gives the best overall advantage.

Main memory utilisation
When reading from magnetic tape, the input area in main memory must be large enough to accommodate the largest block size encountered on the file concerned. If the run requires that an output file be created there must also be an output area large enough to accommodate output blocks. To obtain maximum efficiency it is standard practice to timeshare the reading of a block from the input tape and the transfer of a block to the output

tape; and this implies doubling up on the input and output areas as shown diagrammatically in figure 98.

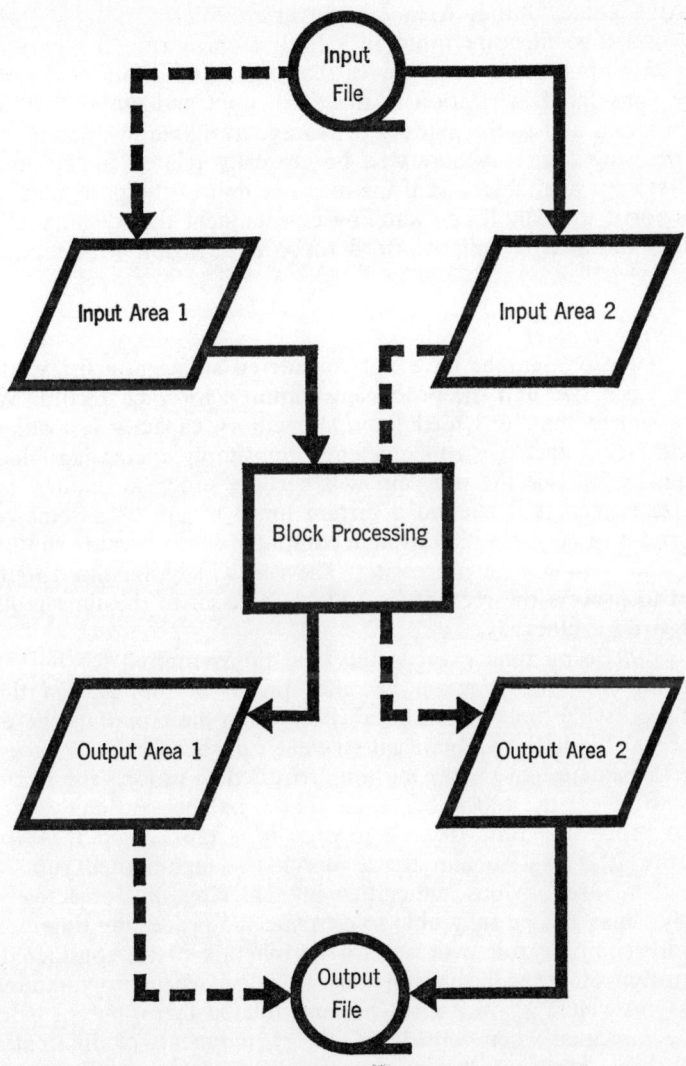

98. Doubling input and output areas for time sharing

This arrangement enables the input tape to be kept running, first loading one input area and then the other. As soon as one area is filled the program edits the data and stores it away in an output area. The output file is written first from one output buffer and then the next, so

283

that the input, output, and processing operations are overlapped. In the diagram above the operation to read data into Input Area 1 is overlapped with the output from Output Area 1. Similarly operations involving Input Area 2 and Output Area 2 are overlapped.

If several tape files are input or output from a run the investment in input and output areas must be further increased. Thus the block size must be considered in relation to the total input and output requirements of the program and to the amount of storage available in internal memory. The processing efficiency may well be critically related to the amount of internal storage available, and if the memory unit of the particular machine is small (or if in a multiprogramming environment the memory allocation is limited) the analyst cannot afford to be extravagant with input/output areas.

Input/output balance
The foregoing paragraphs have not considered sufficiently the relationship between block size and the processing required for each record. We have so far assumed that, provided internal memory capacity is available, we can continue to increase run efficiency by simply increasing block size. However, in practice the program will become processor-limited once the block size is increased beyond a certain limit. Figure 98 assumes that an editing run is to be performed using a computer which permits simultaneous reading and writing with processing. The run will be balanced if the time required to process the records in a block is equal to the time required to read or write a block.

If the processing time exceeds the tape time required for both reading and writing then the program becomes processor limited. On the other hand if processing time is considerably less than the tape time the program is peripheral limited. To obtain an efficient run therefore it is necessary to be able to estimate the processing time required to process the records in a block, and to set the block size such that a balance is achieved between the total processing time (for all records in a block) and the tape time. In practice it is best to aim for a marginally tape-limited run.

Now there are obvious difficulties in achieving this objective. Firstly the analyst may not be fully able to estimate the processing time: certainly in a more complex run with several peripherals to be considered a full consideration of this concept can only be given when programming work is in progress. Here again it must be remembered that different block sizes might be indicated when considering the requirements of different runs.

Also there may be hardware restrictions limiting the amount of simultaneity obtainable with particular tape decks; e.g. the number and arrangement of input/output channels.

Finally it must be stressed that where programs are being run in a multiprogramming environment, it is not possible to predict the effects that the running of various programs will have upon one another. The programs will be under the control of an operating system which will optimise the use of the machine according to the nature of the various programs, priorities that have been set for them, and events and conditions arising at run time. Thus no advantage is obtained by attempting to balance every run to the last instruction cycle. Nevertheless the general principles outlined above should be considered when obtaining block size.

Block size and tape utilisation
For a file of a given number of fixed-length records, the actual length of magnetic tape required is dependent upon the size of the records and the number of records per block. Each block requires a length of tape for the associated inter-block gap; thus the more blocks the more inter-block gaps and the greater the length of tape required for a given number of blocks.

The actual area of tape needed for a given amount of data depends upon the packing density of the particular tape system, but column 4 of figure 99 shows for different block sizes the capacity of a reel of tape with a packing density of 556 characters per inch. It clearly shows that the reel capacity is limited where the block size is small.

Where a file is likely to approach the full capacity of a reel of tape, the analyst will probably prefer to increase the block size to prevent unnecessary use of a second tape for the file. Of course where a multi-reel file is involved extra tape time is required to deal with the file, and extra operating time will be involved in loading files at run time. Furthermore additional tapes will be needed for successive generations of the file, thus involving a higher stock level for the tape inventory, and additional costs in running the tape library.

The tape time needed for reading, writing, or rewinding a magnetic tape file is governed by the speed of the tape deck, and the size of the file. Figure 99 indicates sample times with a 60Kchs tape deck using different block sizes. Again we see that operations are more efficient where the block size is largest.

Tables similar to that shown in figure 99 are generally provided by computer manufacturers as part of the supporting technical documentation accompanying their hardware/software systems.

Varible length blocks
Variable length blocks are only worth using where some advantage is obtained from having a variable record format as described previously in 4.5.2. The usual approach to this problem is to set the block size as high as practicable to obtain general running efficiency, and to attempt to create blocks which equal or most nearly equal this size. The input/output areas in main memory must be large enough to accommodate the largest block that will be encountered. Therefore, if the block size is allowed to vary widely within this upper limit the input/output areas will be under utilised.

Generally it is not worth while splitting records across blocks to obtain a fixed block size while using variable length records, because it introduces unnecessary complications into the logic of the housekeeping routines.

285

The handling of variable length blocks in any case does not present any great difficulty in itself.

BLOCK SIZE	READ/WRITE TIME FOR A MILLION WORDS (minutes)	REWIND TIME FOR A MILLION WORDS (minutes)	REEL CAPACITY (million words)	TIME TO READ/WRITE BLOCK (milliseconds)
10	24.60	6.85	.35	14.8
15	16.93	4.77	.50	15.2
20	13.10	3.72	.64	15.7
25	10.80	3.10	.77	16.2
30	9.27	2.68	.89	16.7
40	7.35	2.16	1.11	17.6
50	6.20	1.85	1.30	18.6
75	4.67	1.43	1.67	21.0
100	3.90	1.22	1.96	23.4
125	3.44	1.10	2.18	25.8
150	3.13	1.02	2.36	28.2
200	2.75	.91	2.63	33.0
250	2.52	.85	2.83	37.8
300	2.37	.81	2.97	42.6
400	2.17	.76	3.17	52.2
500	2.06	.72	3.31	61.8
600	1.98	.70	3.41	71.4
700	1.93	.69	3.48	80.9
850	1.87	.67	3.56	95.3
1000	1.83	.66	3.62	109.7

Hardware Characteristics of the Tape System :
(i)   Packing Density = 556 characters per in.
(ii)   Interblock gap = 0.75 in.
.(iii) Tape speed = 75 in. per second

99. A timing table for a magnetic tape system

When variable length blocks are initially created on an output tape, the output program continues to place records into the output area until the area remaining is too small to accept the next record. A word count or character count is made for the block as it is created, and the count is usually output as the first record in the block when it is written.

Thus when a variable length block is read into memory on a subsequent run, the block count indicates the relevant boundary of the block in the input area. The selection of individual records within that boundary is then made as previously described in 4.5.2.

4.5.4 Timing and evaluation of magnetic tape procedures

Purpose of timing estimates

We have previously discussed some general principles to be observed when designing magnetic tape systems. The best solution to any particular file processing problem is one which achieves the objectives of the application, but which minimises the cost in terms of operating time, processing time, storage media requirements, and peripheral utilisation. There will inevitably be many solutions to any particular problem, and each solution will suggest a different run structure, and different file and record formats. A method of timing and evaluating these different proposals is therefore needed in order to assess the relative advantages of each.

The analyst may work through several proposals before deciding upon a preferred solution to any problem, and it is necessary for him to have a standard approach to each evaluation in order to be sure that he is comparing like with like. For simplicity, we have assumed that the analyst is obliged to design his procedures to suit the requirements of a particular hardware configuration. The problem may then be approached by considering the following factors for each run:

1. The likely volumes of data to be handled during each run, defined in terms of the record and file formats.
2. The resultant peripheral activity required during each run.
3. The total elapsed peripheral time required in each run, taking into account the timesharing capabilities of the particular hardware configuration.
4. The processor time required for the arithmetic and logical operations.
5. The overall running time for the run according to whether processor time exceeds elapsed peripheral time.
6. The operating times for setting up and taking down the run.

Timing estimates based on these factors are used to evaluate design proposals, rather than to create run time schedules. There are many factors which can affect the overall running of job on a day to day basis (e.g. variations in data volumes, or, in a multiprogramming system, the nature of other programs being processed concurrently). The approach suggested in the following pages should present timing estimates expressed in minutes, with reasonable accuracy and without undue complexity.

General timing method

In the first instance it is necessary to consider each program individually:

it is unlikely that it will be possible in the early design phase of a system to assess timing of programs in a multiprogramming environment. In fact it is usual for run time scheduling to be done as a tactical exercise in the day to day management of the computer. The systems analyst should therefore aim at designing efficient programs that comply with general requirements of multiprogramming as previously described in sections 4.4.8 and 4.5.2. The stages in determining the efficiency of individual programs are described below.

Peripheral time

The running times for each peripheral unit used by the program must be found, and these times can be separately considered for magnetic tape units and slow peripherals (e.g. non-magnetic devices such as card readers, card punches, paper tape readers, paper tape punches, and line printers).

Magnetic tape operations can be evaluated by identifying the record and block structure of the files concerned and considering this factor along with the read/write time, reel capacity, and rewind speed. Programs using utility software can usually be assessed from information presented by the computer manufacturer (e.g. timings for sort programs can be extracted from specially prepared tables).

Slow speed peripherals have predetermined operational speeds expressed for example as, lines per minute, or characters per second. The volume of data to be handled by the peripheral has to be estimated, and is used to determine the overall time for individual peripherals.

A more detailed guide to peripheral timing is given later in this section.

Total elapsed peripheral time

Most large computer processors have the speed and power to drive all their peripheral units simultaneously on a timesharing basis; where this is so the total peripheral time is equal to the time absorbed by the longest-running peripheral.

Some smaller processors are restricted in that they can only drive certain combinations of peripherals at any one time, and it is in this situation where some difficulty can be experienced in making timing assessments.

The analyst should be aware of the timesharing characteristics of the configuration with which he is working, and should, if necessary, add to the overall peripheral time a further time allowance to cover time-sharing limitations. Manufacturers of computers provide manuals to enable users to ascertain characteristics relevant to their configuration.

Input/output channels

Another hardware factor limiting full simultaneity is the number of input/output channels; it is not uncommon for two or more tape decks to share a common channel. Where this is so the decks concerned are unable to operate simultaneously, and read and write operations may be held up waiting for one or other of the decks to complete a transfer of data. Usually these limitations can be avoided at run time by allocating decks to programs in such a manner that delays are not incurred.

Data conditions

Peripheral simultaneity may sometimes be limited by data conditions in a particular program, for example, where a slow speed peripheral such as a line printer is used in a main file updating program to print out details of records that are changed in the run. These situations are best avoided by adhering to the principles advocated in section 4.4.8, i.e. avoid using slow peripherals in runs that have predominantly fast input and output.

Where this practice cannot be avoided the analyst has to estimate the penalty incurred by the activity of the slow peripherals at stages during the run. In this example, an estimate of the printer activity can be made for a given hit rate of amendments being applied to the main file, and thus for a given unit of data on the main file it is possible to assess the extent to which magnetic tape input/output operations will be delayed awaiting the completion of a printer operation.

Processor time

The overall processor time is dependent upon the characteristics of the particular computer, but can be considered as consisting of the following factors:

1. The time needed to complete the sequence of instructions for processing data in the program (referred to below as program time).
2. The time required by any supervisory program for operating the user program within the computer system (supervisory time).
3. The time needed by the processor to transfer data into the store from individual peripherals (transfer time).

The first of these factors is an estimate made by considering the total number of instructions that would be required in processing the data. When multiplied by the average instruction execution time this gives the overall computing time. Naturally this sort of estimate is difficult to make because most programs consist of different loops and subroutines that are entered according to conditions arising in the data itself. A sensible way to approach this problem is to identify the different routines in the program, and then arrive at an estimated average execution time for each routine.

Assess how often each routine is to be entered and arrive at the total program time. In many cases only the programmer will be able to supply this information.

Supervisor time for a particular user program is related to the time needed by the operating system (or executive program) to control and monitor activities relevant to the user program. These activities are related to peripheral operations and can be based upon the time required to

289

initiate a transfer of data and check that the transfer has been correctly completed. This time is often expressed as a constant for each transfer operation and the supervisor time is derived by multiplying this constant by the number of transfers made, e.g. record blocks on magnetic tape.

The transfer time is the time needed by the hardware to accept and deal with each element of data (e.g. character or word) from a peripheral device. Different peripherals have different transfer times and manufacturers supply tables to enable such times to be derived. In many cases the transfer time is expressed as a percentage of the individual peripheral times.

Thus overall processor time can be estimated by the following:

Program time + supervisor time + transfer time = processor time.

Overall running time for a program

The running time is quite simply determined by assessing whether the processor time exceeds the total peripheral time: if it does then the running time is equal to the processor time; if on the other hand total peripheral time exceeds processor time then the running time is equal to the peripheral time.

The operating times (some guidance for these is given in the following section) are next added to the running time to arrive finally at the overall running time for the program.

The general timing method can be summarised as follows:

1. Determine the individual peripheral times.
2. Calculate the total elapsed peripheral time.
3. Calculate the processor time.
4. From 2 and 3 derive the running time.
5. Add operating times to arrive at overall running time.

Peripheral times and operating times

To complete the timing estimates as indicated in the preceding section, it is necessary to specify the formats of all records used within the system, and to estimate the average volumes of the various record types incurred during a typical processing period. This must be accomplished for all data media used in the system; e.g. card files, paper tape files, magnetic tape files, output printer formats, etc. In most batch processing systems there are few problems in developing these specifications, but it is advisable to construct the file specifications in a standard manner as suggested below.

Paper tape files

Paper tape may be used both for input and output files. All that is needed is to obtain a clear specification of the size of the records in terms of characters and an estimate of the number of records of each type in the file concerned. Figure 100 shows a sample specification for an order transaction file.

```
RECORD SPECIFICATION — DAILY ORDERS (PAPER
                    TAPE INPUT) FILE.

HEADER SUB-RECORD                              CHARACTERS
                    Customer Number                 6
                    Customer's Order Number         6
                    Date of Order                   6
                    Delivery Address Code           3
                    End of Header Marker            1
                                                   ___
                                                   22

DETAIL SUB-RECORD
Repeated on         ⎛  Item No.                    10
average 15 times    ⎜  Quantity                     6
per record          ⎨  End of Item Marker           1
                    ⎜                              ___
                    ⎜                               17
                    ⎜
                    ⎝  End of Order Marker           1
                       Total Average Record Size  =  278  characters

NUMBER OF RECORDS
Average Number of Orders = 8,000 per day
Total Average No. of Characters = 2,224 thousand

PERIPHERAL TIME
At 1000 characters per second = 36 minutes approx.
```

100. Specification of a paper tape input file

The example shows how efficiently input data can be handled on paper tape, but as a caution it is as well to note further the time required for operator handling. Let us assume that as a rule about 100 orders are punched and verified into paper tape as a separate length of tape. With an average of 15 items per order each tape length will consist of about 27,800 characters. At a 1,000 characters per second such a tape length would be read in less than a minute, but it is probably more realistic to consider that $1\frac{1}{2}$ minutes are needed on average for each reel to be loaded and rewound. To read a complete day's order data therefore requires $80 \times 1\frac{1}{2} = 120$ minutes.

Large input runs of the type described in this example are always limited by the comparatively slow speed of the paper tape reader. As a result large input files on slow peripherals are not usually used with main updating programs, but are dealt with in separate input runs.

Where paper tape is used as an output medium the same principles can be used for making timing estimates; and it is worth noting also that paper tape output is best reserved for use where output volumes are relatively low and do not therefore affect the efficiency of magnetic tape operations in the run.

Card files
Punched cards may be used as an input file, or be punched on line as an output file. To ascertain peripheral times it is necessary to estimate the total number of cards to be read and/or punched and calculate the peripheral time according to the speed of the device concerned.

Sometimes it is necessary to specify the field structure of each card before making an estimate of the number of cards needed. For example, in figure 101 a spread card is utilised and it is necessary to consider the detailed card format before arriving at an estimate of the average number of cards per input transaction. This example uses the same data as in figure 100 and affords an interesting comparison between card input and paper tape input.

Operator handling times for card readers are not significant for small batches of cards, but for large input files, say over 5,000 cards, can add as much as 5 to 10 per cent upon the overall time for the job.

Printer output files
Line printers have certain characteristics which if considered when designing output formats, can minimise the overall peripheral time needed during routine operation. For example, most line printers that utilise a print barrel, can be operated more efficiently where a limited subset of the characters are used for printing. Such a condition usually applies where a sub-set of characters having contiguous positions on the print barrel is chosen. The reason is that vertical spacing operations can be conducted while the unwanted print characters are passing the print position, it being not feasible otherwise to continue directly printing characters while the paper is being moved vertically.

By the same principle, the efficiency of the printing run can be

292

improved by minimising the amount of vertical paper movement required; e.g. if a single line space is chosen between successive lines of the tabulation the total time required in printing will be less than if a double space had been selected. Efficiency can also be improved by use of a paper tape loop (see section 3.2.6).

ORDER CARDS FORMAT

FIELD NAME	NO. OF COLUMNS	COMMENTS
Customer Number	6	Columns 1 to 12 of each card of an order
Customer Order No.	6	
Date of Order	6	Columns 13 to 21 of First card only
Delivery Code	3	
Item No.	10	Repeated for each order as follows:
Quantity	6	

		1st Card	Other Cards
		22 to 37	14 to 29
		38 to 53	30 to 45
		54 to 69	46 to 6l
			62 to 67

Card Number	1	Column 80

VOLUMES

8,000 order per day with 15 items per order on average

= 4 cards per order

= ·32,000 cards per day.

PERIPHERAL TIME

At 600 cards per minute = 44 minutes

At 1000 cards per minute = 32 minutes

---

101. Specification of a card input file

Printing speed characteristics can be ascertained from the technical literature of the computer manufacturer. Generally, this information will be presented in tables prepared by the manufacturer to allow timing estimates to be made. An example is shown in figure 102.

Tables of the type shown in figure 102 are most useful for making

293

simple estimates of the peripheral activity needed for straightforward listing jobs; the analyst simply ascertains the volume of output required in terms of print lines and calculates the overall time by looking up the chart. However, for printing onto preprinted forms where irregular spacing and paper throwing operations are required, a different approach is needed. The analyst must first prepare a definition of a typical output form and produce a timing estimate for printing one form—bearing in mind the characteristics of the line printer. A sample definition is shown in figure 103. The manufacturer will again provide timing specifications for spacing or throwing operations which can be used to arrive at reasonable estimates for printing particular form types.

Printer Listing Speeds with 1300 line per minute printer				
No. of Lines	Time in Minutes (Single Spacing)		Time in Minutes (Double Spacing)	
	48 characters	64 characters	36 characters	64 characters
100	0.1	0.1	0.1	0.1
200	0.2	0.2	0.2	0.2
400	0.3	0.3	0.3	0.4
700	0.5	0.6	0.6	0.7
1,000	0.8	0.9	0.8	1.1
2,000	1.5	1.9	1.6	2.1
10,000	7.7	9.5	7.9	10.6
40,000	30.6	37.8	31.5	42.3

102. Timing table for a line printer

Magnetic tape files

To obtain the peripheral time needed to read or write a magnetic tape file, one has to consider the size and number of records in the file and the way in which the records are organised into blocks. This aspect of file organisation has already been described in sections 4.5.2 and 4.5.3. Having ascertained the file structure the read/write time can be obtained by considering the hardware characteristics of the particular magnetic tape system. Referring back to figure 99, it can be seen that read/write times are given for various block sizes using a particular hardware system. Actual file times can be derived from this table by multiplying figures in the far right hand column by the number of blocks in the file, or from the second column according to the total file length in words. The file specification given in figure 104 has been evaluated using the figures from the table in figure 99.

Where files consist of variable length records/blocks the estimated data volumes should be based upon average occurrences.

DEFINITION OF ADVICE NOTE FORMAT

Line printer at 600 lines per minute

Form Depth = 11 in.

Character Set = All 64 Characters

Time in milliseconds

Name and Address	4 lines	
Heading Data	1 line	
Item Lines (10 average)	10 lines	
Total Line	1	
	16 lines (single spacing)	= 1600

Throwing

(From N & A. to Heading = 1 in. or 8 lines)	=	40
(From Heading to 1st Item = ¾ in. or 6 lines)	=	20
(From 1st last Item to Total = ¾ in. or 6 lines)	=	20
(From Total to Name & Address = 3 in. or 24 lines)	=	120
A. Single Form	=	1800

Volume = 5000 Advice Notes per Day

Peripheral Time = $\dfrac{5000 \times 1800}{60,000}$ = 150 minutes

103. Definition of a print output file. Relevant times for paper throwing have been assumed.

Operating times for magnetic tape files consist of the time needed for putting files onto tape decks and unloading them on completion of the job. Sometimes the operator has to fit or detach write permit rings. These operations can usually be conducted fairly quickly and it is probably best to allow say one minute per tape reel. In multiprogramming operations reels for one job can be prepared or taken down while some other job is in operation.

STOCK MASTER FILE

DATA ELEMENTS	FORMAT	CHARACTERS	WORDS
Word Count	binary	—	1
Part Number	BCD	12)	4
Depot Number	BCD	2)	
Description	BCD	20	5
Cost	binary	—	1
Purchase Tax	binary	—	1
Total Stock	binary	—	1
Total Shortages	binary	—	1
Total Orders on Suppliers	binary	—	1
Demand this month	binary	—	1
Average Monthly Demand	binary	—	1
Total Value of Stock	binary	—	2
Economic Re-Order Quantity	binary	—	1
			20 words

4 depots x 15,000 part numbers x 20 words = 1.2 million words

Block size = 25 records x 20 words = 500 words

Read/Write Time = 2.5 minutes approx.

Rewind = 0.9 minutes approx.

104. Magnetic tape file specification for timing estimate

4.6 DIRECT ACCESS PROCESSING

4.6.1   Direct access file storage principles

In Part 3 of this book we have already discussed the hardware characteristics of direct access storage devices; it has been shown that such storage systems present distinct advantages over magnetic tape systems in that it is possible to address data directly without having to search serially through the storage system to locate items desired. The time required to extract a particular item is relatively short, usually measured in milliseconds, and the access time is not necessarily dependent on the location of the item previously addressed.

With a direct access backing store the user has the potential to develop systems involving a greater degree of man-machine interaction—for example, it is possible to post random transactions to update records on a file, and to satisfy random inquiries for information from the file with a response time measured in seconds.

Such facilities are desirable both in some commercial applications and in scientific and technical research situations. But it must be stressed that these applications also require suitable remote input devices (perhaps involving communications equipment) and fairly complex software has to be provided within the framework of a timesharing computer system. If one studies the requirements of any commercial or industrial organisation it will generally be found that certain aspects of their business are suited for this treatment, but many other problems can best be met by more conventional batch processing procedures. In practice many applications are of a mixed nature perhaps requiring files to be updated in batch processing mode, but permitting some form of fast response to deal with exception conditions or remote inquiries. A multiprogramming computer might well be called upon to operate batch processing jobs as background work to a fast response system (see also section 3.6—operating systems).

The right solution to a processing problem can be best obtained by considering the economic factors relevant to the application concerned. Thus although some form of direct access storage is required for real-time or fast response processing, it is not essential to always utilise such storage devices in this manner. Direct access devices are used in both random and batch processing modes, and can be used for storing input files, transaction files, master files, history files or output files; they also provide an excellent medium for storing program libraries, or executive software such as operating systems.

A direct access backing store can be utilised in several ways, but initially we shall describe the basic modes of operation, and in later sections will consider these in more detail to indicate how and where they can be best adopted.

Modes of processing
There are three methods of processing with direct access stores:

1. Serial processing
2. Sequential processing
3. Random processing

Serial processing implies that a file is processed by reading from the first physical location of that file on the medium, through to the last location

K*

for that file, taking each physical block in turn by strict sequence of the location addresses. This is the processing method used with magnetic tape: to access any particular record the whole tape file has to be searched from the beginning until the desired record is located. As we have seen the characteristics of magnetic tape dictate that transactions are batched and sorted to sequence before being applied to their relevant master file, and a new master file is created on each updating run. It would be possible to use direct access media in this manner but it is wasteful and generally undesirable to do so.

Serial processing is however used in situations where it is required to read and/or amend, every record on the file; e.g. perhaps in reporting from a file or in a file maintenance routine.

Sequential processing takes place when, for example, a transaction file sorted to sequence is applied to a master file. The program selects master file records according to the keys of the records appearing in the transaction file, but only relevant master file records are examined, and the record retrieved at any particular moment may not necessarily be physically contiguous to the record previously addressed.

Random processing implies that transactions may arise in any sequence and access is therefore required to any record on the file without regard to the record previously addressed.

The seek area concept
Before going deeper into the subject of processing modes let us first consider the way in which a direct access store is used. The storage medium is not considered as a number of contiguous blocks of data, instead the file is considered to consist of *seek areas*. A seek area is a unit of storage which can be searched to access data without physical movement of the medium, e.g. without transporting a magnetic card in a magnetic card file or without moving a read/write head across a disc.

For example a disc file might consist of six separate magnetic discs with the read/write heads arranged to form cylinders of information as previously shown in figure 40. The information within each cylinder can be treated as a seek area since once the heads are positioned to cover a particular cylinder, access to data in that cylinder can be achieved without incurring the additional time for head movement as part of the access time. (Switching between heads on different bands is relatively insignificant.)

With a magnetic card file a similar principle applies, switching between bands on a particular card can be readily achieved while the card is being transported around the drum to be read/written. However if data on another card is required additional time is incurred while the card concerned is selected from the appropriate magazine and transported to the drum.

The seek area is a logical concept based upon the hardware structure of the particular storage device. This concept is fundamental to storage and retrieval of data from the device, because the system designer has to design his files and his procedures to minimise the amount of head movement incurred during file processing.

A bucket—the unit of data retrieval
Each seek area is subdivided into logical units generally referred to as *buckets*. Sometimes a particular storage device has a fixed bucket size but usually the bucket size can be varied by the user provided a constant size is adopted within each particular file. A bucket consists usually of one or more blocks of information, up to, say, a maximum limit of sixteen blocks.

The bucket is the basic unit of transfer between a direct access store and the central processor, each bucket can be used to contain a number of file records according to the size of the records and the size chosen for the bucket itself.

Basic principles of file organisation
We have now seen that regardless of the particular hardware mechanism used, the systems designer may visualise a direct access store as consisting of a number of seek areas of predetermined size. These seek areas may be allocated to particular files, and within each file area a specified bucket and record organisation is applied. Within a particular file each bucket is normally given a logical bucket number, these numbers are assigned in ascending sequence from one seek area to the next to cover all buckets allocated to the file.

Files may be depicted by file maps and an example is shown in figure 105. Here a file covering twenty seek areas is depicted and the bucket numbering for the complete file is shown running from zero to 479. It is assumed in this example that the device comprises 200 seek areas each comprising 96 blocks.

Note that the example incorporates a file specification nominating starting and ending addresses for both the seek area and the block within each seek area. In practice it is usually possible, and sometimes necessary, to allocate storage so that separate blocks of a particular seek area are assigned to separate files. Also it may not always be feasible to allocate consecutive seek areas to a file thus necessitating the allocation of more than one area to the same file.

Storage control and file protection
The control of storage allocation and the safeguarding of files is usually achieved under the aegis of executive software; e.g. by an operating system. An area of the store itself is usually allocated to contain various tables and file descriptions that enable the executive software to administrate and control the utilisation of the device and to communicate with the user's

299

programs in opening, closing and scratching files. Details of these functions are given later in section 4.6.10. In the following sections we return to consider further the types of files and to develop a more detailed discussion of three basic processing modes.

Bucket packing density

A bucket may hold one or more records of either fixed or variable length, and records are always placed into buckets so that they can occupy consecutive locations starting from the left hand end of the bucket. Thus unoccupied locations are available at the right hand end as shown in figure 106.

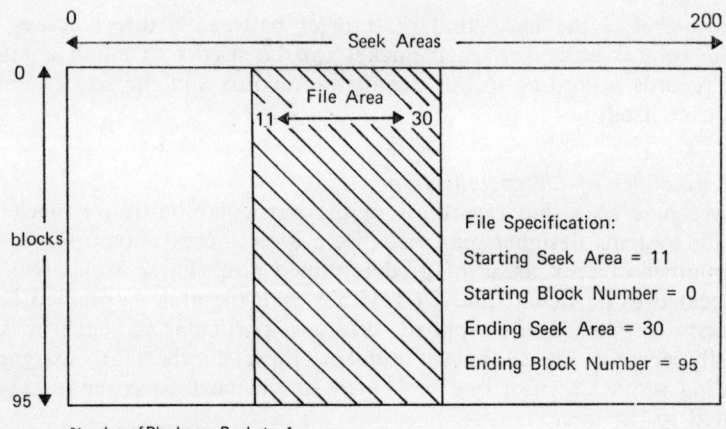

Number of Blocks per Bucket = 4

Number of Seek Areas = 20

Number of Buckets in File Area = 480

Bucket Numbering Table							
Seek Area	Buckets	Seek Area	Buckets	Seek Area	Buckets	Seek Area	Buckets
11	0 to 23	16	120 to 143	21	240 to 263	26	360 to 383
12	24 to 47	17	144 to 167	22	264 to 287	27	384 to 407
13	48 to 71	18	168 to 191	23	288 to 311	28	408 to 431
14	72 to 95	19	192 to 215	24	312 to 335	29	432 to 455
15	96 to 119	20	216 to 239	25	336 to 359	30	456 to 479

105. File map showing logical bucket numbering

The distribution of records over a file area as a whole, is dependent upon the manner in which the file is created; this topic is discussed in detail in subsequent sections. The bucket packing density is a measure of storage utilisation and is expressed as a percentage of storage space used.

The example in figure 106 shows a block size of 512 characters, a record size of 128 characters and a packing density of 75%.

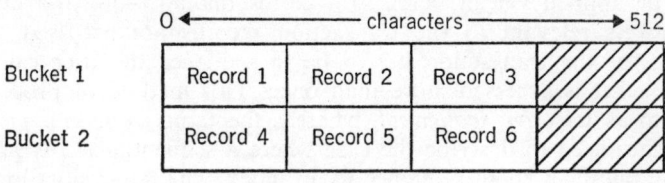

0 ←——————— characters ———————→ 512

Bucket 1 | Record 1 | Record 2 | Record 3 |

Bucket 2 | Record 4 | Record 5 | Record 6 |

106. Bucket packing

4.6.2 File types and processing modes

Serial files

A serial file is one in which the records are stored one after another in successive locations within each bucket up to the maximum number of records of that size allowed, and every bucket in that particular file area is filled. The records may or may not be sorted into key sequence. Transaction files may be held in this manner both before and after being sorted to key sequence; output files also may be held in serial form. Access to serial files is made bucket by bucket in strict sequence of the logical bucket numbers, until the end of the file is encountered.

The bucket packing density in a serial file is usually near to 100 per cent.

Sequential files

Sequential files are characterised by having records within sequence at the bucket level, but within buckets the records may not be in key sequence. Thus the highest significant key in a particular bucket will be lower than the lowest significant key in the following bucket as shown in figure 107.

| Bucket n | Key = 119 | Key = 117 | Key = 118 | Key = 116 | Key = 115 | |
| Bucket n+1 | Key = 125 | Key = 129 | Key = 121 | Key = 122 | Key = 126 | |

107. Organization within a bucket

To address a particular record stored in a sequential file, it is necessary first to access the appropriate bucket, read the contents into main memory, and then search for the desired record. When the record has been updated the bucket is then written away to the direct access memory once more.

Sequential files are usually processed by means of a stored index which

301

provides a cross reference between the key number of each record and the corresponding logical bucket number.

Thus transactions might be selected one by one from a serial transaction file and be applied via an index to a corresponding sequential main file. Only buckets relevant to the transaction records are accessed, and by arranging for the transaction file to be in sequence the user can ensure that no bucket is accessed more than once. This method of processing is referred to as *selective sequential,* whereas the term *sequential processing* is sometimes used to describe the case where a sequential file is processed bucket by bucket without reference to an index. The latter situation would apply where the hit rate between transaction records and main file records is very high.

A sequential file could also be accessed serially, examining each record one by one; e.g. this technique could be used when examining a file to select records containing data elements exhibiting certain logical relationships, or when amending some field as part of an ad-hoc file maintenance routine.

Sequential files can also be processed in random fashion, unsorted transaction records being applied to the index to select buckets containing relevant records. Here there is a possibility of requiring to access the same bucket several times and of having to move up and down the file from one seek area to another. A sequential file will usually be updated in this manner only if the transaction hit rate per seek area is very low.

The subject of indexing is discussed in greater detail in section 4.6.3, and in later sections the subject of the hit rate (or file activity ratio) is explored in greater detail.

Random files
In random files records are assigned to buckets in a random manner, but by some logical process which can be repeated to retrieve particular records required on demand. The process usually entails the manipulation of the record key in order to create an address for a bucket; thus provided the key is known the bucket required can be accessed and searched to find the corresponding record. The technique of creating bucket addresses from keys is usually called *address generation.*

Unless the keys of the records are purposely designed, it is likely that some difficulty may be experienced in developing an address generation technique relevant to a particular problem. The systems analyst rarely has the opportunity to nominate the record keys that his system will operate upon.

Ideally the processing technique should present 100 per cent utilisation of the file area available, and should generate unique addresses for each record, but in practice a far less efficient situation usually must be accepted. Address generation techniques often assign a large number of records to certain buckets, whilst under utilising other buckets—therefore some facilities must be incorporated for allotting overflow locations and for retrieving data from such locations. The subject of address generation is discussed in detail in 4.6.6.

A random file can be processed in true random fashion allowing for single transaction records, or small batches of transactions, to be applied to the file directly as they arise. Generally such techniques will permit a faster response than random processing against a sequential file, because it is not necessary to apply the transactions by looking up an index. Random files are therefore used in real-time situations, where the traffic is relatively low but where quick response is essential. If the file activity is high then it is likely that a sequential file would be preferred.

Reporting from a random file is most easily accomplished where exception reports or selective reports are desired. Sequential reports cannot be obtained so readily, without sorting the file.

4.6.3 Indexing methods for sequential files

Self indexing

To explain indexing it is probably best to consider first the method known as 'self indexing' although in practice it is seldom possible to use this method for accessing large files. In self indexing each record has a key which directly relates to a logical bucket number and often thence directly to the record number within the particular bucket. It is an excellent method for storing and retrieving data where a file consists of fixed length records and where there are few gaps in the sequence of key numbers used for records.

The keys have to be such that by fairly simple arithmetic they lead to actual storage addresses in which the records can be arranged sequentially on the storage device. If an example is considered it can be seen that the method leads to very efficient utilisation of the storage medium, and minimises seek times and problems of overflow.

Let us assume that there are 10,000 records each of 40 words and that they are arranged five to a bucket, thus giving a total file size of 2,000 buckets. The key numbers of the records run consecutively from 1 to 10,000, and a section of the file might therefore appear as shown in figure 108.

Bucket No.

B23	R111	R112	R113	R114	R115
B24	R116	R117	R118	R119	R120
B25	R121	R122	R123	R124	R125
B26	R126	R127	R128	R129	R130
B27	R131	R132	R133	R134	R135

108. Storage allocation for self-indexing

If we take record R119 its address can be located simply by dividing the record key by 5 ($119 \div 5 = 23$ remainder 4). Now add one to the quotient thus derived ($23 + 1 = 24$) and this gives the bucket number containing the required record. The remainder is then used to refer to the record position itself within the bucket, in this case the fourth position.

This is a good method where the analyst can be sure that record keys will always be allocated consistently within this sort of framework. Where this is so additions and deletions are easily applied to the file. It is however a wasteful method if the key numbers are sparsely populated within the overall limits of the key ranges.

Partial indexing

Because conditions are rarely as simple as those described above, it is usually necessary to create index tables on the direct access device itself, in order to refer the keys of transactions directly to the file storage areas where the relevant main records are stored. This system of addressing is referred to as *partial indexing*. The key of each transaction record is used to search through a table which contains an entry for each key value, each entry giving the bucket address of the desired main record. The desired bucket is then read into main memory the required record is accessed, processed as necessary, and the whole bucket written back to the direct access device once more. Often there may be several records stored in a bucket, but because the records are allocated to buckets sequentially, the index will need show only the key of the highest record in each bucket.

Now a file might consist of several seek areas each containing perhaps several thousand records, and a method is therefore required to organise the index so as to minimise the time spent searching for desired records. It is quite usual to have an initial seek area index which simply indicates the highest and lowest key number in each of the seek areas. At the beginning of each seek area there would then be a bucket index giving the highest key in each subsequent bucket—in practice the bucket index for a particular seek area could be spread over several buckets at the beginning of the seek area.

The seek area index might be small and could be brought into main memory and retained for the duration of an updating run. Indeed the bucket indexes could be held in similar fashion, but if the file is a large one the relevant bucket indexes will probably be called selectively into main memory as sorted batches of transactions are processed. Let us examine a simple example of a bucket index:

Assume that a seek area has capacity for 5,000 variable length records of average 100 words in length. In this example assume also that the bucket capacity is 2,000 words, thus on average 20 records could be stored in a single bucket, and the seek area must contain at least 250 buckets. The data records are stored sequentially in buckets but within buckets may be in random order, an index has therefore to be constructed to show the highest key number that is stored in each particular bucket, and this index must have capacity for 250 cross references (5,000 ÷ 20) between key numbers and logical bucket numbers.

If the keys numbers are, say, three words in length, and a further word is necessary to store each logical bucket number, then the bucket index for the seek area must contain 1,000 words. Such an index could be stored in the first bucket of the seek area as represented in figure 109.

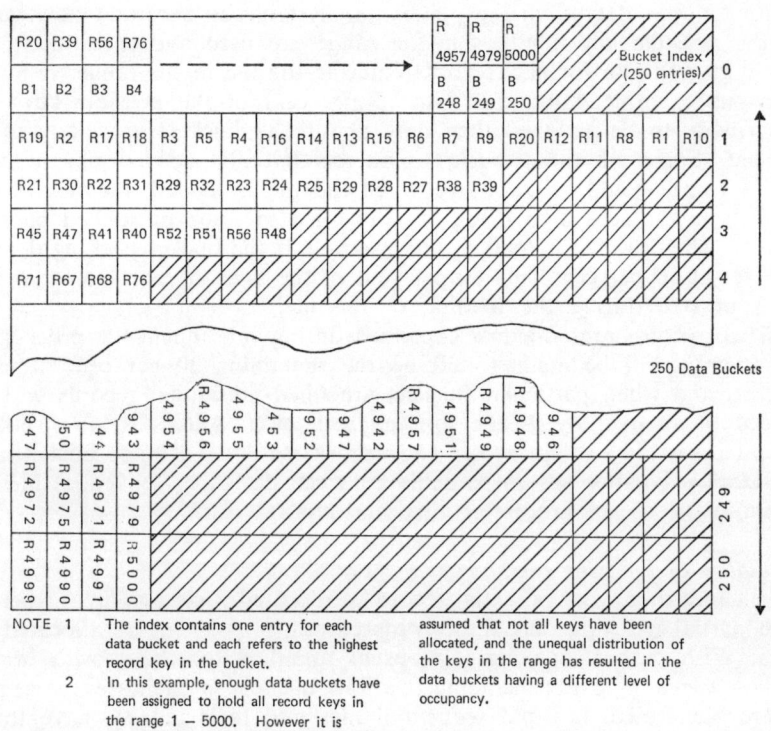

NOTE 1 The index contains one entry for each data bucket and each refers to the highest record key in the bucket.

2 In this example, enough data buckets have been assigned to hold all record keys in the range 1 — 5000. However it is assumed that not all keys have been allocated, and the unequal distribution of the keys in the range has resulted in the data buckets having a different level of occupancy.

109. Sequential file with bucket index

Packing density and bucket overflow

We can see in figure 109, that this particular index contains reference only to records that exist on the file (e.g. record R57 is not stored). Where gaps occur in the sequence of the keys, there is space in the storage area to accommodate them should it be necessary to add them later. However, in practice, it is likely that considerably less than 100 per cent of the possible keys might be allocated, and that in consequence as a deliberate policy rather less than 250 buckets need be allowed to receive the whole file. Since it is not usually possible to predict which of the keys will be ultimately used and which will be redundant, there must remain the possibility that at some future stage an attempt will be made to load more

305

records into a bucket than can be accommodated and an overflow condition arises.

Overflow is dealt with in more detail in the next chapter, but let us first note possible methods of dealing with a sequential file where overflow is likely.

At the time of loading the file and creating the initial index (these functions can usually be achieved by standard software provided by computer manufacturers) the systems analyst has to ascertain how many of the possible key numbers in the range are used and to estimate the likely growth rate for new records entering the file in the range. If in the previous example, not more than 75 per cent of the possible keys are ever likely to be assigned then only about 188 buckets are likely to be needed. Some of these buckets can be initially defined as *overflow buckets* and would be kept free when the file is loaded, the rest of the buckets in the seek area being initially filled to some prescribed packing density when the file is loaded. For example if the bucket packing density was set at 70 per cent then 30 per cent of the area of each bucket would be kept free during the loading of the file.

When records are added or expanded during subsequent file processing, an attempt will be made to fill up the remaining 30 per cent of each bucket, and when particular buckets are filled additional records will be placed in an overflow bucket on the seek area. Access to an overflow record takes longer than access to a record stored in a home bucket, and therefore a balance has to be achieved when designing sequential files to optimise the use of processing time and storage capacity.

Summary of indexing methods

In self indexing every potential record is allocated a storage area, whereas with partial indexing only the records currently assigned are allocated an area. With partial indexing it is usual to allow for the growth of the file by allocating areas including overflow buckets in each seek area, and where file growth is rapid sequential files and their indexes have to be re-organised frequently.

Self indexing provides faster access to records because there is no need to search index tables or data buckets to retrieve particular records. On the other hand, records must be fixed length, and few gaps must be included in the key sequence if storage utilisation is to be economic.

Partial indexing allows variable length records and although file structures appear complex at first sight, the use of standard software to create files and indexes, and to update and re-organise them reduces many of the problems involved.

4.6.4 Overflow of sequential files

As we have seen overflow of sequential files occurs when an attempt is made to assign more records to a bucket than there is capacity within the bucket. This event can occur because records have been added to buckets since the file was originally created and/or because certain records have increased in size during subsequent processing. When overflow arises it is necessary to store one or more of the records in other buckets, and it is better in the long run if special overflow areas are set aside for this purpose. Preferably these overflow areas should be in the same seek area so that it is not necessary to incur additional head movement on the storage device concerned when retrieving overflow records.

Many of the problems associated with overflow are catered for by manufacturers software, but the user may have to write his own software or at least have the opportunity to select a particular mode of software operation to meet his particular problem. There are at least two basic methods of handling bucket overflow, and many refinements and conditions exist for these methods. The two methods discussed here are known as *chaining* and *tagging* respectively.

Chaining

In chaining, overflow records are written to overflow buckets, but the address of the overflow record is retained as part of the last record in the home bucket. Thus in the simplest case the program retrieves the home bucket and examines the records to see if the desired record is present or whether an overflow address is given for the key concerned in the last of the home records. If the latter condition exists the program retrieves the appropriate overflow bucket to obtain the desired record. Naturally some programming convention should exist for storing and unpacking overflow addresses from home records.

Now in certain circumstances overflow can occur in the overflow buckets themselves, and if the *chaining* principle is followed to its logical conclusion long chains can develop requiring several bucket access operations before obtaining the desired record. Where such conditions are likely to exist *tagging* is the better method.

Tagging

Tagging entails placing a stub record in home buckets for each overflow record, giving the keys of the record and its overflow address. Thus there is a separate stub for each overflow, which normally provides a pointer directly to the desired overflow bucket.

The example in figure 110 shows that three records have been stored in a home bucket, and two records have been placed in an overflow bucket. Continuing with that example, assume that a further record is to be placed in the home bucket, but that on this occasion there is not room even to store the tag. A likely solution is to design the software so that Record 3 is also stored in an overflow area and a tag is created both for it and the new record to be stored, both of the records being assigned to overflow locations.

Obviously bucket overflow can occur in the overflow area itself and it is best if the software can simply assign a further bucket within the overflow area and ensure that tags in the home buckets are correctly maintained. In every case the tag in the home bucket should point directly to the address containing the particular overflow record.

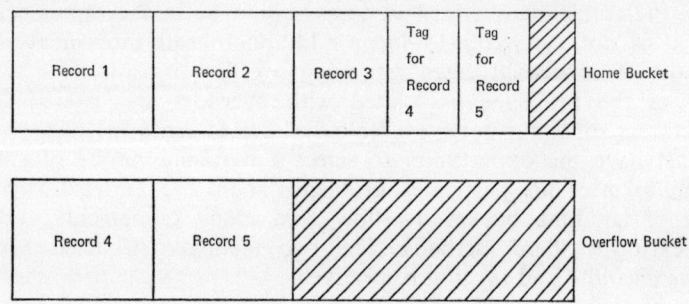

110. Tagging showing home and overflow buckets

In severe cases overflow records could completely fill the assigned overflow area, and most software systems incorporate warnings to indicate the imminence of this condition whilst maintaining master overflow areas for use by all files on the storage device in order to safeguard users against loss of file data. Such master areas are usually on special seek areas reserved for this purpose, and when a file is suffering from this degree of overflow, the processing of the file takes more time since additional seek times are incurred.

To the systems analyst working with a particular software system, there may not be a great deal of choice for the methods of dealing with overflow, but he will principally have to decide on the size of file areas to be allowed both for home and overflow buckets. He will want to minimise the entire processing time caused by overflow, whilst enconomising the use of storage areas on the device. Furthermore the analyst will need to establish the frequency with which the files are to be reorganised.

File reorganisation

File reorganisation routines are generally provided as part of the utility software supplied by the computer manufacturer. Basically a sequential file is reorganised by copying it to another storage area which is larger than the original area. In so doing the tags are encountered and the corresponding overflow records are retrieved and placed in home buckets. The file reorganisation is carried out to a specified bucket packing density and a new index table is created for the file. In some cases overflow records will be found to be unnecessary, because records have been previously deleted from the file leaving space in the home buckets. This condition alone can be corrected without the necessity of copying the whole file, by simply editing the file in its original file area.

308

Analysing overflow

The systems analyst will require to consider the necessary size of a file area when initially setting up the file and when reorganising it. In so doing he must bear in mind the potential rate of growth both in the number of records stored and their size, so that he can allocate an appropriate overflow area. This area can be of a size relating to the bucket packing density in the file area and the frequency with which it is intended to reorganise the file.

The basic weakness of any calculations to be done in this respect, lies in the initial difficulty in assessing file growth. Observations of existing systems over a period of time can produce reasonable estimates of the number of new records entering a file in a period, and for deletions of redundant records. Where new systems are being set up to handle data not previously processed difficulties may be met.

Often activity is concentrated more highly in certain key ranges of the file and rather than have to intersperse new keys among existing ranges it may be possible to extend certain file areas at intervals in a sequential manner. Again certain seek areas can be given a lower packing density than others.

The problem with variable length records is not so readily tackled. Where the nature of the data is such that some records are likely to expand unpredictably in size, the analyst may have to make predictions based upon observations of records in the file or parts of the file.

As a result of this type of study the analyst will need to derive a figure representing the number of new records entering the file in a certain period. Then knowing the number of records in the file at the outset, and their average size, he can calculate the degree of overflow likely to be experienced with various packing densities. The rigour with which he tackles these calculations may be tempered by an understanding of the limitations of the basic data. Most systems analysts will settle for a method which produces a reasonable estimate of potential overflow, bearing in mind that software is generally available to analyse file organisation and provide information and experience to enable a suitable routine for file reorganisation to be established.

4.6.5 Timing characteristics of sequential files

Processing methods

There are three methods for processing sequential files:

Sequentially:
This implies that each bucket is selected one by one in bucket number sequence, and is processed against a transaction file already sorted into record sequence. Every bucket has to be read and any bucket in which a record is amended has to be written back to the file. Calls are made from time to time to the overflow area to extract buckets containing records represented by tags in the main file area. By this method all overflow records are eventually examined.

309

Where an amendment to the records in a bucket results in bucket overflow, the next appropriate overflow bucket has to be obtained, updated and written back to the storage device.

Selective-sequentially:
With this method the transactions file is pre-sorted and is applied to an index during an updating run to select buckets corresponding to transaction records only. Wherever possible the initial seek areas index for the file is stored in main memory at the outset, and the bucket index for each seek area should be located in memory, while that seek area is being processed. Relevant seek areas are accessed in sequence, and within these areas only buckets containing hit records are read, updated, and written back to the file. All hit records in a particular bucket are updated while that bucket is in memory. Thus not all seek areas, or buckets within seek areas, are processed, and activity may be confined to particular areas of the file according to the nature of the data. Each time a transaction is matched with a tag in a home bucket, a call must be made to the overflow area.

Random processing:
In this method transactions are applied as they arise, and by implication are not in sequence or batched for long runs. The file index has to be used to obtain the bucket address for each main file record required. Thus head movement on the storage device cannot be minimised since successive transactions may require access to different seek areas and buckets up and down the file area. If the file has several levels of index it may not be practical to hold a secondary level in memory since it will not be possible to predict the next seek area required. The number of seek areas accessed will depend upon the nature of the data and the volume of transactions arising in a given period. The number of data buckets read and written can be assumed to equal the number of transactions arising, and in fact several calls may be made to the same bucket. Overflow processing is the same as for selective-sequential processing.

Timing estimates
Before examining the effects of these methods in further detail, let us say that the choice of processing method is largely dictated by the nature and objectives of the processing run, and not purely by file processing economics. A system that necessitates some fast on-line response to transactions will probably need a random processing approach. It may be possible in such a context to collect small batches of transactions on-line and presort them before the file processing run, but this will be governed by the frequency with which transactions arise and the period of response that can be tolerated by the users.

In practice therefore the choice between random and selective-sequential processing is not governed strictly by the simple formulas given below, but there may be cases where the hit rate is such that the advantages of one method or the other is marginal.

Simple sequential processing is most obviously used where the volume of transactions is such that a very high proportion of buckets in the main file are going to be accessed.

Let us then look in detail at the basic factors affecting file processing times, for the time being ignoring overflow.

(a) *Sequential processing:*
Timing estimates are given by considering the following: —

    (i) Every seek area is accessed requiring
          ($Sn \times St$) milliseconds.
        Sn = number of seek areas.
        St = minimum seek time.
   (ii) Every bucket has to be read therefore we must add
          ($Bn \times BRt$) milliseconds.
        Bn = number of buckets.
       BRt = bucket read time.
  (iii) Every updated bucket has to be written therefore add also
          ($BHR \times BWt$) milliseconds.
        BHR = bucket hit rate.
       BWt = bucket write time.

Most of these factors can be derived from the file specification or the equipment specification, except for BHR which must be estimated according to the number of buckets in the master file, the number of records per bucket, and the number of transactions to be applied in the run. Computer manufacturers will usually provide charts enabling users to plot the BHR given the factors just mentioned. It should be stressed that for most timing estimates it is merely necessary to know whether, say, 20 to 40 per cent of the buckets are likely to be hit. In practice conditions certainly vary considerably when programs are running and activity is often higher in certain file areas than others.

(b) *Selective sequential processing:*
Timing estimates are given by the following: —

    (i) Selected seek areas are accessed requiring
          ($Sh \times St$) milliseconds.
        Sh = number of seek areas hit.
       St = minimum seek time.
   (ii) Buckets corresponding to hit records have to be read from the device and then written away, therefore we must add
          BHR ($BRt + BWt$) milliseconds.
  (iii) To these factors we must add another factor I, sum of the individual times required when searching the index. This depends upon the number of levels of index, whether all or portions of it can be stored in memory, the searching technique adopted, etc. For example, assume that 1,000 transactions are to be applied to a sequential file and each search requires an index of, say, one bucket to be retrieved, if the bucket read time is 30 milliseconds, I consists as a minimum of (1000×30) milliseconds.

(c) *Random processing:*
Timing estimates can be given by the following: —

    (i) Almost every transaction will require a new seek area, although it is

311

possible that successive transactions will hit the same seek area. This is somewhat dependent upon the number of transactions and the number of seek areas involved. We assume here that every transaction requires a new seek area to be selected.

$(T \times St)$ milliseconds.

T = number of transactions.

    (ii) For every transaction a bucket must be read and written back to the file, therefore add

        $T (BRt + BWt)$.

    (iii) Then I must be calculated and added. This will probably entail more time than in the selective-sequential method since it will not generally be advantageous to store portions of the index in main memory.

Effects of overflow on file processing times

The timing of overflow processing presents several difficulties, and can best be estimated in the light of particular facilities incorporated in the software used. One of the main factors concerns the availability of suitable buffer areas for holding both home and overflow buckets in main memory. We shall see below that a considerable penalty is incurred if only one buffer is available to be shared alternately by both home and overflow buckets.

There are two principal overflow activities that consume time:

1. The need to retrieve a bucket from the overflow area upon hitting a tag in a home bucket during a file processing run.
2. The need to write a new record to the overflow area, when overflow in a home bucket is caused by adding or expanding a record during a file processing run.

As previously stated overflow buckets can be obtained with greatest efficiency if they are stored in the same seek area as the corresponding home buckets. It is assumed in the following text that this condition is present; if this is not so additional time for head movement is incurred to access the seek area concerned.

For example, when a file overflow area itself contains bucket overflow conditions, time will be wasted while access is made to master overflow areas provided on the storage device.

Dealing with existing overflow records

When a file is updated sequentially or selective sequentially recourse to the overflow area is necessary whenever a transaction record corresponds to a tag in a home bucket. If a separate buffer is provided for overflow buckets, then the program simply calls the required overflow bucket into memory, updates the record contained therein, and writes the bucket away to the overflow area once more. Meanwhile the home bucket, previously stored in main memory, can be retained until any subsequent transactions relating to it have been dealt with.

Assume on the other hand that only one buffer has been provided for the main file, and that a tag has been encountered in a home bucket after having already updated one or more records in the bucket. The following actions are then necessary:

1. Write the home bucket back to the storage device.
2. Read the required overflow bucket into main memory and update the relevant record.
3. Write away the overflow bucket to the storage device.
4. Read the original home bucket again, to carry on processing relevant transactions.

Thus two additional peripheral operations are required for each tag, as compared to the method using two buffers.

It should be noted that the records in the overflow area are not likely to be in sequence because they will have been accumulated in the overflow area during several processing runs; therefore no advantage can be obtained by retaining an overflow bucket in its buffer until a further tag is encountered.

Dealing with new overflow records

Timing here again depends upon the software approach adopted; a method for handling overflow is described below and we shall consider the implications of this approach.

Let us assume that the overflow area is filled sequentially starting from the first overflow bucket through to the last overflow bucket. Thus at any instant there will be a particular bucket assigned to receive new overflow records; we shall call this the *current overflow bucket*. The address of the current overflow bucket can be kept on the storage device as part of the file control data needed for handling the file, thus programs using the file can obtain this address to retrieve the current overflow bucket during the run. The current overflow bucket can be updated with further records, and when it is full the next overflow bucket is selected and the program must update the current overflow address on the file control area.

With this method therefore it is necessary to read the current overflow bucket and write it back to the device each time a new overflow record is created. Again if only one buffer is provided additional time is needed to write away and retrieve the current home bucket.

Each time that a new overflow bucket is needed additional peripheral operations are necessary to fetch, update, and write away the new address of the current overflow bucket. The frequency of this operation depends obviously upon the number of records that can be accommodated in each overflow bucket.

Summary

Timing estimates relating to overflow for sequential files can be made by first assessing the number of transactions that are likely to hit tags in the main file area. This will be related to the degree of overflow existing at the time, and the worst conditions can be expected just before the file is due for reorganisation. This number can be multiplied by the number of read and write operations necessary to deal with the overflow records, according to the number of buffers provided as described above.

To the time thus derived add the time for creating new overflow records. This time is calculated from an estimate of the number of new records that will need to be written to the overflow area in the run, which in turn is based upon the number of transactions to be added, and the existing bucket packing density for the file. Bear in mind that a proportion of these new overflow operations will require the selection of a new overflow bucket and require therefore additional operations to update the current overflow address in the file control area.

4.6.6 Random files and address generation

Address generation princples
Random files are employed where transactions arise singly or in small batches, and have to be dealt with immediately. At any moment, any of the records in the main file may need to be addressed, regardless of the key value last processed. The retrieval technique employed usually depends upon deriving a logical bucket directly from the key of the record being processed. Some arithmetic procedure is used to manipulate the key of a record to produce the address of the bucket containing the record.

Taking a simple example; assume that records have keys six digits in length lying in the range 100,001 to 199,999; and that of the numbers in this range 55,000 only have been allocated as record keys, other numbers at present remaining unused. Suppose also that a file area is available to receive these records, packed two per bucket, in bucket addresses 40,000 to 69,999, thus allowing a maximum of 59,998 records to be stored. The following technique might then be used to generate storage addresses:

Multiply record key by 2
drop the least significant digit
add 20,000 to the number thus derived.

The problems presented by this technique are demonstrated by applying this algorithm to a group of consecutive numbers.

Possible record keys	×2	Drop last digit	Add 20K
137110	274,220	27,422	47,422
137111	274,222	27,422	47,422
137112	274,224	27,422	47,422
137113	274,226	27,422	47,422
137114	274,228	27,422	47,422
137115	274,230	27,423	47,423

Here, five consecutive numbers have generated the same bucket address, and therefore, if these numbers are all allocated as keys there will be an overflow condition in respect of three of the records. Therefore this technique is only acceptable in practice if the records are evenly distributed over the range of possible keys.

Ideally, address generation should permit a unique address to be created for every record simply by manipulation of the key, this process to be achieved in such a manner that the operation can be repeated for any key in the series, according to a predetermined formula. Furthermore the process should result in an economic distribution of the records over the available file area; i.e. the packing density should approach 100 per cent utilisation of the storage area.

These conditions are not easily obtained in practice; some buckets are under utilised whereas others are assigned too many records. In the example used above, 55,000 records are assigned to a storage area capable of holding 60,000 records, giving approximately 92 per cent utilisation of the area concerned. However, as we have seen overflow can occur in many of the home buckets.

The basic steps in address generation can be considered as:

1. Manipulate the record keys to produce results with an even distribution over a desired range of numbers.
2. Expand or compress the first results obtained to correlate with a range of addresses available.
3. Add or subtract a base number of these intermediate results to line up with actual bucket numbers.

Sometimes the systems analyst may be able to develop the key numbers as an integral part of his project, but he invariably has to develop a technique which will enable existing codes to be continued. Therefore it will be necessary to deal with ranges of numbers which contain alphabetic codes, or ranges in which the distribution of existing keys is not uniform.

Overflow in random files

With random files overflow can occur when the records are initially loaded into the file area, and also when additional records are created during updating or maintenance runs. With sequential files we have seen that a special group of buckets are assigned to receive overflow records on each seek area. This is not practicable with random files; instead it is usual to assign overflow records to any bucket that may have capacity at the time. Thus those buckets which would otherwise remain unused, due to the limitations in the address generation method, will eventually be given overflow records. It would be very satisfactory if overflow records were assigned to such buckets only, but this will not be possible in most situations. Home buckets will also be loaded with overflow records, thus potentially creating further overflow situations.

The extent to which overflow develops in a random file, depends upon the distribution of the keys throughout the range to be handled; the effects of overflow should always be mitigated by assigning records to the same seek area as their home buckets. Again it is best if tags are used as pointers to retrieve overflow records, rather than long address chains.

The address generation system employed should be designed to offset as far as possible, the bias encountered in the distribution of the record keys. Ideally it should distribute records evenly to each of the buckets in the file area, but although a random distribution of the keys may be attained, it is not possible that all buckets will be equally occupied.

4.6.7 Inverted files

Descriptive record structure

The file structures we have examined differ considerably in the way individual records are located within a file, but have an essentially similar record: for example, a personnel record has as its unit an employee; fields or data elements which together describe the unit represented by the record: for example, a personnel record has as its 'unit' an employee; and the various fields within the record, such as salary, job description, department all describe the individual concerned. With fixed length records, each unit, each employee in our example, will have the same set of attributes associated with it, and the different values give to these fields or attributes constitute the unique description of the unit. With variable length records, only those attributes applicable to a given unit are associated with it, and some method of recognising which attributes are present in each record is adopted, as described more fully in section 4.5.2. Both fixed and variable length records of this type may be said to describe a unit, and may be called *descriptive records*. Descriptive records are used in most data processing applications, in which the main requirements are for the recording and analysis of large volumes of essentially similar records.

Problems of information retrieval

A common requirement in any information system is for the extraction of records from a large data file which satisfy certain criteria. For example, a requirement in a personnel records system may be to extract all systems analysts earning over £1,500 a year. This is a very simple example, and using descriptive records, the requirement would have to be satisfied by examining all records in the file, looking at the job description and salary fields, and extracting all records which satisfy the conditions. This implies processing a file *serially,* regardless of whether it is stored on a direct access device or not. Obviously, if it were possible to access records satisfying given conditions without having to examine all records in a file, the benefits of direct access processing could be extended to this type of information retrieval requirement.

Inverted file structure

The information retrieval problem outline above has as its fundamental concept the matching of records to attributes. In an inverted file, instead of holding data as lists of records, data is held as lists of *attributes*. A separate list is opened for each possible value of each attribute which can be identified: for example, in a personnel application, *systems analyst, programmer, operator* are values of the attribute *job title*, and *under £1,000, £1,000-£1,500, £1,501-£2,000* are values of the attribute *salary range*. Into each list is entered a reference number for each unit, e.g. 'employee' which satisfies that attribute. A separate index list is also kept which cross-refers the file's unit reference number with an external identification of the unit, e.g. name or employee number. A simple form of inverted file is shown in figure 111, which also illustrates how data is added to the file. Additional attributes may also be added by extending the attribute lists when more attributes or values need to be identified.

Information retrieval from inverted files
An information retrieval requirement which relates different attribute values, e.g. 'systems analysts earning over £1,500', can be satisfied, using inverted files, by merely accessing the relevant attribute lists, without having to search through all records. An inverted file has already pre-classified units by attribute value, and the satisfying of an information enquiry is performed by creating a derived attribute list containing unit record numbers which satisfy the criteria specified. A simple example is shown in figure 112.

Hardware and software considerations
Inverted files occupy a considerable volume of storage, since separate lists have to be maintained for each attribute value, and the number of attributes and associated values in a complex data file may be very large. Since the justification of using inverted file techniques is the rapidity and simplicity with which information retrieval problems can be handled, direct access storage devices are an essential hardware requirement for inverted files. The handling of inverted files will be performed by means of specialised software, which will provide facilities for loading data into the file, extending attribute lists, and producing derived attribute lists in response to information queries. In addition sophisticated analysis techniques often enable inquiries of a considerable complexity to be processed.

318

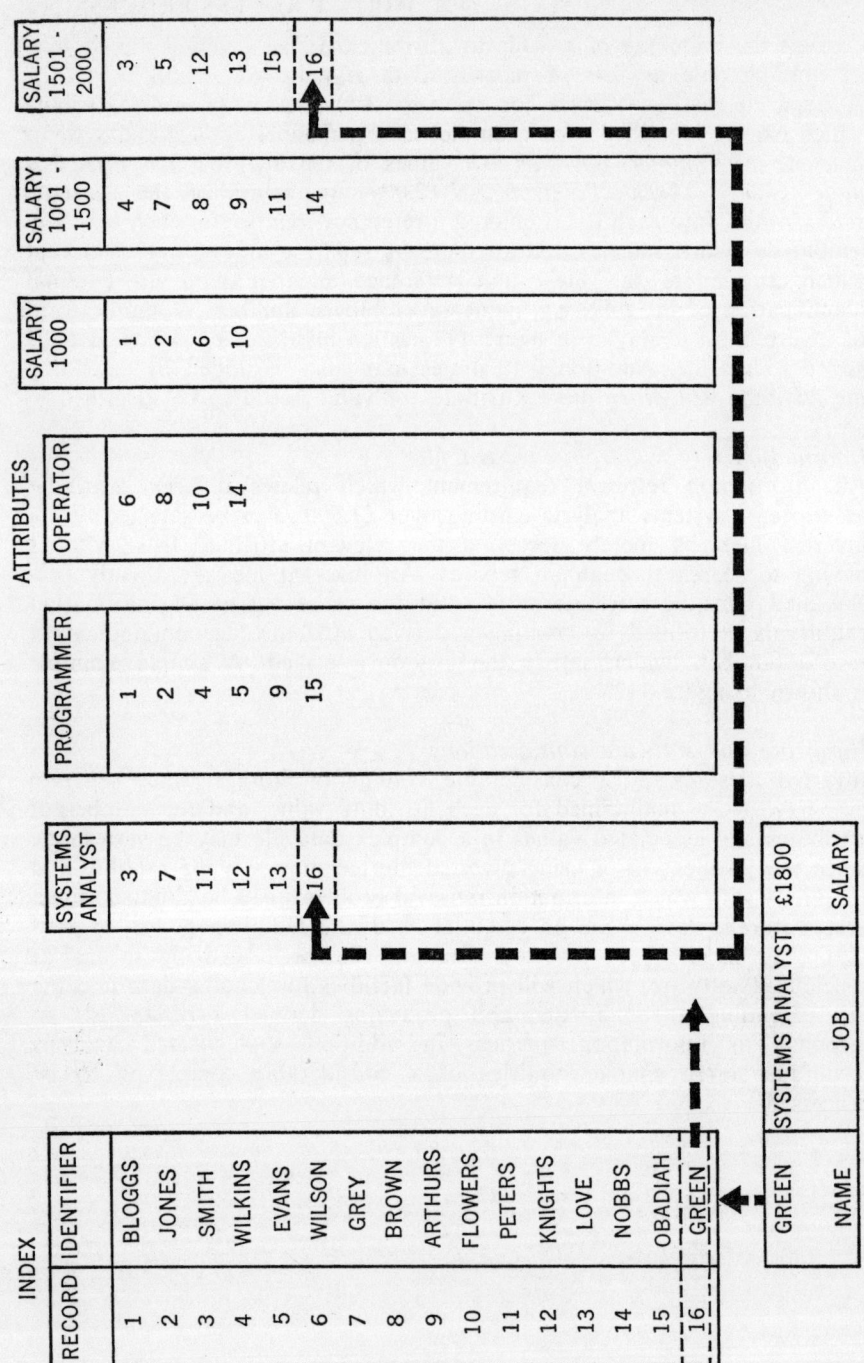

111. Adding a new record to an inverted file

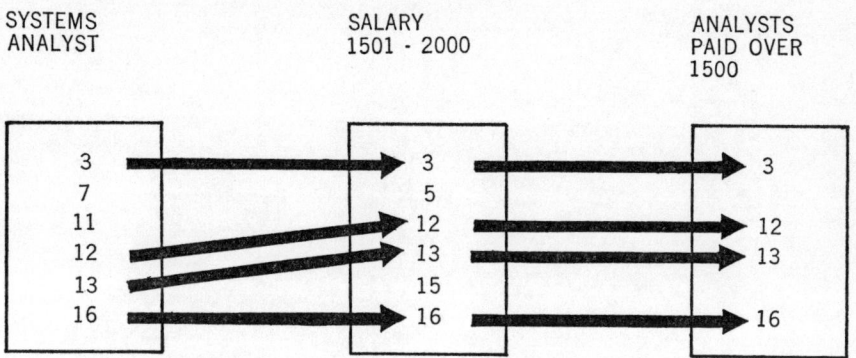

112. Inverted file: creation of a derived attribute list

PART 5
Systems Implementation

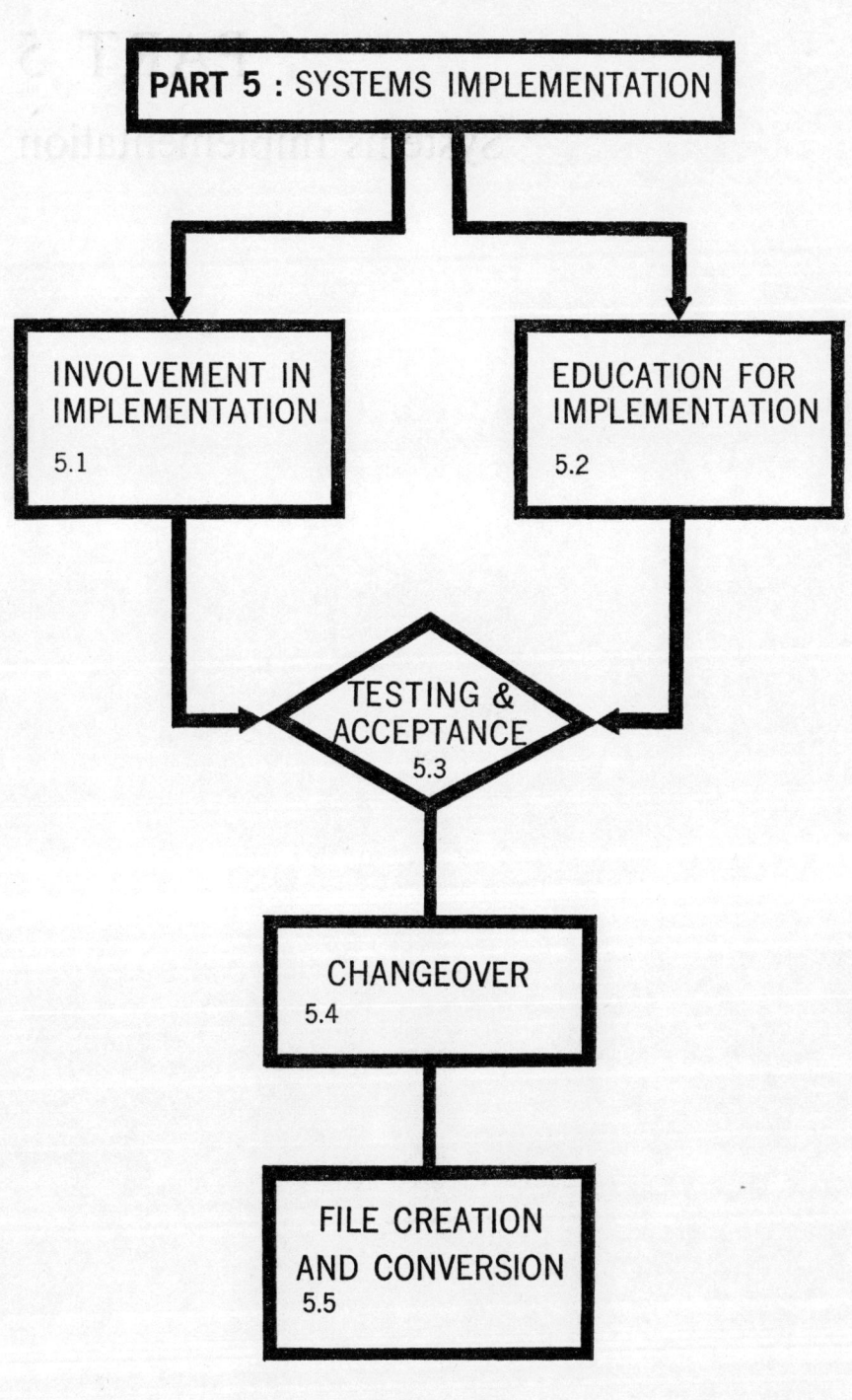

PART 5 : SYSTEMS IMPLEMENTATION

INVOLVEMENT IN IMPLEMENTATION
5.1

EDUCATION FOR IMPLEMENTATION
5.2

TESTING & ACCEPTANCE
5.3

CHANGEOVER
5.4

FILE CREATION AND CONVERSION
5.5

5.1 INVOLVEMENT IN IMPLEMENTATION

5.1.1 User participation

General

One of the guiding principles of systems analysis is that the keystone of any system is the attitude of the *people* whose daily lives are affected, either as users of a system's output, or as creators of the system's input. In short, if the user isn't happy, the system will not work, or will work inefficiently. User participation in all aspects of a system's development has already been discussed and we must now consider it as an essential part of the next stage in systems work, the system's *implementation*. If the preliminary stages of systems analysis and design have been carried out with good user co-operation, the analyst will have built up a sound and happy relationship. This relationship is of vital importance at the implementation stage, for now the faith of the user in the analyst's professional ability will be tested. The user will have been satisfied that the system will work on paper. The task of the analyst when implementing the system is to ensure it works in practice, and to ensure that the system does in fact satisfy the user. The analyst must bear in mind that however clearly his system has been defined for the user in an agreed systems definition, when the system 'goes live', even if it conforms in every way to the written specification, the user will find unforeseen difficulties. 'I never realised that...', 'I thought it would mean...', 'I never expected Bloggs to have to...', are not unusual opening gambits in a frustrated user's vocabulary when he is faced with the actual functioning of an implemented system. The aim of user participation at the implementation stage is therefore to overcome these difficulties, and gain for the system in practice the acceptance it has gained during the development stage.

Implementation stages

Considerable effort is required from users at three main stages of system implementation. The user must be involved with the supply of test data, and must agree both the expected results and the results based on the use of the test data. The user must be involved in planning for pilot running and parallel running; and the user must be involved in the education and training of himself and his staff where this is necessary. User participation will of course be maintained throughout the changeover to a new system so that there is minimum disruption of existing procedures and a smooth transition to new ones.

5.1.2 Project teams

General

The formation of Project Teams as a systems design philosophy has already been discussed in section 2.1.6. The normal function of a project team when designing a system is to involve the user in creating a system to fit his needs. Once the system is completed, the user should feel a sense of satisfaction in a large part he is able to identify with 'his' system and feel personally involved in its success. However, involvement must continue during the stages of implementing a system, and the concept of project teams is equally valid at this period of systems work.

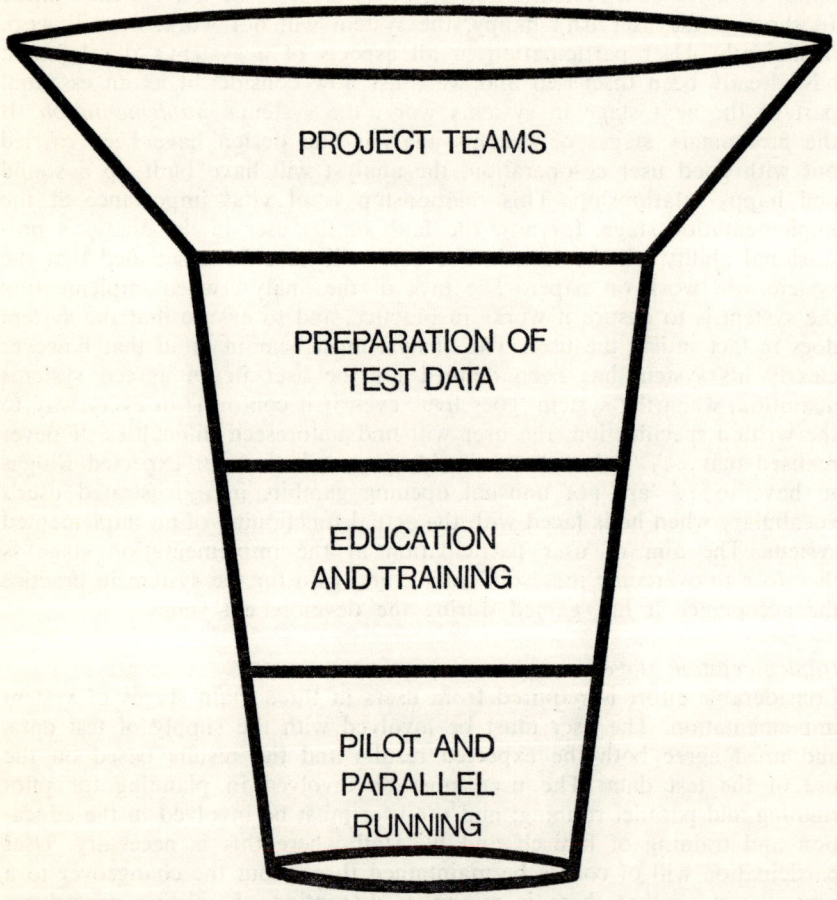

114. Major phases of user participation

Types of project

Project teams consisting of the systems analyst responsible for implementation and user representatives will be established to cover the main stages of systems implementation. These include:

324

1. A team to define and plan user education in new procedures.
2. A team to control systems testing including the provision of test data and approval of test runs.
3. A team to plan and control the changeover to the new system.
4. A team to define and control initial file creation procedures.

These various activities may themselves need to be broken down into separate individual projects, depending on the complexity of the system being implemented. Succeeding sections below discuss these various stages in more detail.

Special considerations
Project teams at the implementation stage of a new system have a different function from teams set up to design a new system. The function of implementation teams is to monitor progress, and to check that the system as designed is possible in practice. The implementation stage involves getting things done: people trained, forms printed, systems and programs checked, files created and converted. It is thus particularly important to make sure that realistic time tables are set and maintained, and that individual responsibilities are clearly defined and understood.

5.2 EDUCATION FOR IMPLEMENTATION

5.2.1 Planning education

General
One of the most important stages in the implementation of a new system is the education of users in the operation of the system. The extent of education and training necessary will obviously depend on the complexity of the procedures being introduced. In planning a programme of education the analyst must establish

1. The categories of staff requiring training;
2. The time scale over which training is to be completed;
3. The types of training to be given.

More detailed discussion of points 3 will be found in section 5.2.2. This section examines the staff to be trained and the problem of timing.

Staff categories
In the course of designing a system the analyst will have defined all the various procedures covered by the new system, and should thus have identified all categories of staff involved. A list of people involved in implementing a new system may include some or all the following:

L

Senior management
Line management
Clerical staff
Data preparation staff
Computer department staff (work assembly, operators)

All these people will be involved more or less directly in the working of a new system. However, the analyst must not forget the fact that other people may well be affected indirectly by a new system. For example, company auditors, who will have agreed any new system, may have to modify their audit functions to cope with different printout formats and clerical operations: consideration must therefore be given to any formal re-training or other explanation of the new procedures to staff involved. Another category of people often forgotten when a new system is introduced is the computer department staff who will be processing the computer runs. Operating instructions should have been prepared and agreed, but some effort spent on giving background information about a new system to the data processing staff is often helpful in ensuring that the significance of the system is understood. Operators who understand and are interested in what they are doing will be of invaluable help in making sure that the computer operations work smoothly.

Time scales
The timing of training sessions must be carefully planned by the analyst. If too long is left between training staff in new methods and the date at which they actually start to use them, they will tend to forget what they have learned and lose interest in the new procedures. It is important not only to teach new methods, but to stimulate interest and enthusiasm in the new system: both these factors can be dissipated if too much time is left between training and implementation. On the other hand training must not be rushed and skimped through lack of time. The analyst must also avoid the obvious errors of arranging training at peak holiday periods, or at times when a department's work loads are at a maximum. For example, the wrong time to start training a salary clerk is at the time of a company's annual wage review!

5.2.2 Types of training

General
This section describes some of the types of training which may be used at the implementation stage of a new system. Where a computer is being installed for the first time, training will also be required for existing staff who are to become operators, programmers and other specifically data processing employees. The training of data processing staff is not considered here: such training will not normally be the responsibility of an analyst responsible for the design of a new system, but will be planned for by

the user with the manufacturer providing the initial equipment. In this section we are specifically concerned with staff who will have to work with a new system, as explained in section 5.2.1.

Full-time courses

A full time course implies that staff concerned are taken completely off their normal work for the duration of the course. Such a course will thus normally be held away from the usual place of work to avoid the inevitable distractions which arise if staff can still be contacted. Full time courses can be either non residential or residential.

A residential course can be held either in an organisation's own training establishments or in a hotel or conference centre. The advantage of such and environment is that it ensures complete detachment from the normal working conditions. A group of people placed together in this way and subjected to a fairly intensive programme of lectures and discussions generate a very good 'learning climate', which should if skilfully handled by the course managers extend well beyond the fixed study sessions into the more informal relaxation periods. However, such a course must never be allowed to degenerate into an excuse merely to 'have a good time': the primary purpose of the exercise will be to instil new skills in as efficient and rapid a manner as possible, and the pace of the course must be intense and exacting.

The same principle applies to a full time non-residential course. However, where such a course takes place in normal 'office' conditions, even though at a different locality from the students normal place of work, it will not be possible to work at the same pace as at a residential course.

A full time course is the best method for teaching completely new skills, where these are substantially different from any already possessed by the staff concerned, and where they involve a degree of sophistication and complexity needing formal instruction and practice. A good example might be teaching staff how to use a remote terminal for inquiring processing, involving the manual dexterity for handling a machine, and the ability to formulate logical enquiries in a format acceptable to the system.

Part-time courses

It may not always be possible to release staff for full time course attendance. Part-time courses can be substituted in these cases, where staff attend at morning or afternoon sessions, doing their normal jobs at other times. Part time courses are suitable where the degrees of complexity of the new techniques are not excessive, and where the new skills are in some part related to old ones. If anything too radically different is attempted, a part time course will not be satisfactory. Inevitably concentration is lessened if staff are thinking about their daily work, which will always remain uppermost in their minds. By their nature, part-time courses must be held near the normal place of work and the temptation to rush off to one's desk at intervals, or for urgent messages to interrupt a lecture, will often prove irresistible. However, provided some discipline over these distractions is exercised, a part-time course can be a successful method of compromising between the need to train and the need to allow normal work to continue.

Week-end courses

Another method of allowing normal work to continue at the same time as training is to organise week-end courses. However, people naturally enough do not like having to sacrifice spare time, and in order to gain acceptance a week-end course normally has to be held in an attractive environment away from normal work places. A certain amount of formal lectures may be given, but again for psychological reasons the best approach is to organise the time as informally as possible, with discussion groups and seminars. This approach is most suitable for appreciation and background courses rather than formal instruction in new skills.

Lectures

Instruction can be given by a series of lectures given at regular intervals attended by all or a large part of the staff concerned. This approach involves minimal interference with the normal work flow of a department, since work is only suspended for the duration of the lecture. This approach is very useful for extending background and general knowledge. However, it is not possible to teach detailed skills involving practice and example by this method, since the time will not normally allow this. Single lectures can form a useful part of a training program by giving staff background information, for example about computers generally and the overall system in which their own activities play a part. This means that when more detailed instruction on a new procedure is given it will be more intelligible and therefore more interesting.

On the job

Most cases of training in new procedures involve the teaching of a relatively simple procedure, followed by practice until the new procedures become completely assimilated. The actual instruction stage must be done at a formal training session. New routines can be practised as part of normal working as well as at formal sessions. On the job training can be fitted into normal work in several ways. If a new method for dealing with existing input in office procedures is involved, each batch of input can be treated first in the old system and then followed by repeating operations by the new method. A second method is for work processed by one section of the department in the old method to be passed on to a section undergoing training to process in the new method.

At the initial training stage the results of the new method will probably not be utilised further, although there will be checks to see that procedures are being applied correctly. When some skill is achieved the same procedures can be used for parallel or pilot runs, as explained in section

328

5.4.2. The 'new' output will be used to complete the cycle of systems testing at all subsequent levels.

Simulation

In some cases it may not be practical or possible to use actual data on which to train staff. This situation will normally arise when a completely new system is being introduced, so that there is no current system providing basic data on which to perform operations. An example of this would be the introduction of the Post Office Giro system, or a company deciding to set up a centralised spares depot to replace a dispersed system. The situation may also arise in which live data cannot be used because of its confidential nature, for example in a payroll system. In these situations the analyst and the user departments will have to devise simulated data for use as training material. Such test data must follow the principles of all system test data, in that it must be comprehensive and cover all possible types of situation which might occur in real life. In addition, the quantities must be sufficient to allow for plenty of practice. One of the aims of any test must be to see how long the operations can be performed by staff before fatigue of one kind or another sets in. As an example, the colour of print on a form can in some cases cause eye-strain when forms are used in quantity. A 'simulated' test on a small number of forms will not highlight this factor, whereas a normal workload would.

Scheduling

It is not usually possible to shut down a department's normal activities for any length of time in order to train staff. Training must therefore be planned so that the department can continue to function while part of the staff are taking part in on or off job training. If possible the time chosen for training should coincide with a period when the department's work load is known to be below full capacity. Where this is not possible, then spare capacity must be arranged, either by agreeing overtime working, by reducing the service provided by the department (delaying the production of the department's output, deciding not to prepare a monthly return for a given month, increasing the time taken to turn round a document), or by engaging temporary staff. All these measures sound drastic, and may well raise objections by the department concerned or by others affected. But the justification for the expense involved is the same as the justification for spending money on the new system in the first place, and training is as important a part of a new system as programming and computer time. Once some spare capacity becomes available, the department can be divided into sections, so that while any sections are undergoing training, the remaining sections can continue to operate existing functions.

5.3 TESTING AND ACCEPTANCE

5.3.1 Test data

General

The importance of test data in systems design cannot be too strongly emphasised. The earlier an error in a system can be identified, the less trouble, time and cost will be required to put right, and good test data is the surest way of spotting errors early. The function of test data is two-fold. It provides programmers with a basis on which to develop their programs, and the results produced by these programs can be compared with the expected results prepared at the time the test data was created as a confirmation that the various runs are working correctly. Test data also plays an important part in the acceptance by a user of the validity of the procedures in a system. At this stage the testing is of clerical and allied procedures, volumes and time scales: the nature of the data used for testing will thus be of a somewhat different nature to program test data. These types of test data are described in this section, and sections 5.3.2 and 5.3.3 discuss the procedures for program acceptance and procedure acceptances.

Program test data

A careful program specification will contain directions to the programmer to enable him to cater for all conditions which the program is likely to encounter. This will include both normal conditions and all conceivable exception conditions. Once a program has been written, the programmer will need to test his program to see whether it produces the required output. In order to do this the program must be supplied with input data which will simulate all the contingencies for which the programmer has provided routines, both the normal and the exceptional. At the preliminary stages of testing a program, the programmer may well devise his own test data. But this procedure is dangerous, since the programmer will tend to create data to test for conditions he knows he has catered for and will not normally think up 'unexpected' conditions. The systems analyst must therefore supply the programmer with data which is specifically designed to contain examples of every type of data including all conceivable errors. At the same time expected results from the test data must be prepared by the analyst. This will include samples of all printout expected, including control totals, record counts and error messages where applicable. Data should be in sufficient quantities to test volume situations, for example if a printout is to show brought forward and carry forward totals where details continue onto more than one page, sufficient details must be supplied to cause this condition. Some volume situations may have to be simulated in other ways. For example if it is expected that data will be such as to require more than one reel of magnetic tape, it would be uneconomic to

330

supply test data to fill a normal reel: a full reel can be simulated by using a short tape for test purposes and supplying enough data to fill this short tape. Similarly 'paper low' conditions in a printer must be simulated by feeding in small quantities of stationery.

The data as presented to a program must be in the form which the program expects to find it, on paper tape, cards, magnetic tape or other storage medium. Where data is required on paper tape or cards, the data can be created either by punching from live source documents, or by inventing artificial data. Where live data is used the analyst must satisfy himself that it is fully representative of all expected *and* unexpected conditions. He may have to add suitably doctored data to cover error conditions and other unusual events which will not occur in normal data. Where a program expects data to be on a backing store, steps must be taken to create a suitable test file. The analyst may be able to specify the contents of a test file by filling in special forms showing the contents of each location on the file: special purpose software is sometimes available for creating test files in formats so specified. If no such software is available, it may be necessary to specify a special test file creation program. Normally, however, part of any new system will include a program for file creation. Once the program is working it can itself be used to create test files for subsequent input to other programs in the suite. Utility software will be available to print out the contents of test files and these should correspond with the analyst's specification of the test file contents.

The principles for creating test data for use with new programs apply equally to the testing of specific applications of general purpose software used in a system. Many systems will make extensive use of such routines as sort programs, report generators, information retrieval programs and data management software. In each of these, requirements are specified by means of a parameter set defining the particular run requirements. Test data will be used in exactly the same way as with program testing to ensure that particular parameter sets produce precisely the required output.

Procedure test data
The purpose of procedure testing is to ensure that all the non-computer parts of a new system are satisfactory, as well as making sure that all the various individual programs in a system work together as a complete suite. Since procedure testing involves manual procedures, such as input document coding, control checking and use of computer output, test data must simulate actual operating conditions rather than be specially designed to cover every abnormal occurrence. As far as possible, 'live' data should be used in normal quantities. Section 5.4.2 describes pilot schemes and parallel runs, in which live data is used to test a system before full implementation. Another source of test data is historical data, in cases where documents are filed and stored for some time. Where such real data cannot be used for any reason, for example if the system has com-

pletely replaced all documents, or if it covers procedures for which there is no existing parallel, the analyst will have to create suitable test data. Unlike program test data, the emphasis here will be on quantity, so that sufficient data is available to provide a realistic test of timings, volumes, and the ability of staff to handle quantities. Reference to this has already been made in section 5.2.2 on types of training, where test data is also required to enable staff to practise new procedures.

5.3.2 Program acceptance

It is important that some formal procedure for accepting that a program is completed and working satisfactorily is set up during implementation of a new system. One of the commonest causes of delays and frustration at the implementation stage of a new system is the incorrect functioning of the programs in the system. Too often, the enthusiasm of a programmer with his program, on the basis of a single correct printout from limited and inadequate test data can mislead the analyst into plunging into parallel runs and even full production running, with disastrous results the first time some untested loop of the program is entered. Equally, the programmer can be infuriated by a stream of requests from the analyst to modify details in the program or to cater for new conditions which the analyst has suddenly discovered. These situations can be avoided by careful program specification, clear documentation and properly prepared test data and expected results schedules. Any program changes must be agreed with the same formal procedure as the original specification. Acceptance of the program by the analyst will involve checking all output from the test data supplied with expected results schedules. Any discrepancy must be examined: this may be due to errors in the test data schedules rather than the program. In the course of writing and testing a program, certain minor changes may be made in the original specification: for example, additional error messages may be included to cater for hardware failures. In some cases details of switches to be set externally may be left to the discretion of the programmer. All these details must be recorded and checked by the analyst before the program is accepted. Once a program has been formally accepted, it is important that no changes are made to the program without a formal amendment procedure. It is always tempting for a programmer or analyst to add embellishments to an accepted program. Normally these may not apparently affect the operation of the program, but unless the acceptance procedure is repeated for a modified version the possibility of subsequent error cannot be ruled out.

5.3.3 Procedure acceptance

General

Acceptance of system procedures by the user of a new system will not normally need the same formal steps as program acceptance. The satisfactory operation of new procedures involves the smooth functioning of many inter-related steps, and includes the successful training of the user's staff as well as the correct functioning of the computer procedures. Procedures will be tested during training of staff, and during the operation of pilot schemes and parallel running described in section 5.4.2. However, in the course of these preliminary stages, the users of a new system must be satisfied that the major points in the system are acceptable. Among these will be included:

1. Acceptance of input documents.
2. Acceptance of input controls.
3. Acceptance of validation procedures.
4. Acceptance of system output formats.
5. Acceptance of system timings.

Input documents
The user must be satisfied that any coding documents or other inputs to the system he has to prepare conform to original designs, and are in every way acceptable as working documents, with special regard to colour, size, ease of coding, legibility. If documents are prepared for subsequent direct input to a machine, such as a document reader or mark sense card punch, it must be confirmed that the document handling procedures do not damage the document so that it cannot be read mechanically. Where documents are to be used as punching forms it must be confirmed that the data preparation staff can use the forms for this purpose.

Input controls
The user must confirm that any controls on input documents and subsequent data input to the computer do in fact provide adequate safe-guards in practice. When relevant, auditors or other authorities must also confirm their acceptance of control procedures.

Validation procedures
Where the new system is being used to validate data errors, the user must be satisfied both that the system correctly performs the specified validation, and that any error correction procedures are adequate and can be operated effectively.

Output formats
The user will have already agreed the formats of all output from the computer system. At the acceptance stage it must be confirmed that the output produced in 'live' conditions is acceptable.

System timings
A crucial feature of the acceptance of any system by its user is the ability for proposed timetables to be maintained. Nothing is more guaranteed to lose a user's confidence than the failure of system timings. Timetables

333

will be tested on actual volumes of data during pilot and parallel runs, and the user must accept the feasibility of the system's timetables at this stage.

5.4   CHANGEOVER

5.4.1   Changeover planning

General

Before a new system is implemented in full, the new procedures must be thoroughly tested. Individual parts of a system will have been checked and accepted, as described in sections 5.3.1, 5.3.2, 5.3.3. Staff will have been trained in new procedures, as described in section 5.2.2. The actors have learned their parts, the scenery is ready: but before the play is really launched, a dress rehearsal is necessary. The various stages leading up to the first night require careful planning. In section 5.4.2 two methods of changeover to a new system, pilot schemes and parallel running, are described in some detail. In this section some of the general points to be considered in changeover planning are described.

Continuity

Where a new system replaces some or all the functions of a previous system a major implementation problem will be the smooth transition from old to new. Among factors to be considered must be:

1.   Compatibility of new and old inputs and outputs;
2.   Volatility of movements;
3.   Necessity to meet deadlines.

These points may be explained in more detail by giving some examples. Compatibility of data and generated information will be a major factor in the changeover of management information systems. In particular, data required for presenting company accounts may change its nature, and the method of presenting accounting information may be changed. It will thus be necessary, when planning the changeover procedures, to close the old system and open the new one, making sure that the closing and opening positions are reconciled to the satisfaction of accountants and auditors. In such a case, it may be possible either to implement the system on a phased basis, each account being implemented separately, or the entire system may be adopted at the same time. A phased changeover will normally only prove feasible if each phase is independent of other phases, since if there is inter-relation of data, the problems of relating new and old may well prove excessively complex.

Different continuity problems will arise from differences in volatility of movements. Some systems involve large volumes of data being handled daily, the results of new data on the systems output being relatively significant, for example in a stock control system with a large daily turnover. These may be contrasted with relatively slow moving systems, for example the updating of personnel records. A high volume system in which updating occurs frequently, or even in real time, will require an instantaneous changeover, the new system becoming effective immediately it starts operating, otherwise essential procedures such as stock replenishment may well fail to occur when required. With a slow moving data flow the new system may be introduced over a longer period provided that the new procedures provide information which is as up to date as required. The problems of compatibility and volatility affect the third main factor to be considered in achieving continuity during system implementation. Some systems require certain outputs to be achieved without fail by certain fixed dates. For example, if a payroll is delayed large numbers of people are going to complain in no uncertain terms. Changeover planning must take into account any urgent deadlines. It may in fact be possible to alter deadlines during a changeover, but this must be planned for in advance. People may accept a small delay in pay day for the changeover period, but they must know about it and if necessary be compensated for it. The same applies to the production of output: some urgent reports and analyses may be delayed during a changeover, but any delays must be planned and those affected consulted in advance.

Timing
The planning of a changeover to a new system requires careful consideration of time tables. As already described in section 5.2.1, a new system should if possible be introduced at a time when the normal work load of the organisation is at a minimum. For example, in many ways the worst time to introduce a new payroll system is at the beginning of the tax year, when a pay department is involved in the clerical function of new tax codes, special end of year analyses, etc. In the same way, a changeover should avoid holiday periods if extra work will be required from staff. However, where a new system will involve a reduction of staff the best time for changeover would be a period where normal staff turnover minimises the problems of redundancy.

5.4.2 Pilot schemes and parallel running

General
As described in 5.4.1 one of the major problems in implementing a new system is that of ensuring a smooth transition between old and new. In some situations it is impossible to have a full 'dress rehearsal': a new system must take over completely from the old without any intervening stages. However, it is usual to test out all procedures in a live situation

335

before the complete system is implemented. Two methods by which this testing can be performed are by pilot schemes or by full parallel running.

Pilot schemes

A pilot scheme involves choosing some subset of the total area in which a new system is to be implemented, and operating the system within that subset initially. For example, the new procedures can be confined to a particular geographical area, or to a department, or to a particular class of transaction. The criteria for choosing the unit for use in a pilot scheme are those of completeness. Within the area chosen all likely problems and events of the full system must be present. Indeed, it is best to choose a particularly difficult or awkward area. For example, in choosing a pilot scheme for a payroll application, a department known to employ staff with complex pay structures or a high turnover should be chosen in preference to a more 'normal' or less volatile department. When the procedures implemented during the pilot scheme are fully tested and operational, the new system may be safely extended, either immediately to all other areas of application, or on a phased basis by a gradual extension to other units.

Parallel running

Parallel running of a new system involves operating the new procedures while the existing ones continue unchanged, until the new procedures are working satisfactorily and can fully replace the old ones. Parallel runs may be truly parallel, in that the new procedures operate simultaneously with the old on the same data, or may be staggered, so that the new procedures operate on data already processed by the old. Parallel running normally involves considerable strain on resources, since staff are operating a double work load. It is thus important for the time spent on parallel run to be minimised as far as possible by careful preparation, education and system testing.

5.5 FILE CREATION AND CONVERSION

5.5.1 Initial file creation: data collection

General

One major technical problem in the implementation of a new system is that of setting up data files. Many systems involve the collection of movements to data files, the updating of a file with movements, and the subsequent reporting from updated files. Once the system is in operation this pattern can repeat itself indefinitely. However, the heart of the system is the basic data file around which the activity takes place: the system

336

cannot start operating effectively without its files. Creating initial files is thus a first step in such a system, a step which may involve a separate system study and suite of programs of its own, even though it will only be used once, in contrast to the continuing repetition of the stages of the full system. This section describes the problems of collecting data for an initial data file where there is no existing set of data which can be used for direct conversion into the initial file. Section 5.5.2 describes the creation of data files by means of file conversion techniques.

Parallel data collection
Any system based on file updating will have the facility for adding new data to a file; indeed data must be 'added' as a first step, before any subsequent movements can be accepted to modify the original data. One method of creating an initial data file, is to treat all data as 'addition' movements for an initial period. For example, in setting up a payroll file, all employees can be considered as new starters, and the new starter procedure used until all records have entered the new system. This method can be used during the parallel run stage of a new system, the old system continuing to produce output until all records have been entered in the new system. During this operation it is important to remember that once a record has been added to a file, all subsequent movements to that record must be recorded in the new system, so that the new record is kept up to date. Some data files are relatively volatile, that is to say the period between adding and deleting a record on the file is relatively small, for example on an outstanding order file in an organisation with rapid turnover. In this situation it may not be necessary to transfer existing data into the new system at all. All new data is added as it comes in, and the old file is allowed to disappear as movements affecting it are used to delete records. Any records still outstanding after a fixed period may be transferred to the new system if necessary; these will normally be a relatively small proportion of the original total.

Where the new system is collecting data of a completely new type, so that there is no existing data to be captured, the situation is relatively simple, since all data must be treated as additions to the basic files.

Phased file creation
In section 5.4.2 the concept of pilot schemes was discussed. In the same way that a system can be implemented in stages by extending the areas covered on a phased basis, so can initial files be created to cover areas of data. For example, a personnel records system can be implemented by taking on the records of individual departments in an organisation in stages. This approach is suited to situations where sections of data can be treated as separate units, and where any movements entering a system can be easily identified as belonging to records on the new or old systems. The same approach could also be adopted on any file, by arbitrarily deciding to take into the new system some subset of records identified by

a range of keys, for example all accounts with a reference number in a given range may be taken into the new system first.

Dummy file creation

On some systems it may be possible to know before the system is implemented some details about the data to be held on the system's file. For example, in a stock control system a record will be required for each item held in stock, and details of all possible items may be pre-defined. In this case a record may be created for each such item. This record must be updated with 'movements' to give the initial values of variable information such as stock levels. In some situations even the 'variable' items may be pre-determined. It may be decided to start such a system off with the stock levels at a pre-determined level, and steps may be taken to make sure that at a given date the items are physically checked and make sure that they are at the given pre-determined level.

Special exercises

In some cases it may not be possible to use any of the above methods for creating initial files. The data volumes may be too great to be dealt with adequately by existing staff on a parallel run or phased basis, or the volume of movements may be so great as to make any initial file out of date more quickly than it can be updated by existing staff. In such situations it may be necessary to employ additional staff solely to create the initial file. Special programs may be necessary to accept input data in a different format, since the file creation operation may be performed more efficiently by devising special purpose input documents. It will still be necessary for any movement data being captured to be recorded if it occurs after the initial data is recorded. This may be done by staff operating the new system on a parallel basis.

Coding

One major problem at the stage of data collection for an initial file will be coding the new data. A new system will in many cases involve applying new codes to such basic data as staff numbers and account numbers. The transition problems can be eased in several ways. If possible, new codes should be adopted before the conversion to a new system takes place, thus eliminating the problem at the file creation stage. In many cases, however, this will not prove possible, either because the new codes will not be compatible with the old system, or because of the work load it could involve. An alternative is to create an index of old and new codes. At the initial take-on old codes can be accepted: once all data is captured on an initial file, the index may be used to convert the old codes to their new values. This approach may also be used for coding of individual elements of data applying to each record, where there is a close correspondence between old and new. Codes where there is ambiguity in any conversion may be translated into a special symbol and converted to the new value as a separate exercise.

5.5.2 Converting existing files

General

Section 5.5.1 described some of the problems in creating an initial data file in a system where no previous files exist in a form which can be directly input to a new system. However, many new systems replace systems in which part at least of the data required is already held on some medium capable of direct input to a computer, for example in a conventional punched card system. Where data is present in this form the problem of file creation may be eased considerably by use of file conversion techniques.

Record structures

When designing the formats for records on a new system the analyst will have as his primary aim the creation of an efficient new system, without regard to any existing record structures. Thus any existing data files will not necessarily be compatible with the new record formats. In the simplest use of file conversion, it may be possible to read existing files onto the medium of the new file and perform a direct editing of the old record into the new format. A first complication to this simple approach may be incompatibility of codes. This can be overcome by using conversion tables, if there is a direct or close co-relation between old codes and new ones. Where there is no direct co-relation the 'old' codes may be retained, and modified later by an updating of the converted file with new values.

Simulated updating

In many cases the new record format will not be directly compatible with any existing data formats. In particular, when converting from punched cards to magnetic storage media, a common problem is record size: several cards may be required to create a single record on tape or disc. Where the new system includes an updating program in which the new values for individual data elements are input as a series of amendments on a movements file, simulated updating may be used to create the initial file. This means that the individual cards from the old file can be edited into amendment formats rather than directly into the full record. These amendments can be used to create an initial file by using the new system updating program.

Use of software

Many manufactuers provide various forms of data management software which enable users to create and manipulate data files by means of flexible parameter sets. Initial file creation may be simplified by judicious use of such software. For example, where multiple old records are needed to create a single new record, separate files for each card type may be created and then combined into a single record by means of a series of data management programs.

339

Partial file creation
A new system may well require data elements which are not recorded on any existing files. In these circumstances it may be possible to create an initial file from existing data for those data elements common to the old and new systems. The missing elements can then be added either in series of updating runs at the parallel run stage of a new system, or as a special exercise of data collection as described in section 5.5.1.

PART 6
Documentation Standards

PART 6 : DOCUMENTATION STANDARDS

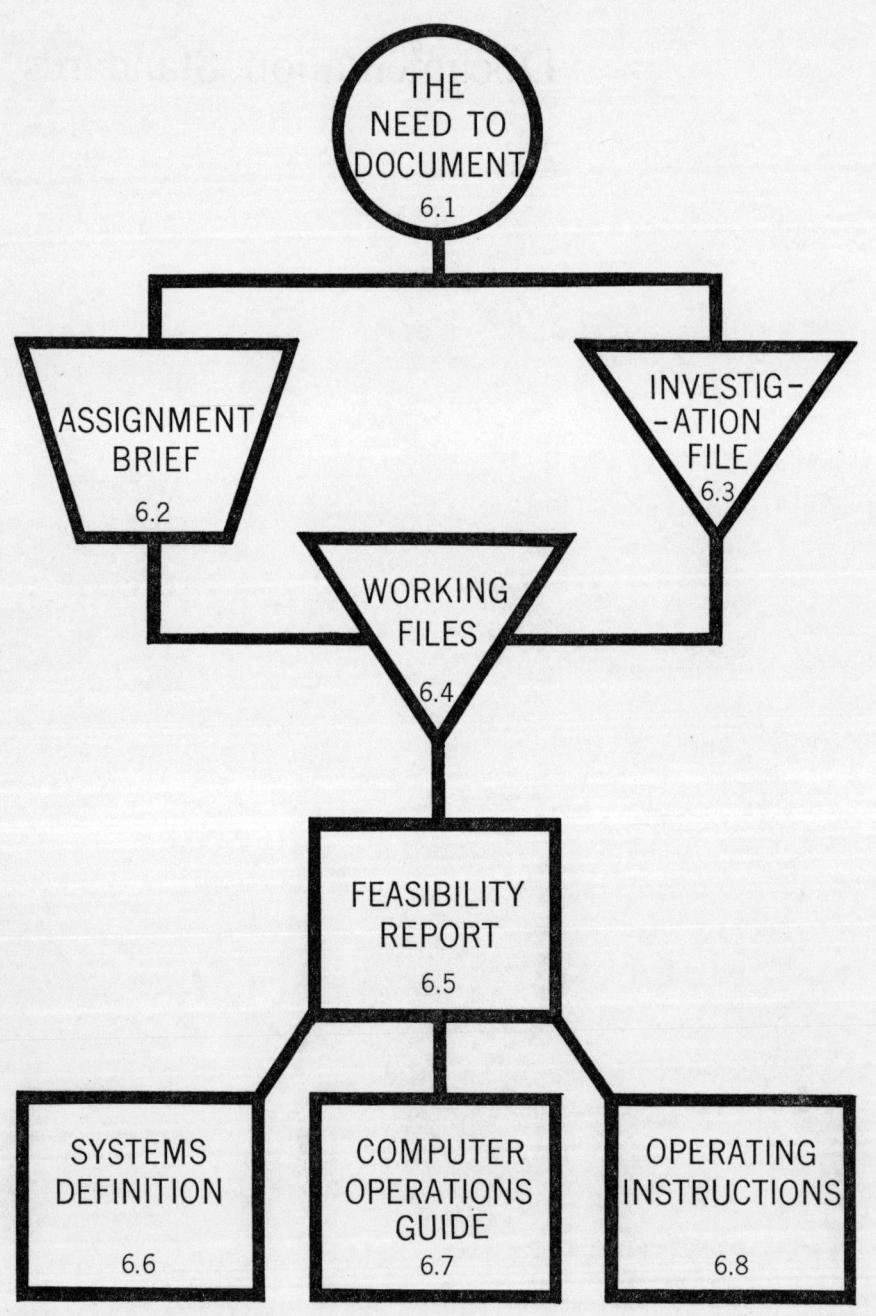

6.1 WHY DOCUMENTATION IS NECESSARY

Many systems projects have long lead times from the initial investigation through to the final live operation of the system. Some projects may take, say, 10 man/years, to develop and there may be many individuals involved in the design and implementation. It is essential for communication between people to be conducted without ambiguity, in order that valuable time and resources should not be squandered on fruitless activities. Good documentation standards provide for the establishment of precise goals for systems and programming staff, and enable the progress of work to be monitored by both line management and data processing management.

Above all else, the adoption of good documentation reduces the dependence of the systems department on individuals and safeguards against staff changes during the development period. In addition, it must be remembered that no system will run for ever without amendments being necessary (see Part 7), and if documentation is weak such amendments will be difficult to make. System amendments are often associated with changes in company policy or organisation structures, and are usually required to be undertaken as quickly as possible.

Documentation is a necessary evil; there is no doubt that many man hours can be expended in committing all details of a system to paper. It is important to get the right balance and to prevent documentation from becoming an obsession: if standards are too onerous the analysts and programmers in a department will not adopt them and will tend to make their own way.

There is no one and final standard for all systems work, and each organisation develops its own standards manual to suit its particular way of working. In principle all standards manuals are similar; but the documentation in an organisation having separate programming and systems departments, for example, will be different from one in which systems staff do all their own programming.

Documentation is important all through the various stages of systems activity, and can be related to the completion of certain events in the life of a project as shown in figure 116.

The documents shown represent a formidable amount of work, yet in practice it is sometimes found that one systems analyst will have the whole responsibility for this load, apart from conducting all the other activities associated with his project. If the project is a simple one he may be able to fulfil this task, but in other cases the amount of work to be done should be recognised and the job should be shared by a number of analysts working in a team.

In subsequent sections of Part 7 of this book, the documents listed in figure 116 are dealt with in greater detail. The principles expressed in these sections will have general application in all data processing departments, and form a suitable framework for developing a standards manual.

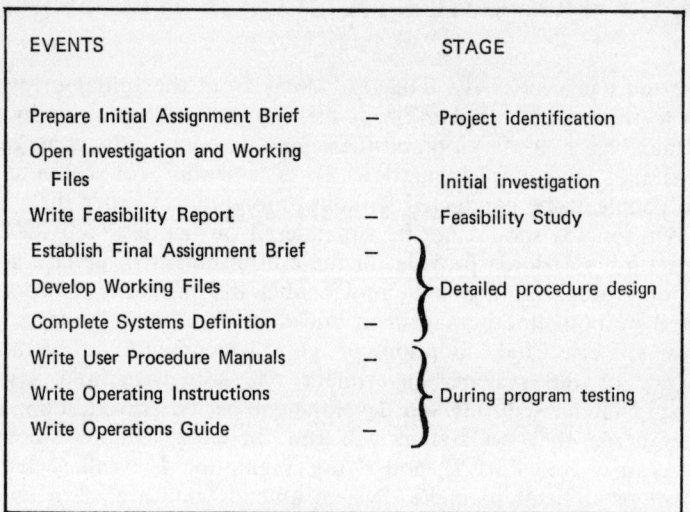

EVENTS		STAGE
Prepare Initial Assignment Brief	–	Project identification
Open Investigation and Working Files		Initial investigation
Write Feasibility Report	–	Feasibility Study
Establish Final Assignment Brief	–	
Develop Working Files		Detailed procedure design
Complete Systems Definition	–	
Write User Procedure Manuals	–	
Write Operating Instructions		During program testing
Write Operations Guide	–	

116. Documentation in systems development

6.2 THE ASSIGNMENT BRIEF

The purpose of the assignment brief has been explained in section 2.1 of this book, and the text which follows is a summary of information given earlier.

The assignment brief is a very important document, since its objective is to enable line management and data processing management to agree the bounds of a particular project. An assignment could cover a very simple task, such as the amendment of a program to effect a change in an output report, or could call for a detailed investigation into some major operation within the organisation.

The assignment brief for a major project is developed in two stages:

1. At the initiation of a systems investigation a preliminary brief is drawn up to provide guidelines for a feasibility study.
2. After the feasibility report is reviewed, the final assignment brief is prepared to specify the parameters for the detailed systems study.

Essentially the brief must be a short document since it has to be studied and approved by management at the highest level in the organisation. It serves to identify the particular problem area, and should be formally approved before relevant stages of the systems work are attempted.

Such a document can be initiated by a senior executive in the company, or it may be prepared by a systems analyst at the request of such an executive. The assignment brief is a request for systems activity, and usually has to be approved by the data processing steering committee.

6.3 THE INVESTIGATION FILE

The investigation file is set up to record the existing structure of the departments subject to investigation and to identify particular functions or procedures under review. The file is best kept in one or more large ring binders so that notes, memoranda, source documents and relevant papers can be collated as required. The analyst will have to take care that the compilation of this file does not become a fetish; the experienced man will develop a feel for what is important and what is not (section 2.2.4 gives detailed advice for this activity) but the following items form a useful check list for anyone conducting this task:

1. Compile existing documents and written notes to show organisation structures, and identify departments and procedures concerned.
2. Refer to copies of existing systems documentation or procedure manuals.
3. Seek existing copies of relevant organisation charts, but check details with current practice.
4. Obtain up to date terms of reference of departments.
5. Ascertain numbers and categories of staff and compile notes concerning their functions.
6. Collect samples of transaction documents.
7. Chart existing procedures to confirm uses made of documents, entries made, by whom and when.
8. Record details of working papers and files used in the current system.
9. Interview staff operating the current system and record details of their formal and informal roles.
10. Compare actual procedures with the existing manuals or specifications and record significant differences.
11. Write notes concerning any weaknesses and strengths of existing systems, e.g., suitability of staff, needs for training, organisation problems, transport and communication problems, and so on.
12. Make notes for future reference in the design stage indicating possible solutions, or identifying major factors to be considered.

This file must generally be compiled quickly and with attention focussed on the problem defined in the original assignment brief. All papers should be dated and copies of documents should be checked with the people concerned in the user department.

6.4 WORKING FILES

The objectives of the working file

Reference has been made to opening working files at the beginning of the design phase of each assignment. This section sets out guide lines on setting up and organising these important files. The objectives of the working files are:

1. To provide, on file, in a clear, orderly and accessible manner all relevant information required to enable another systems analyst to take over the assignment, quickly pick up the threads, continue with the remaining work and bring the assignment to a successful conclusion.
2. To provide a reference book and check list for the systems analyst engaged on the assignment.
3. To build up a stage by stage record of the progress of the assignment from which the final systems documentation (systems definition) can be readily prepared.

To facilitate reference to working papers it is recommended that the working file be divided into sections and that papers relating to each section be filed in a separate file cover or folder. The contents of these sections may differ from one organisation to another, but the recommendations given below (which are in fact based on standards in use in a systems department using a high level programming language) should suit most situations.

Contents of working files

Section 1 A copy of the assignment brief and any general correspondence on the assignment. A copy of a feasibility report when and if one is produced.

Section 2 A programme of the significant stages of the assignment including dates by which the various stages should be completed. At the beginning of an assignment this programme will be based on estimates agreed with the user. However as the assignment progresses and stages are completed the original programme will be amended, and possibly extended, so that it becomes not only a schedule of future work to be planned and completed but also a reliable record of the progress achieved as at any particular date.

Section 3 Final output. This section contains details of final output reports required from the system including the content, format, frequency and distribution. Where practicable specimens should be included in the file and these should be signed by the user to signify his agreement to the contents, format, etc.

Section 4 Data. This section will contain:

346

1. Data element list
2. File layouts
3. Code lists

The *Data element list* will include the name of each data element, its size and source. This list, which will be compiled progressively as the assignment proceeds, must be detailed and should show:

1. Full descriptive name of data element.
2. Length of the data element; i.e. in words or characters
3. Source, i.e. file reference.
4. Storage units (e.g. £p, tons-cwt.-qrs., cwts.-qrs.-lbs., etc.).
5. Internal format, i.e. character, binary, left or right justified, zero or space filled, etc.
6. Code list references.

File Formats—these will cover all file formats including input files (e.g. card or paper tape formats), main and intermediate files on magnetic tape or disc. The formats should be correctly drawn on graph paper or special forms and include a list of data elements and their addresses.
The information in this section should include full details of file sizes (i.e. number of records and average size of records) and the sequence in which the file is to be maintained (e.g. part no., type no., etc.).

Section 5 Computer run structure. This section of the working file will include:

1. A run flowchart which identifies each computer run and includes a descriptive narrative of each run.
2. A full and detailed run by description which should be sufficiently detailed to enable another systems analyst to continue the project without difficulty.

Section 6 Run development. It is recommended that this section of the working file be further sub-divided and that separate folders be provided for each job (i.e. suite of computer runs). Each sub-section will include:

1. Diary of events in development and testing programs.
2. Source programs, object program listings.
3. Test data.
4. Test results (e.g. reports and file prints, etc.).

Section 7 Tested versions of object programs (e.g. on paper tape) and relevant operating instructions. These will be filed separately for each job, and will be built up progressively as jobs are tested successfully and will be held ready for handing over for operating.

Section 8 Minutes of project team meetings. Separate files or folders will be maintained in this section of the working file for the minutes of meetings of each separate project team set up in developing and implementing a particular assignment.

347

6.5 FEASIBILITY REPORTS

Purpose of feasibility reports

A feasibility study is an investigation into the ways and means of solving a particular business problem, the solution of which may or may not involve the use of a computer. At the end of the study a feasibility report should be produced defining the problem, showing possible methods of solving it, and providing recommendations for the choice of method, based on the resources required for running the system and the costs and benefits of the proposals. A feasibility study could well take several weeks or months for completion. On the other hand, it may be possible for an answer to a problem to be produced within a few days. The development of the feasibility report depends largely upon the nature of the problem, but it should include the following headings:

1. Introduction
2. Problem definition
3. Possible solution
4. Nature of outputs
5. System evaluation
6. File information
7. Input information
8. Run structure
9. Programming required
10. File creation
11. Implementation plan
12. Recommendations

The objective of the feasibility report is to provide management with a sound basis on which to make decisions about future systems work. It therefore needs not only to demonstrate general methods of solving the problem, but also to give estimates of the benefits to be obtained, the resources that will be required, and estimates of the development time. Each of the headings referred to above is discussed in further detail below.

Contents of a feasibility report

Section 1—Introduction: The introduction should be a brief summary of the problem with reference to the initial assignment brief.

Section 2—Problem definition: The problem definition is one of the most important parts of the feasibility report; it must state clearly the nature of the problem to leave no doubt about the extent of the area under review. It should attempt to define the problem in language suitable for line managers and executives in the departments concerned, and specialised data processing terminology must be avoided.

A feasibility study is required for major projects only, e.g. an investigation into an entirely new problem area, perhaps involving the complete operation of a department. Here considerable time might have to be spent in understanding the nature of the problem, and in determining the objectives of the system to be developed.

To attempt to define such a problem the systems analyst must know what the terms of reference of the department are and what sort of objectives they seek to meet in doing their day to day work. It is important when defining the problem that the systems analyst is not constrained by his knowledge of the hardware and software resources currently available.

The treatment of problem definition is very much related to the type of work being undertaken, but it is important when conducting any feasibility study to make certain that there is a mutual understanding of the problem with management.

Section 3—Possible solutions: In some cases it may be necessary to come up with just one sensible and practicable solution of a problem. On the other hand in a major feasibility study it is usually important to make management aware of different methods that could be adopted which may present different benefits, but each having different cost factors to consider.

In presenting the solutions it is important to access the relative merits of each method. This can be done by estimating the relevant costs incurred in development and in live running, and presenting them along with the statements of the operational benefits of the system. The lead time for developing each solution should also be presented, so that readers can appreciate when the operational benefits are likely to accrue. For example, let us assume that an investigation is required to ascertain the best means of order processing in a warehouse operation, with the added requirement that this system needs to be compatible with related stock control, provisioning, and accounting routines. Two methods of tackling this problem may present themselves:

1. Institute a daily routine in which orders received in the order department are written to magnetic tape and allocated against the stock files to generate despatch documentation and update stock figures and accounting routines.
2. Have the stock file available on a large disc store with access for allocating orders on-line by means of teletype terminals in the order department.

It may be that management have expressly asked for method 2 in order to have the ability to make random enquiries of stock positions and to meet urgent demands for information for consumers, as and when requests arise. As a result of his study the systems analyst may realise that to fulfil this objective would take up to two years and he may therefore feel justified in recommending an interim solution on the lines of method 1 above. In making this recommendation he would probably have to point out that method 1 would produce tangible operational benefits in a shorter time, that its development costs would be lower than the full on-line system, and that it would provide valuable experience for the organisation in controlling data and file maintenance problems before operating a full on-line system. The report should also indicate the operational benefits which would be foregone in the meantime should the management decide to go for the on-line solution.

Section 4—Nature of output: The objectives of any assignment can to some extent be represented by the nature of the output reports required from the system. This does not imply that output reports have to be designed down to the last detail at the feasibility stage, but does imply that considerable thought should have been given to the data elements required in output reports, the type of response required from the system, the type of device to be used in output (e.g. line printer, teletype terminal, VDU terminal) and any software constraints there may be. If the output can be shown as a print format then it is worth while including in the feasibility report samples of these formats prepared as computer output and therefore looking exactly as they will appear when the system is in operation. The most important thing, however, is to ensure that the output is specified in sufficient detail to identify the functional requirements of the users concerned, and to establish the feasibility of providing information to meet these needs.

Section 5—System evaluation: This section of the feasibility report should deal in some detail with the benefits of the solutions proposed in Section 3 of the report. From this section, management should be able to assess the overall resources for operating a system when it is live and should also be able to develop a clear picture of the effectiveness of the system. It is often useful at this stage to highlight any major difficulties that may have to be overcome in creating or running the system. Particularly it is necessary to assess the facilities required by line departments to support the running system.

Benefits may be considered under some of the following headings: improved operational efficiency, better customer service, improved quality of information, more timely information, reduction in overheads, reduction in stocks, improved cash flow, increased profitability, improved communications, and so on. It is necessary to justify any such claims and where possible to quantity them, but care should be taken not to create false hopes. Integrity is the key note, because at this stage the analyst is establishing relationships with line management which will be the background to all future work.

It will be necessary also to include a statement of estimated costs under the following headings:

Development costs	Systems activity
	Software development
	Data preparation
	Clerical and other supporting activities
	Computer equipment utilization
Operational costs	Computer time
	Data control and preparation facilities
	Stationery and documentation
	Capital equipment costs

If during the conduct of the feasibility study the systems analyst sees that the feasibility report will cover ground which is outside the bounds set in the assignment brief, then he must report back to management for guidance immediately this situation is apparent. This reporting back should generally result in a re-appraisal of the original assignment brief or proposals to establish a new brief to cover development of additional stages of the problem area.

Section 6—File information: This section of the report is designed to indicate the affects on file structures and can probably be considered under the following headings:

1. The effects of the proposals on existing file structures.
2. Details of new files to be set up.

Recommendations about changes to existing files can be made in narrative form with detailed layouts where necessary. The report should consider the effects on data volumes in the files concerned: for example, the effects on record size, and the number of records in the file. These points must be made in sufficient detail so as to gauge the effect on processing times in the system under consideration, and any side effects on existing systems. Particularly, the study should indicate the effects of such file changes on existing programs, and cross references can be made from this section of the report to Section 9 which deals with programming requirements.

For new files it will be necessary to indicate the data elements to be stored and the source of these elements. The file storage medium must be specified and also an indication of record sizes and the number of records must be included. If it is necessary to explain the file structure, file lay-outs can be incorporated but the emphasis in this section is on defining the file medium and file sizes, so that timing estimates can be made and file storage requirements evaluated.

Section 7—Input information: After specification of the file information, consideration should be given to the transaction types needed to maintain the file data with an identification of the source of such transactions and the volumes of data expected. If there are particular problems in the collection of the data it will be necessary to identify them in this report.

Generally the systems analyst should make recommendations to overcome any such problems and should give proposals concerning the methods of data collection to be employed. The report should describe the volume of data to be collected, the frequency of its collection, and the data preparation and data control requirements.

Section 8—Run structure: This section should provide general run flowcharts of the computer procedures needed in maintaining and updating files and in reporting from them. It is to be expected that at this stage the run structure may be provisional, but the analyst should attempt to show the basic functions and processes that take place. The purpose of this section is to identify the major programs required so that programming estimates can be developed, and to give an overall picture of the processing time required. The run flowcharts should be accompanied by an explanatory narrative.

Section 9—Programming required: The object of this section is to identify new programs needed, to show where existing software can be used, and to develop an overall picture of the manpower resources needed to develop new programs. The programming requirements can be considered under the following headings:

1.   New software required (e.g. new general purpose programs).
2.   New programs required for operation in this particular system.
3.   Programs required in file creation or development.
4.   The effects upon existing programs (e.g. amendments made necessary by changes in existing files).

Section 10—File creation: In this section the analyst should indicate appropriate sources for the collection of data needing to go into new files. If necessary he should specify particular validation requirements or stages of reconciliation necessary for file creation. Computer run flowcharts with narrative showing phases of file creation can be included where necessary, and cross references can be made to Section 9 above if it is necessary to highlight any particular programming requirements.

Section 11—Implementation plan: The implementation plan should be outlined in sufficient detail to give a clear idea of the various stages involved in bringing the system into operation. It should highlight any major events, particularly those which will require resources either from within or outside the data processing department. The successful completion of this section of the feasibility report will depend, therefore, on the degree of detail contained in previous sections of the report and recorded in the assignment working file. It must be the aim of the systems analyst to include in this section a reasonably accurate estimate of the time it will take to complete the implementation of the project.

Section 12—Recommendations: This section of the report should provide practical suggestions for action based on information in previous sections of the report. For example, it may recommend that the complete system is put into operation as a series of well defined stages with a clear indication of the value and benefits to be gained from each separate stage, or phase of the system. To some extent this section will merely repeat essential comments contained in previous sections but it should be concise and aim at giving the reader the necessary facts to enable him to understand and appreciate the systems analyst's final assessment. Where the systems analyst has put forward more than one proposal in Section 3 (possible solutions) then recommendations are necessary to give assistance in selection a particular solution.

Summary of the approach to feasibility studies
The feasibility report arising from the feasibility study is intended for management concerned with policy in the problem area. The report must identify the problem, make practical solutions for dealing with the problem and identify the overall resources required. It should be considered as a document which will provide the steering group with sound recommendations which will enable that group to make decisions resulting in the proper control of the development of systems. The time scale required to implement the system must be given and a clear indication of the benefits to be obtained from the system is essential.

Finally, the feasibility report must be capable of being considered side-by-side with the original assignment brief and therefore the recommendations made must be within the bounds set by the assignment brief.

6.6 THE SYSTEMS DEFINITION

Purpose of systems definition
This document is usually developed from the systems analyst's working files, and is the final statement of the computer and clerical procedures forming the new system. It is used primarily to specify the files, documents, programs, and procedures needed for the system when it is in live operation, and may also specify routines and procedures to be adopted to create files prior to their live operation.

Generally the systems definition is used as a formal specification to obtain final acceptance of the principles and detailed procedures that will appear in the final system. The following representatives may be asked to sign their formal acceptance of the system:

1. Line departments concerned with running or setting up the system.
2. Data processing management, e.g. systems manager, programming manager, and computer operations manager.
3. Departments receiving output from the system, or supplying input data, or other supporting services.

4. The manager of the internal audit department or auditor's representative.
5. The chief accountant or his nominee.
6. Representatives of the data processing steering committee.

In practice the approval by line management of the system may have been secured through the normal workings of project teams engaged in the development of the system. In the same way the feasibility report, if one has been previously produced, will have served to obtain approval at the executive or policy making level in the organisation. Where this is the case the final signature on the systems definition may appear to be a rather hollow ritual. In practice however the document still serves a valuable function in that it is useful as a reference manual for all those concerned with running the system; particularly for the computer operations department who will have to maintain and sustain the procedures perhaps for many years while the system is in routine operation.

Contents of systems definition

Introductory pages The cover and the title page of the systems definition should state: the project name and identifying project file number, the date of production of the definition, the systems analyst's name, the names of the divisions or organisations using or operating the system, and the names of the departments or divisions to whom charges for operation of the system are to be attributed.

There should be an authorisation page for the signatures of those approving the implementation of the job. This page is followed by a contents list, and an amendment control record to log the receipt of all amendments made subsequently.

Section 1—Glossary

Most business organisations have a tendency to develop their own jargon to describe phenomena or concepts current in their particular industry or trade. The analyst will inevitably find himself using these terms in his everyday work, and will start to write them into his reports and specifications unselfconsciously. A clear and concise definition of these terms is an extremely useful aid in understanding an organisation and its procedures, and should be provided when relevant at the beginning of the systems definition.

Section 2—Aims of the system

This should be a simple and concise statement of the scope of the system, and the benefits to be obtained from its implementation. If this subject has been well covered in the original assignment brief then a simple cross reference to the documents concerned can be provided to avoid extensive repetition of the existing text.

A page or two should be sufficient to complete this section (each item should be dealt with succinctly), and the following topics should be covered:

354

1. The major procedures covered by the new system.
2. The departments and divisions concerned with the project.
3. Major computer files used or maintained by the system.
4. The benefits to be provided by the system when it is operational under the following headings:
   1. Financial advantages
   2. Improved manpower or resource utilisation
   3. Improved management control
   4. Better information quality and availability.

Section 3—Systems description

Computer procedures The first part of this section should contain a run flowchart of the computer procedures, to show the various program suites that make up the entire job and to identify the individual programs. A narrative should accompany the flowchart to identify the files used and to describe the general activities that take place in each program. If this has been previously covered in a feasibility report, then the analyst can simply insert the relevant section of that report into this section of the systems definition, having made any alterations that might have arisen in the detailed design phase.

Clerical procedures If the clerical procedures are fairly simple it might be appropriate to describe them in detail in the systems definition. If, on the other hand, they are extensive, they should be dealt with in separate procedure manuals developed in detail for the user departments concerned. In the latter case a concise description of the clerical procedures should be given in the systems definition sufficiently to identify the operations that closely support the computer routines, and to identify the nature and role of the input documents or output reports. Charting techniques similar to those in figures 15, 16 or 17 can be used to provide charts supporting the narrative.

Data preparation and data control These procedures should be developed in full as part of the systems definition, particular emphasis being given to the controls used to verify the operation of the data preparation and collection routines. Cross references may be made to other sections of the systems definition, e.g. to source document formats, record and file layouts, code lists, etc.

The written procedures should specify how source documents are collected and batched, when they are to be collected, the means used to transport them, and methods of data preparation and control. All transaction types must be identified and the daily/weekly/monthly volumes quoted.

Section 4—Source documents

All new source documents should be shown as samples, and be identified by name and form number. A brief statement of the purpose of each

355

document should be provided along with a description of how it is completed. Reference can be made to the clerical procedures described in Section 3.

Section 5—File specifications

All files used in the computer system must be specified in detail including all files held on backing storage, and punched card and paper tape files used as input.

The separate record types used within each file must be defined and file layout diagrams showing the arrangement of data elements within records should be provided. In some cases these diagrams can be drawn up on special charts, but they can be represented just as well on squared paper. An example is shown in figure 117.

0	1	2	3	4	5	6	7	8	9	10
	IDENTIFICATION No.		AGEING 1 2 PAY SUB	DATE YY MM DD	▽ INVOICE No.		DESCRIPTION		▽▽▽	CUSTO

117. File layout diagram

The individual data elements in the file must be defined; a simple list as in figure 118 can serve this purpose.

The file name must be specified and the sequence of the records within the file. The retention period must also be defined. File names should be included in the specification, and the format and content of any special header or trailer labels used in the system. If the file conforms with some predetermined format standard, e.g. a computer manufacturer's housekeeping software, then this should be stated.

Section 6—Output specifications

Samples of printouts should be drawn on planning charts as shown previously in figure 74. It is essential to give every printout an identifying name and preferably an identifying number. All fields printed in the report should be named and maximum and minimum sizes of fields specified. The desired lateral and vertical line spacing can be shown on the chart. The heading lines and total lines should also be specified. Some narrative should accompany each sample stating the purpose of the report, the type of stationery to be used, the number of copies required, their routing, frequency, and distribution.

356

DATA ELEMENT	SIZE	FORMAT	SIGN	RANGE
Customer Name	16 ch	A/Numeric		
Address	48 ch	A/Numeric		
Customer No.	6 ch	Numeric		
Credit Code	1 ch	Numeric		
Credit Limit	1 word	Binary	+ only	0 to £20K
Document No.	6 ch	Numeric		
Product Code	3 ch	Numeric		
Unit Cost	2 word	Binary x 10^3	+ only	
Qnty Due	1 word	Binary	+ or -	
Extended Value	2 word	Binary x 10^3	+ or -	

118. Data element list

Section 7—Implementation procedures
This section should identify all the major events that must be completed to get the system running. Particularly it should specify the procedures necessary to create files, placing emphasis on such important matters as:

1. Identifying sources of data.
2. Specifying validation requirements in taking on the data.
3. Coding and/or editing needed.
4. Highlight heavy loads in say data preparation.
5. Identify and specify any special programs required.

Plans should be defined for conducting the changeover to the new system, to show what resources are likely to be needed in pilot running or parallel running with the existing system. A statement must also be given to show what steps must be taken to verify the accuracy of the system as it goes live and to reconcile with existing systems where necessary.

Section 8—Equipment utilisation
The purpose of this section is to identify the equipment resources needed to run the system when it is operational, and to set up files initially. Such resources may include data preparation equipment as well as computer hardware. These time estimates are fairly easy to make for daily, weekly, monthly routines and can be made by referring to run flowcharts included in Sections 3 and 7.

357

M

Section 9—Program specifications
This section of the definition is probably one of the most important: it should be completed in detail and must be updated even for relatively minor changes. This will probably be one of the most important records that will need to be retained throughout the development and operational life of the system—for example, in order to amend programs when the system is changed to meet new management requirements.

The precise layout of a program specification prepared by an analyst will to some extent depend on the standards of specification required by the programming unit to whom the specifications will be given. However, all specifications must contain full details of the requirements from a program, prepared in a clear and unambiguous manner and providing sufficient detail for the programmer to prepare the required program. The use of flowcharts and decision tables in providing a link between specification and program has been discussed in sections 3.4.1 and 3.4.2. The main points to be included in a program specification are outlined below.

Specification contents
1. *Introduction* This will contain a summary of the program requirements and outline the overall system of which the particular program is to form a part.
2. *Print layouts* All printer output will be shown on sample layout charts, annotated where necessary, as in Section 6 of the system definition.
3. *File specifications* All input and output files will be described in detail, as in section 5 of the system definition with lists of data elements, their locations within records, the types of storage medium to be used, and file labelling conventions. In addition, the analyst should where possible indicate expected file sizes, expected frequency of variable data elements and any other information available concerning volumes.
4. *Procedure description* All procedures to be carried out by the program must be described in detail. These will include:

> Start procedures
> Main procedures
> End procedures
> Dump and restart requirements

The section on main procedures will be the chief section, and must include details of all requirements, such as validity checks, control totalling, maximum and minimum sizes of all value fields and accumulations, any rounding conventions, all error detection and reporting procedures, arithmetic and logical operations on data

358

elements, opening files, closing files, dump and restart routines, run-time options and operator messages.

5. *Test data*   The analyst must prepare detailed test data, and expected results from his data against which the program can be checked to ensure it has worked correctly. The procedures for establishing test data and using it in program acceptance are given in sections 5.3.1 and 5.3.2.

6.7   COMPUTER OPERATIONS GUIDE

This document will be used by the computer operations staff, and where necessary data control staff, in the preparation of a job. It is usually a description of the job and of the associated setting up procedures. The following items should be considered for inclusion.

1. A title page giving the job name, job number, charging code, the name of the analyst, and a contents list.
2. A brief description of the job in narrative with a run flowchart. A job will often be some subset of the runs forming a complete system. The description can be very brief with a cross reference to the relevant systems definition.
3. A list of all contacts in user departments with references to type and source of input data: schedules should show the time of collection and methods of transportation.
4. Actions required to prepare each run; e.g. changing run-time parameters. Where relevant, instructions should be included for preparing specific run time sheets for operators, e.g., to notify switch settings to effect options in processing operations.
5. Details of file control methods. All files in the job must be identified and instructions for retention of file generations should be stated.
6. Distribution instructions for all reports, and error lists produced by the system. The number of copies, methods of copying, and a circulation list should be included.
7. Methods of applying quality control checks to the job. This will include examination of print formats, checking record counts, block counts, control accounts, etc. Tolerance limits can also be established for error reports.

6.8 OPERATING INSTRUCTIONS

Operating instructions are intended to act as a precise procedure for the computer operators running the job in the computer room. To make for maximum efficiency the instructions should be capable of use without changes for each particular run, but if necessary a separate job sheet can be provided to give specific run time instructions for any particular day.

A job will generally be regarded as a distinct set of runs that are operated together as a logical unit. For example, a daily invoicing routine, a weekly sales analysis, or a monthly statements routine.

There will usually be a separate folder of instructions for each job, with one page to describe the detailed instructions for each run in the suite of programs. A sample page is shown in figure 119.

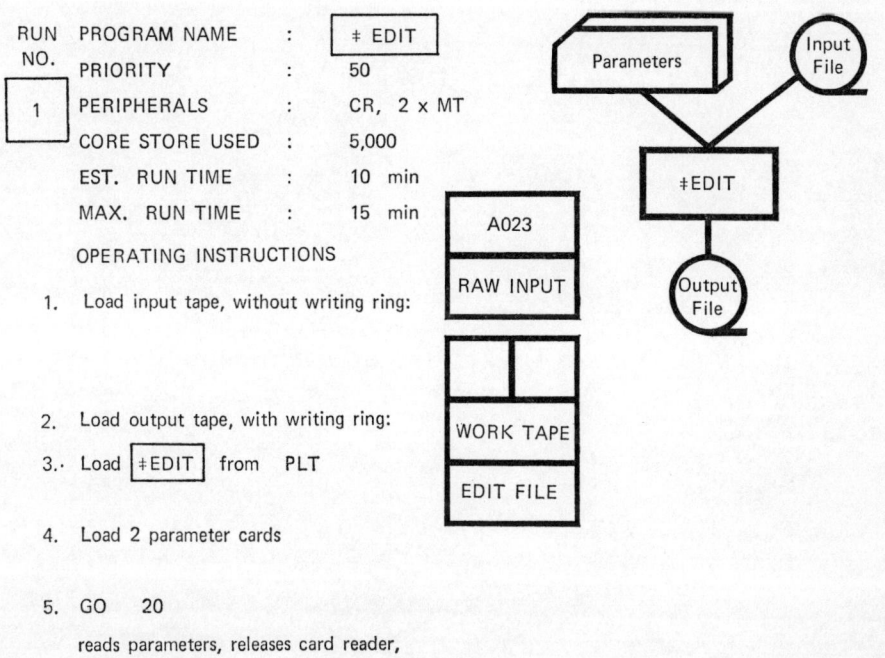

RUN NO. [1]

PROGRAM NAME : ‡ EDIT

PRIORITY : 50

PERIPHERALS : CR, 2 x MT

CORE STORE USED : 5,000

EST. RUN TIME : 10 min

MAX. RUN TIME : 15 min

OPERATING INSTRUCTIONS

1. Load input tape, without writing ring:

2. Load output tape, with writing ring:

3. Load ‡EDIT from PLT

4. Load 2 parameter cards

5. GO 20

   reads parameters, releases card reader,

   reads input tape,

   writes output tape,

   releases both tapes,

   DISPLAY-COUNT OF TAPE RECORDS READ nnnnn

   DISPLAY-COUNT OF TAPE RECORDS WRITTEN nnnnn

   HALTED-END OF PROGRAM

6. Go to run number [2]

EXCEPTION CONDITIONS

1. Tape failures — GO — this causes DISPLAY, followed by HALTED-RUN ABANDONED; then re-run, using different decks.

2. Other failures — Log and abandon; go to run number [END]

119. Sample page of operating instructions

361

PART 7

Systems Maintenance

PART 7 : SYSTEMS MAINTENANCE

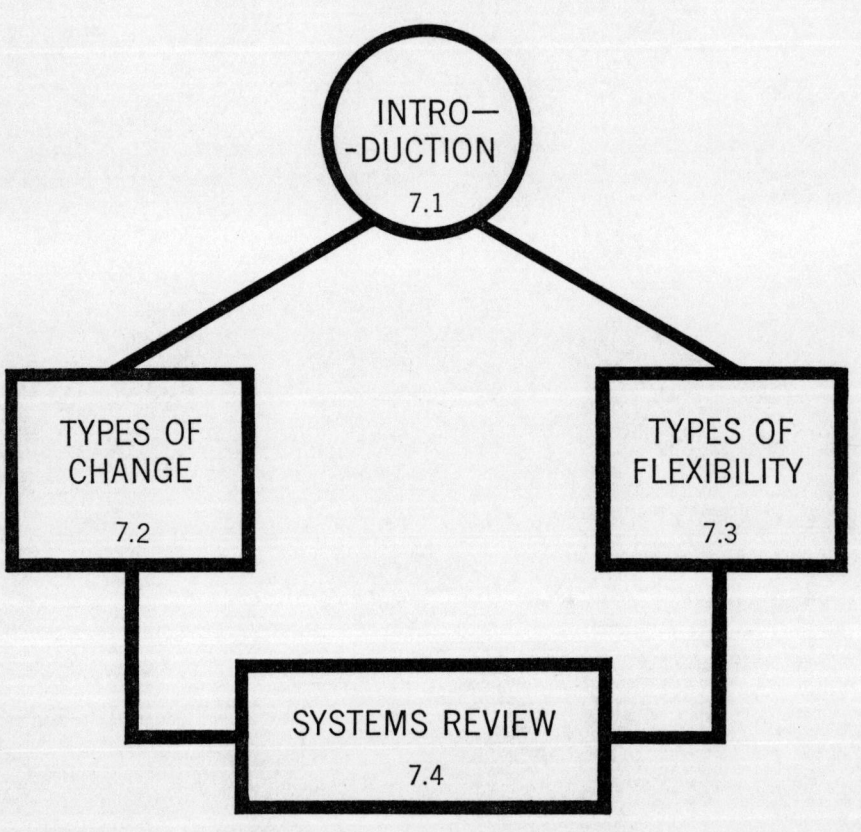

7.1 INTRODUCTION

So far in this book we have discussed all events which lead up to the completion of a systems task. We have considered the quality of the analyst, the tools he has at his disposal, the techniques he requires when designing a system, and the problems he has to tackle when implementing the new system. One may imagine the tyro analyst, the day after his first system has gone 'live', heaving a sigh of relief, pouring out his celebration drink, and turning with a fresh and invigorated gaze at his next assignment brief. If he imagines that the system he has just completed will require no more attention, that it will continue immutable and perfect until the end of time, he will be mistaken, probably sadly. No system, however thoroughly designed, however carefully tested, however impressively documented, will ever last unchanged. Indeed, as we have emphasised in earlier sections, the good analyst will, at all stages, plan for change. Flexibility in input and output formats, in record structures, in data flow and validation, all are important at the development stage, since the analyst can, while the system is being designed, make modifications with the minimum disturbance to work already completed. Thus a good system will already have built into it the seeds of change, for modifications simple to make at the development stages will be equally simple after implementation. In the same way, comprehensive and accurate documentation is in a way an invitation to change, for change is simplified if there is a clear and accurate statement of what happens now. Systems change for many reasons: some of them are discussed in section 7.2; section 3.7 outlines some of the techniques of flexibility which enable the analyst to make changes with the least possible trouble; and section 7.4 outlines a way of keeping an eye on an existing system to make sure it continues to function efficiently and effectively.

7.2 TYPES OF CHANGE

General
There are many reasons why systems change, and it would be impossible to give a complete list of all likely events, particularly since a great number come back to that most vulnerable link in any system, the human being. However, causes of change can be grouped into two main categories, those external to the system, and those internal to the system. Among external causes will be changes to equipment available, both hardware and software; changes to the organisational structure controlling the system, such as mergers; changes of product line and changes as a result of

government action. Internal changes include changes of attitude or responsibilities of those operating the system and changes in volumes of data processed. How these various broad categories of change affect systems are discussed below.

Equipment changes

As time progresses, computer manufacturers develop and extend their range of equipment. A system must always be designed to utilise equipment available at the time of implementation: but the analyst must always remain alert to the possibilities of new developments, and aware of the potential gains to be achieved by using new devices in an existing situation. For example, the development of optical readers and audio devices and the potential development of equipment able to read ordinary handwriting or recognise sounds made by the human voice offer exciting possibilities in the field of data collection. Data communications networks and visual display units are other devices which are becoming increasingly available and may well alter the conventional pattern of even such well-established systems as payroll and stock control. The analyst must of course also be aware of improvements to existing devices: faster tape units, faster card readers and line printers, may all effect considerable economies in an existing system without necessarily changing it in any way.

Software development is another area in which changes may well be required in an existing system. The development of operating systems is a powerful example of the abilities of software to extend the power and capabilities of a computer system, and the full use of an operating system may well involve changes to existing procedures. Developments in information retrieval and analysis software, report generators, inquiry languages, may all call for revision of existing methods for extracting information from an existing system. In all these examples of hardware and software improvements, it is likely that any initiative for change will come from the analyst, although the line manager operating a computerised system should also be aware of developments and how these may affect his efficiency.

Changes in organisation

Any organisation using a computer for some or all of its functions is likely to change its structure in reacting to changing markets, changing research, and changing situations of all kinds. It is clearly impossible to foresee all possible changes of this type when a system is designed, but the analyst must be aware of the possibility that changes may occur. A company payroll system must be capable of standing up to the strains of a merger, with new employees requiring changes to pension fund calculations, types of deduction and conditions of service. A stock control system may well have to be modified when the organisation's product range changes; an accounting system may need to be modified if new accounting codes and control systems are introduced. Another type of change which can be imposed on a system as a result of external changes are all those due to government action: new taxes, new regulations, currency decimalisation, new official statistics, will all require the attention of the systems analyst. Unlike the equipment changes, initiative for system changes in these circumstances will normally come from those in day to day charge of operations using the system: but the analyst too must be aware of imminent changes, and plans must be made well in advance to cater for them.

User attitudes

The types of change so far discussed are imposed on a system as a result of changes external to it, but a major source of change to a system will be as a result of the attitudes and experience of those involved in the day to day operation of the new system. Inevitably, the way the system operates will change: individuals will always think of a way to modify and improve an existing method. Input documents may be controlled differently; dates when data is processed will change; forms will be modified. The data itself may change. Some items will become redundant, while others, initially less significant, will become important, and perhaps need to have an extended code structure, or a different system of recording. Changes may result not only because of such apparently minor changes to detail, but because an imaginative manager may see opportunities for extracting additional information from data already recorded in a system, either by extending the data elements recorded, or modifying existing elements, or changing output formats. It is impossible to forecast every new requirement: but a system designed for change will grow and develop as the people using the system become increasingly aware of the power of computers and the increased control offered by well-designed systems.

Volumes

A further important source of change to a system is change in volumes of data processed. In some situations it will be possible to make a reasonable forecast of the change expected in data volumes over a period of time, and changes necessary to cope with them can be built in at the start of the new system. But changes cannot always be foreseen, and as volumes change, so details of the system may need to be modified. Updating frequency may need to be increased; it may be necessary to switch from a batch processing of movement to the random processing of a direct access file.

Where many systems share a computer, increases in data volumes of unrelated systems may mean a revision of timetables and a change in system details. The user and the analyst must watch the volumes being processed and ensure that changes can be reflected by suitable system modification.

7.3 TYPES OF FLEXIBILITY

General

Earlier sections of this book have emphasised the need for flexibility in systems design. This section outlines some of the main techniques available to the analyst for building flexibility into a system, enabling it to cope with changes arising out of the various situations described in section 7.2.

Data flexibility

No analyst and no user can be certain that all possible data elements likely to be required have been recorded in a new system. For those elements which have been included, there is no certainty that the formats which have been designed for them will remain fixed; code structures, number of characters, validity checks, all may require changing as the system is used in practice. Thus data flexibility must be two-fold. The facility for adding new data elements to record structures, and deleting elements no longer required, must exist. It must also be possible to expand and extend existing data elements, where necessary modifying associated input documents, updating procedures and validity checking. If possible rigid and inadequate codes should be avoided. If there are 24 categories of contract, a code in the range A–Z may seem sensible: but when the accountants produce a requirement for another dozen categories the system may well begin to creak. Of course, flexibility must not be taken to extremes, and it would be absurd to carry within a record large blank spaces waiting to be filled. However, an intelligent consideration of the likelihood of expansion of data elements will help to avoid headaches in the future.

Program flexibility

Any change to a system will nearly always involve some changes to the detailed operations performed by one or more programs in the system suite. If such changes involve extensive re-writing of an existing program, with all the attendant time-consuming and expensive stages of program specification and testing, a valuable change may in fact never get implemented, or the delays involved will cause frustration and loss of confidence by the user.

Where possible programs in a suite should be made adaptable. This can be done by specifying particular detailed requirements to a program by means of system parameters or by writing the programs in a modular way, so that changing specific modules will not affect other parts of the program. The use of standard software often provides a considerable amount of flexibility, since software packages are designed for differing applications and hence can accommodate system changes to a given application.

368

Procedure flexibility
A good system requires all users to exercise certain disciplines in the procedures operated. Indeed in many cases a new system is justified solely on the grounds that order replaces chaos, quite apart from the additional gains of computerisation. Any system will therefore lay down procedures to be followed, in matters of data collection and input, data control and validation, and output control procedures. But although standards must be adhered to, a system so rigidly designed that it cannot accommodate procedural changes when these prove desirable and necessary will not prove a success. For example, a system may be designed to operate on the turn-round document system for updating procedures. But in some circumstances rigid adherence to input from turnround documents only may impose severe strains. The turnround system may be based on an estimated 20 per cent of records being modified at each updating cycle, but some special event may involve changing all records: this may call for a quite different method of obtaining the required data, and procedures should be able to accommodate such a requirement.

Documentation
Section 6 on documentation standards has emphasised the importance of clear and accurate documentation at all levels of system design. The existence of good documentation is an essential element in system flexibility because if documentation accurately describes an existing situation, the effects of any modification can be seen clearly, and modifications can be made quickly and accurately.

Once modifications have been agreed, they must pass through the same formal acceptance and documentation procedures laid down for the original system, so that documentation remains up to date. If a system is in operation for some time it is very likely that new staff and new analysts will be involved in effecting changes: documentation then plays an essential part in communication, giving all concerned the complete understanding of an existing situation essential before any attempt can be made to change it. Good documentation provides the essential continuity over the period of existence of a system, so that however different a system may become, there is always a clearly documented link back to its original state.

7.4 SYSTEMS REVIEW

'The system analyst's job is to make sure that a system works in the best possible way.' This sentence began our book: and it is suitable that we should be reminded of it in the concluding section. Systems are implemented for users, not machines, and this principle must be constantly kept in mind at all stages of development, implementation and, finally, after the new system begins its operational life. We have already discussed the importance of close co-operation with users of a new system during the design and implementation stages, particularly the use made of project teams. This close co-operation must be maintained after a system has been implemented. A useful technique is to carry out a regular review of the way the system is operating, perhaps at six-monthly or yearly intervals. Users and analysts should examine all aspects of the system, and any suggestions for modification and improvement can be discussed. Apart from such formal review procedures, the mutual confidence which will have been built up between analyst and user during development and implementation should mean that the user will have confidence at all times that the computer system he uses is in fact *his* system, giving him the effectiveness and control he expects: this, after all, is the true justification for the use of computers.

BIBLIOGRAPHICAL NOTE

It is usual in a work of this kind for the authors to append a bibliography listing suitable books for further reading. In this case, however, we feel we can do no better than refer readers to the International Computer Bibliography published by The National Computing Centre Limited. It is obtainable from most libraries and provides a thorough and complete review of a very large number of books on computers, suitably grouped and cross-referenced under appropriate subject headings.

INDEX

This index shows key words in context and, as well as giving the page number of the reference, shows the title of the section in which the reference occurs. This will allow those looking for a reference to search the index in the following way:

Find the key word, listed in the centre of the page in alphabetical sequence;

Establish the relevance of the reference by looking at the context of the key word;

Establish the level of treatment by looking at the section reference (printed in italics);

Turn straight to the appropriate reference after checking the page number.

		Key word	*Page No.*
4.6.1 direct access processing	direct	access file storage principles	296
	multi	access *3.6 operating systems*	173
5.3.2 program acceptance		acceptance procedure	332
	procedure	acceptance *5.3 testing and acceptance*	332
	program	acceptance *5.3 testing and acceptance*	332
	testing and	acceptance *5. systems implementation*	330
	random files and	address generation *4.6 direct access processing*	314
	run-time	amendment *3.5.6 program handling software*	161
2.3 analysis of recorded data		analysis	75
2.3 analysis of recorded data		analysis as a separate stage	72
	investigation and	analysis of existing procedures *2. investigation and analysis*	47
2. investigation and analysis		analysis of recorded data	72
	stages of systems	analysis *1. the role of the systems analyst*	27
	qualities of the systems	analyst *1.1 the systems analyst*	21

371

INDEX

	Key word	Page No.
the role of the systems	analyst *Part 1*	19
4.1.1 importance of a clear objective	approval of output design	209
user	attitudes *7.1 types of change*	367
systems	applications *3.5.5 data management software*	160
4.3.4 system controls and auditors initial	approval methods	234
3.1.2 control logic	arithmetic unit	87
3.1.5 data operations	arithmetical operations	91
compilers and	assemblers *3.4.4 languages*	140
2.1 establishing system objectives	assessment of benefits and costs	40
preliminary	assignment brief *2.1.2 identifying areas for systems study*	38
6. documentation standards the	assignment brief	344
role of the	audit department *4.3.4 system controls and auditors*	234
system controls and	auditors *4.3 systems controls*	234
foreground and	background working *3.6 operating systems*	164
input/output	balance *4.5.3 block organisation for magnetic tape files*	284
4.4.1 structure of magnetic tape systems	batch processing	237
4.2.3 source documents and data preparation	batch sizes	225
4.2.3 source documents and data preparation	batch totals	225
4.4.2 input transcription programs	batch totals	240
assessment of	benefits *2.1.4 assessment of benefits and costs*	42
assessment of	benefits and costs *2.1 establishing system objectives*	40
2.2.4 methods of fact recording	block diagram (see also flowchart)	65
4.5 the design of a magnetic tape system	block organisation for magnetic tape files	281
variable length	blocks *4.5.3 block organisation for magnetic tape*	285
George	Boole *3.1.5 data operations*	91
packing density and	bucket overflow *4.6.3 indexing sequential files*	305
4.6.1 direct access file storage principles	bucket packing density	300
4.6.1 direct access file storage a	bucket – the unit of data retrieval	299
3.4.7 programming techniques	buffering	151
electronic	calculator *3.8 punched card equipment*	198
eighty column	card *3.7.1 punched cards*	179
magnetic	card *3.3.5 magnetic card file*	123
punched	card equipment *3. data processing equipment*	194
magnetic	card file *3.3 storage*	123
3.2.2 card reader	card codes	96
3.2.3 card punch	card codes	98
3.7.1 punched cards	card layout	180
operation of	card punch *3.2.3 card punch*	99
programming a	card punch *3.7.1 punched cards*	186
3.2 peripheral units	card punch	98
speeds of	card punching *3.7.1 punched cards*	186
operation of	card reader *3.2.2 card reader*	96
3.2 peripheral units	card reader	95

	Key word		Page No.
punched	cards	*3.7 data preparation*	178
verifying punched	cards	*3.7.1 punched cards*	187
punching	cards	*3.7.1 punched cards*	183
4.6.4 overflow of sequential files	chaining		307
types of	change	*7. systems maintenance*	365
5. systems implementation	changeover		334
scheduling of	changeover	*5.4.1 changeover planning*	334
5.4 changeover	changeover planning		334
equipment	changes	*7.1 types of change*	366
7.1 types of change	changes in organisation		366
3.2 peripheral units	character recognition equipment		109
function	chart	*2.2.4 methods of fact recording*	56
2.2.4 methods of fact recording	chart section		65
horizontal flow	chart	*2.2.4 methods of fact recording*	67
vertical section	chart	*2.2.4 methods of fact recording*	67
modulo 11	check	*4.3.1 standard hardware/software*	
		controls	229
4.3.1 standard hardware/software controls	check digits		228
3.4.5 diagnostics	checking		142
error	checking	*3.2.3 card punch*	99
error	checking	*3.2.2 card reader*	97
error	checking	*3.7.4 keyboard to*	
		magnetic tape	193
error	checking	*3.3.5 magnetic card file*	125
error	checking	*3.3.4 magnetic disc*	123
error	checking	*3.3.3 magnetic drum*	121
error	checking	*3.3.2 magnetic tape*	119
error	checking	*3.2.5 paper tape punch*	103
error	checking	*3.2.4 paper tape reader*	101
error	checking	*3.2.6 printer*	105
parity	checks	*3.1.1 information patterns*	84
4.4.2 input transcription programs	check sums		241
4.2.1 impact of output requirements	classification of data elements		219
open and	closed shop	*3.4.6 remote testing*	145
card	codes	*3.2.3 card punch*	98
paper tape	codes	*3.2.5 paper tape punch*	102
paper tape	codes	*3.2.4 paper tape reader*	100
punching	codes	*3.7.1 punched cards*	180
5.5.1 file creation: data collection	coding		338
	coding	*3.4 programming*	135
record type	coding	*4.4.3 sorting and merging*	245
3.8 punched card equipment	collator		199
initial file creation: data	collection	*5.5 file creation and*	
		conversion	336
parallel data	collection	*5.5.1 file creation: data*	
		collection	337
eighty	column card	*3.7.1 punched cards*	179
3.6 operating systems	command languages		165
3.4.4 languages	commercial languages		139
3.4.3 coding	communication		137
3.2 peripheral units	communications equipment		113
1.1 the systems analyst	communication skills		22
3.1 processors	compatibility		89
3.4.5 diagnostics	compiler messages		142
3.4.4 languages	compilers and assemblers		140

373

	Key word	Page No.
source document direct to	computer *3.7 data preparation*	193
6. documentation standards	computer operations guide	359
5.4.1 changeover planning	continuity	334
file	control *3.5 utility software*	157
interrogation and display unit	control *3.5.1 housekeeping packages*	153
operating system	control *3.6.2 command languages*	166
storage device	control *3.9.1 housekeeping packages*	153
4.3 systems control	control accounts	229
3.1 processors	control logic	86
3.2.6 printer	control loop	105
4.3 systems controls	file control and file security	232
maintenance of	control procedures *4.3.2 control accounts*	231
basic peripheral	control software *3.5.1 housekeeping packages*	152
errors and	control totals *4.4 processing magnetic tape*	265
3.1.2 control logic	control unit	87
input	controls *5.3.3 procedure acceptance*	333
standard hardware and software	controls *4.3 systems controls*	226
4.3 systems controls system	controls and auditors	234
3.6 operating systems	conversational mode	172
file creation and	conversion *5. systems implementation*	336
partial file	conversion *5.5.2 converting existing files*	340
5.5 file creation and conversion	converting existing files	339
3.5.4 file control	copying	158
	core storage *3.1 processors*	88
3.3 storage	core storage	116
magnetic	cores *3.3.1 core storage*	116
full-time	courses *5.2.2 types of training*	327
part-time	courses *5.2.2 types of training*	327
week-end	courses *5.2.2 types of training*	328
assessment of benefits and	costs *2.1 establishing system objectives*	40
2.2.4 methods of fact recording	cost table	58
file	creation *3.6.4 file storage*	168
dummy file	creation *5.5.1 file creation: data collection*	338
5.5 file creation and conversion initial file	creation: data collection	336
phased file	creation *5.5.1 file creation: data collection*	337
5. systems implementation file	creation and conversion	336
test	data *5.3 testing and acceptance*	330
procedure test	data *5.3.1 test data*	331
program test	data *5.3.1 test data*	330
initial file creation:	data collection *5.5 file creation and conversion*	336
impact of output requirements on	data collection *4.2 data collection requirements*	219
5.5.1 file creation: data collection parallel	data collection	337
4.2 data collection requirements		
summary of	data collection principles	226
developing	data collection procedures *4.2 data collection*	220
3.5.1 housekeeping packages	data editing software	152
classification of	data elements *4.2.1 output and data collection*	219

	Key word	Page No.
7.3 types of flexibility	data flexibility	368
3.5.5 data management software	data management operations	159
3.5 utility software	data management software	159
3.1 processors	data operations	90
3.3.5 magnetic card file	data organisation	125
3.3.4 magnetic disc	data organisation	122
3.3.3 magnetic drum	data organisation	120
3. data processing equipment	data preparation	178
source documents and	data preparation *4.2 data collection requirements*	222
1.2 the data processing function	data processing department structure	25
Part 3	data processing equipment	75
1. the role of the systems analyst the	data processing function	24
3.4 programming	decision tables	129
job	description *3.6.2 command languages*	165
3.6 operating systems systems	design for operating systems	174
2.1.1 feasibility studies	defining the problem	37
4.4.5 file maintenance	deletion of records	262
4.5 design of a magnetic tape system record	design for magnetic tape files	277
error	detection *3.7.2 paper tape*	190
3.2.11 communication equipment error	detection units	115
3.5.1 housekeeping packages storage	device control	153
3.4 programming	diagnostics	141
block	diagram *2.2.4 methods of fact recording*	65
4.6.1 direct access processing	direct access file storage principles	296
2.2.4 methods of fact recording	directory section	54
magnetic	disc *3.3 storage*	121
interrogation and	display unit control *3.5.1 house-keeping packages*	153
visual	display unit *3.2 peripheral units*	106
3.6.3 on-line and off-line working the	document concept	168
3.7 data preparation source	document direct to computer	193
2.2.4 methods of fact recording	document section	55
3.4.3 coding	documentation	138
3.5.7 program maintenance	documentation	162
7.3 types of flexibility	documentation	369
6. documentation standards why	documentation is necessary	343
input	documents *5.3.3 procedure acceptance*	333
4.1.4 choice of output medium turn-around	documents	214
magnetic	drums *3.3 storage*	120
5.5.1 file creation: data collection	dummy file creation	338
4.1.2 definition of output formats	dummy reports	209
3.5 utility software	dump and restart	163
3.5.8 dump and restart	dumping	163
3.5.6 program handling software	dumping	161
use of	dumps and restarts *4.4.9 dumps/restarts on magnetic tape*	274
4.4 processing methods for magnetic tape	dump routines	273
4.4 processing methods for magnetic tape	dumps and restarts using magnetic tape	272
input	editing *4.4.2 input transcription programs*	243
3.5.1 housekeeping packages data	editing software	152
4.4.3 sorting and merging	editing while sorting	248
planning	education *5.2 education for implementation*	325

INDEX

	Key word	Page No.
5. systems implementation	education for implementation	325
3.7.1 punched cards	eighty column card	179
4.5.4 timing and evaluation	total elapsed peripheral time	288
3.8 punched card equipment	electronic calculator	198
dealing with	end-of-file conditions 4.4.4 file updating methods	258
3.2.3 card punch	error checking	99
3.7.4 keyboard to magnetic tape	error checking	193
3.3.5 magnetic card file	error checking	125
3.3.4 magnetic disc	error checking	123
3.3.3 magnetic drum	error checking	121
3.3.2 magnetic tape	error checking	119
3.2.5 paper tape punch	error checking	103
3.2.2 card reader	error checking	97
3.2.4 paper tape reader	error checking	101
3.2.6 printer	error checking	105
3.7.2 paper tape	error detection	190
3.2.11 communications equipment	error detection units	115
maintenance of an	error file 4.4.6 error reporting	266
4.4 processing magnetic tape	error reporting	263
4.1 establishing the output requirements	error reporting conditions	217
guide lines for	error reporting 4.1.6 error reporting conditions	217
requirement for	error reporting 4.1.6 error reporting conditions	217
3.4.5 diagnostics	error types	141
2.2.4 methods of fact recording	establishment table	57
principals of	exception reporting 4.1 establishing output requirements	215
investigation and analysis of	existing procedures 2. investigation and analysis	47
testing against	expected results 5.3.2 program acceptance	332
internal and	external formats 3.1.1 information patterns	84
methods of	fact finding 2.2 investigation and analysis	47
methods of	fact recording 2.2 investigation and analysis	54
contents of	feasibility report 6.5 feasibility reports	348
6. documentation standards	feasibility reports	348
2.1 establishing system objectives	feasibility studies	37
4.1.2 definition of output formats	field limitations	211
magnetic card	file 3.3 storage	123
types of	file 3.6.4 file storage	168
3.5 utility software	file control	157
3.6.4 file storage	file creation	168
5. systems implementation	file creation and conversion	336
dummy	file creation 5.5.1 file creation: data collection	338
5.5 file creation and conversion initial	file creation: data collection	336
4.3 systems controls	file control and file security	232
5.5.2 converting existing files partial	file conversion	340
5.5.1 initial file creation phased	file creation	337

376

	Key word	*Page No.*
2.2.4 methods of fact recording	file descriptions	62
4.4 processing magnetic tape	file maintenance	260
techniques of	file maintenance *4.4.5 file maintenance*	261
basic principles of	file organisation *4.6.1 direct access file storage*	299
storage control and	file protection *4.6.1 direct access file storage*	299
4.6.4 overflow of sequential files	file re-organisation	308
3.6.4 file storage	file security	171
file control and	file security *4.3 systems control*	232
3.6 operating systems	file storage	168
3.6.4 file storage	file storage utilisation	170
direct access	file storage principles *4.6.1 direct access processing*	296
inverted	file structure *4.6.7 inverted files*	316
3.6.8 systems design for operating systems	file structures	175
4.6 direct access processing	file types and processing modes	301
4.4 processing magnetic tape	file updating methods	250
card	files *4.5.4 timing of magnetic tape files*	292
converting existing	files *5.5 file creation and conversion*	339
inverted	files *4.6 direct access processing*	316
magnetic tape	files *4.5.4 timing of magnetic tape files*	294
paper tape	files *4.5.4. timing of magnetic tape files*	290
random	files *4.6 direct access processing*	302
serial	files *4.6 direct access processing*	301
4.5.2 record design for magnetic tape	fixed length (and variable length) records	279
floating point and	fixed point representation *3.1.5 data operations*	92
data	flexibility *7.3 types of flexibility*	368
procedure	flexibility *7.3 types of flexibility*	369
program	flexibility *7.3 types of flexibility*	368
types of	flexibility *7. systems maintenance*	368
4.1.5 principles of exception reporting	flexibility–program techniques	216
3.1.5 data operations	floating point and fixed point representation	92
3.4 programming	flowcharting	128
computer procedures	flowchart *2.2.4 methods of fact recording*	67
horizontal	flowchart *2.2.4 methods of fact recording*	67
3.6.1 general principles (operating systems)	foreground and background working	164
internal and external	formats *3.1.1. information patterns*	84
output	formats *5.3.3 procedure acceptance*	333
approval of output	formats *4.1 establishing output requirements*	211
2.2.4 methods of fact recording	form descriptions	59
2.2.4 methods of fact recording	function chart	55
input and output	functions *3.6.8 systems design for operating systems*	174
reproducer and	gangpunch *3.8 punched card equipment*	197
3.7.1 punched cards	gangpunching	186
3.4.4 languages	generators	140
3.5 utility software	generators	153
3.2.11 communications equipment	GPO transmission facilities	115

INDEX

	Key word	Page No.
3.2 peripheral units	graph plotter	108
program	handling software *3.5 utility software*	160
3.4.4 languages	high-level languages	138
2.2.4 methods of fact recording	horizontal flowchart	67
3.5 utility software	housekeeping packages	152
2.1 establishing systems objectives	identifying areas for systems study	38
2.1 establishing systems objectives	identifying management requirements	39
education for	implementation *5. systems implementation*	325
involvement in	implementation *5. systems implementation*	323
systems	implementation *Part 5*	321
5.1.1 user participation	implementation stages	323
self	indexing *4.6.3 indexing sequential files*	303
partial	indexing *4.6.3 indexing sequential files*	304
summary of	indexing methods *4.6.3 indexing sequential files*	306
4.6 direct access processing	indexing methods for sequential files	303
5.5 file creation and conversion	initial file creation: data collection	336
3.1 processors	information patterns	82
problems of	information retrieval *4.6.7 inverted files*	316
write (permit or)	inhibit rings *4.3.1 standard hardware/ software controls*	227
3.6.8 systems design for operating systems	input and output functions	174
5.3.3 procedure acceptance	input controls	333
5.3.3 procedure acceptance	input documents	333
4.4.2 input transcription programs	input editing	243
choice of	input medium *4.2 data collection requirements*	220
4.5.1 magnetic tape files – structure	input/output balance	276
4.5.3 block organisation for magnetic tape	input/output balance	284
4.5.4 timing of magnetic tape procedures	input/output channels	288
minimising	input/output time *4.5.2 record design – magnetic tape*	277
4.4 processing magnetic tape	input transcription programs	240
3.4.3 coding	instructions	135
operating	instructions *6. documentation standards*	360
3.2 peripheral units	interface	95
3.1.1 information patterns	internal and external formats	84
2.3 analysis of recorded data	interpretation of recorded data	72
3.8 punched card equipment	interpreter	198
3.5.1 housekeeping packages	interrogation and display unit control	153
structure of the	interview *2.2.3 interviewing techniques*	51
seven hints for	interviewing *2.2.3 interviewing techniques*	53
2.2 investigation and analysis	interviewing techniques	50
preparation for	interviews *2.2.3 interviewing techniques*	50
2.2.4 methods of fact recording	interview section	65
4.6 direct access processing	inverted files	316
information retrieval from	inverted files *4.6.7 inverted files*	317
Part 2	investigation and analysis	35
2. investigation and analysis	investigation and analysis of existing procedures	47
6. documentation standards the	investigation file	345

378

	Key word	Page No.
2.1.6 establishing project teams	involvement	45
5. systems implementation	involvement in implementation	323
on the	job *5.2.2 types of course*	328
3.6.5 job scheduling	job concept	172
3.6.2 command languages	job description	165
3.6.2 command languages	job running	166
3.6 operating systems	job scheduling	171
3.6.2 command languages	job set-up	166
3.7 data preparation	keyboard to magnetic tape	192
4.4.3 sorting and merging	key changes on a main file	250
4.4.4 file updating methods dealing with	key changes during updating	258
sort	keys *3.5.3 sorts*	155
3.5.4 file control	labelling	157
3.4 programming	languages	138
command	languages *3.6 operating systems*	165
commercial	languages *3.4.4 languages*	139
3.4.4 languages general	languages	140
high-level	languages *3.4.4 languages*	138
low-level	languages *3.4.4 languages*	138
scientific	languages *3.4.4 languages*	139
card	layout *3.7.1 punched cards*	180
print	layout charts *4.1.2 definition of output charts*	209
5.2.2 types of course	lectures	328
single-	level storage *3.1.3 core storage, central processors*	89
program	libraries *3.5.7 program maintenance*	162
types of	line printer *3.2.6 printer*	103
3.2.11 communications equipment	line systems	113
3.5.4 file control	loading	159
3.5.6 program handling software	loading	160
program	lockout *4.3.1 hardware/software controls*	228
3.1.5 data operations	logical operations	91
control	logic *3.1 processors*	86
table	look-up *3.4.7 programming techniques*	151
control	loop *3.2.6 printer*	105
modification and	loops *3.4.7 programming techniques*	147
3.4.4 languages	low-level languages	138
3.3.5 magnetic card file	magazine	123
3.3.5 magnetic card file	magnetic card	123
3.3 storage	magnetic card file	123
3.3.1 core storage	magnetic cores	116
3.3 storage	magnetic disc	121
3.3 storage	magnetic drums	120
3.3 storage	magnetic tape	117
keyboard to	magnetic tape *3.7 data preparation*	192
output	magnetic tape *4.1.4 choice of output medium*	215
sort techniques	(magnetic tapes) *3.5.3 sorts*	156

379

INDEX

	Key word	*Page No.*
block organisation for	magnetic tape files *4.5 design of a magnetic tape system*	281
characteristic structure of	magnetic tape systems *4.4 processing magnetic tape*	236
multiprogramming with	magnetic tape files *4.4 processing magnetic tape*	269
record design for	magnetic tape files *4.5 design of magnetic tape system*	277
reporting from	magnetic tape files *4.4 processing magnetic tape*	268
timing and evaluation of	magnetic tape procedures *4.5 magnetic tape system*	287
4.5 design of a magnetic tape system	magnetic tape files – structure and relationships	275
3.5.4 file control	maintenance	158
file	maintenance *4.4 processing for magnetic tape*	260
techniques of file	maintenance *4.4.5 file maintenance*	261
program	maintenance *3.5 utility software*	161
system	maintenance *Part 7*	363
identifying	management requirements *2.1 establishing systems objectives*	39
3.5.5 data management software data	management operations	159
3.5 utility software data	management software	159
3.2.10 character recognition equipment	mark reading	112
3.7 data preparation	mark sensing	191
characteristics of	mark sensing systems *3.7.7 mark sensing*	192
types of	master data *4.4 processing magnetic tape*	250
	memory (see storage)	
4.5.3 block organisation for magnetic tape	memory utilisation	282
sort/	merge generators *3.5.2 generators*	154
3.5.4 file control	merging	158
sorting and	merging *4.4 processing magnetic tape*	244
3.2.11 communications equipment	message switching	114
compiler	messages *3.4.5 diagnostics*	142
3.1.2 control logic	method of operation	88
3.2.10 character recognition equipment	MICR	112
conversational	mode *3.6 operating systems*	172
transfer	modes *3.2.1 interface*	95
3.2.5 paper tape punch	modes of punching	102
3.2.4 paper tape reader	modes of reading	101
3.4.7 programming techniques	modification and loops	147
5.3.2 program acceptance	modification to programs	332
3.2.11 communications equipment	modulation	113
4.3.1 standard hardware/software controls	modulo 11 check	229
3.4.5 diagnostics	monitor prints	144
3.6 operating systems	multi-access	173
4.1.3 approval of output formats	multiple copies	213
timesharing and	multiprocessing *3.1 processors*	93
3.1.6 timesharing	multiprogramming	93
4.4 processing magnetic tapes	multiprogramming with magnetic tape	269
scheduling for	multiprogramming *4.5.1 magnetic tape files – structure*	275
4.4.3 sorting and merging sorting	multi-reel files	250

	Key word	Page No.
establishing systems	objectives *2. investigation and analysis*	37
1.3 stages of systems analysis	objective setting	27
2.2.2 methods of fact finding	observing	49
3.2.10 character recognition equipment	OCR	112
on-line and	off-line working *3.2.11 communications equipment*	115
on-line and	off-line working *3.6 operating systems*	167
3.2.11 communications equipment	on-line and off-line working	115
3.6 operating systems	on-line and off-line working	167
4.1.4 choice of output medium	on-line output	215
3.4.6 remote testing	on-line testing	146
3.4.6 remote testing	open and closed shop	145
6. documentation standards	operating instructions	360
3.6.2 command languages	operating system control	166
3. data processing equipment	operating systems	164
systems design for	operating systems *3.6 operating systems*	174
method of	operation *3.7.3 mark sensing*	192
3.2.3 card punch	operation of card punch	99
3.2.2 card reader	operation of card reader	96
3.2.10 character recognition equipment	operation	111
3.3.1 core storage	operation	116
3.2.9 graph plotter	operation	108
3.3.4 magnetic disc	operation	121
3.3.3 magnetic drum	operation	120
3.3.2 magnetic tape	operation	117
3.2.6 printer	operation	104
3.2.7 typewriter	operation	106
3.2.8 visual display unit	operation	107
method of	operation *3.1.2 control logic*	88
arithmetical	operations *3.1.5 data operations*	91
data	operations *3.1 processors*	90
data management	operations *3.5.5 data management software*	159
computer	operations guide *6. documentation standards*	359
changes in	organisation *7.1 types of change*	366
data	organisation *3.3.5 magnetic card file*	125
data	organisation *3.3.4 magnetic disc*	122
data	organisation *3.3.3 magnetic drum*	120
magnetic tape	output *4.1.4 choice of output medium*	215
on-line	output *4.1.4 choice of output medium*	215
input/	output balance *4.5.1 magnetic tape files – structure*	276
input/	output channels *4.5.4 timing of magnetic tape*	288
approval of	output design *4.1.1 importance of clear objective*	209
appraisal of	output effectiveness *4.1 establishing output requirements*	218
4.2 data collection requirements impact of	output requirements on data collection	219
5.3.3 procedure acceptance	output formats	333
definition of	output formats *4.1 establishing output requirements*	209
input and	output functions *3.6.8 design – operating systems*	174

381

	Key word	Page No.
choice of	output medium *4.1 establishing output requirements*	214
analysing	overflow *4.6.9 overflow of sequential files*	309
packing density and bucket	overflow *4.6.3 indexing sequential files*	305
4.6.6 random files and address generation	overflow in random files	315
4.6 direct access processing	overflow of sequential files	306
4.6.5 timing of sequential files effects of	overflow on file processing times	312
segmentation and	overlays *3.4.3 coding*	137
housekeeping	packages *3.5 utility software*	152
4.6.3 indexing sequential files	packing density and bucket overflow	305
bucket	packing density *4.6.1 direct access file storage*	300
3.7 data preparation	paper tape	188
3.2.5 paper tape punch	paper tape codes	102
3.2.4 paper tape reader	paper tape codes	100
3.2 peripheral units	paper tape punch	102
3.2 peripheral units	paper tape reader	100
5.5.1 file creation: data collection	parallel data collection	337
5.4.2 pilot schemes and parallel running	parallel running	336
pilot schemes and	parallel running *5.4 changeover*	335
5.5.2 converting existing files	partial file creation	340
4.6.3 indexing sequential files	partial indexing	304
user	participation *5.1 user participation*	323
5.2.2 types of training	part-time courses	327
3.1.1 information patterns	parity checks	84
3.1 processors information	patterns	82
4.5.4 timing of magnetic tape	peripheral and operating times	290
3.5.1 housekeeping packages basic	peripheral control software	152
4.4.8 multiprogramming with magnetic tape	peripheral limited runs	270
total elapsed	peripheral times *4.5.4 timing of magnetic tape*	288
3.1.6 timesharing	peripheral transfers	93
3. data processing equipment	peripheral units	94
4.4.8 multiprogramming with magnetic tape	peripheral utilization	272
4.3.1 standard hardware/software controls write	permit (or inhibit) rings	227
5.5.1 file creation: data collection	phased file creation	337
5.4.2 pilot schemes and parallel running	pilot schemes	336
5.4 changeover	pilot schemes and parallel running	335
changeover	planning *5.4 changeover*	334
5.2 education for implementation	planning education	325
graph	plotter *3.2 peripheral units*	108
3.1.5 data operations floating	point and fixed point representation	92
floating point and fixed	point representation *3.1.5 data operations*	92
data	preparation *3. data processing equipment*	178
2.2.3 interviewing techniques	preparation for interviews	50
basic	principles (of coding) *3.4.3 coding*	136
general	principles *3.6 operating systems*	164
3.6.5 job scheduling	principles of scheduling	171
3.5.3 sorts	principles of sorting	154
4.1.2 definition of output formats	print layout charts	209
3.2 peripheral units	printer	103
types of line	printer *3.2.6 printer*	103

	Key word	Page No.
4.5.4 timing of magnetic tape	printer output files	292
3.5.4 file control	printing	158
3.4.5 diagnostics monitor	prints	144
store	prints 3.4.5 diagnostics	145
3.1.6 timesharing	priorities	94
acceptance	procedure 5.3.2 program acceptance	332
5.3 testing and acceptance	procedure acceptance	332
7.3 types of flexibility	procedure flexibility	369
5.3.1 test data	procedure test data	331
validation	procedures 5.3.3 procedure acceptance	333
random	processing 4.6.1 direct access file storage	298
sequential	processing 4.6.1 direct access file storage	298
serial	processing 4.6.1 direct access file storage	297
data	processing equipment Part 3	75
4.4.8 multiprogramming with magnetic tape	processor limited runs	270
file types and	processing modes 4.6 direct access processing	301
4.5.4 timing of magnetic tape	processor time	289
3. data processing equipment	processors	82
5.3 testing and acceptance	program acceptance	332
7.3 types of flexibility	program flexibility	368
3.5 utility software	program handling software	160
3.5.7 program maintenance	program libraries	162
4.3.1 standard hardware/software controls	program lockout	228
3.5 utility software	program maintenance	161
3.6.8 systems design for operating systems	program specification	175
flexibility	program techniques 4.1.5 exception reporting	216
5.3.1 test data	program test data	330
3.4.5 diagnostics	program testing	143
5.2.1 planning for education general	programme	325
3. data processing equipment	programming	128
3.7.1 punched cards	programming a card punch	186
3.4 programming	programming techniques	146
modifications to	programs 5.3.2 program acceptance	332
types of	project 5.1.2 project teams	324
establishing	project teams 2.1 establishing systems objectives	45
5.1 involvement in implementation	project teams	323
storage control and file	protection 4.6.1 direct access file storage	299
card	punch 3.2 peripheral units	98
operation of card	punch 3.2.3 card punch	99
programming a card	punch 3.7.1 punched cards	186
paper tape	punch 3.2 peripheral units	102
tabulator and summary	punch 3.8 punched card equipment	200
3. data processing equipment	punched card equipment	194
3.7 data preparation	punched cards	178
verifying	punched cards 3.7.1 punched cards	187
modes of	punching 3.2.5 paper tape punch	102
speeds of card	punching 3.7.1 punched cards	186
3.7.1 punched cards	punching cards	183
3.7.1 punched cards	punching codes	180
2.2.2 methods of fact finding	questioning	49

INDEX

	Key word	Page No.
4.6 direct access processing	random files	302
overflow in	random files *4.6.6 random files and address generation*	315
4.6 direct access processing	random files and address generation	314
4.6.1 direct access file storage	random processing	298
operation of card	reader *3.2.2 card reader*	96
card	reader *3.2 peripheral units*	95
paper tape	reader *3.2 peripheral units*	100
2.2.2 methods of fact finding	reading	48
uses of mark	reading *3.2.10 character recognition equipment*	112
modes of	reading *3.2.4 paper tape reader*	101
3.2 peripheral units character	recognition equipment	109
2.2.4 methods of fact recording	record description	62
4.5 the design of a magnetic tape system	record design for magnetic tape files	277
5.5.2 converting existing files	record structures	339
4.4.3 sorting and merging	record type coding	245
deletion of	records *4.4.5 file maintenance*	262
fixed length	records *4.5.2 record design for magnetic tape files*	279
variable length	records *4.5.2 record design for magnetic tape files*	279
3.1.2 control logic	registers	87
3.4 programming	remote testing	145
error	reporting *4.4 processing magnetic tape*	263
3.5.2 generators	report generators	154
principles of exception	reporting *4.1 establishing output requirements*	215
error	reporting conditions *4.1 establishing output requirements*	217
4.4 processing magnetic tape	reporting from magnetic tape files	268
dummy	reports *4.1.2 definition of output formats*	209
floating point and fixed point	representation *3.1.5 data operations*	92
3.8 punched card equipment	reproducing and gangpunching	197
dump and	restart *3.5 utility software*	163
4.4.9 dumps and restarts – magnetic tape	restart routines	273
4.4 processing methods for magnetic tape dumps and	restarts using magnetic tape	722
3.5.8 dump and restart	restarting	163
testing against expected	results *5.3.2 program acceptance*	332
3.3.5 magnetic card file	retrieval unit	124
systems	review *7 systems maintenance*	369
dump	routines *4.4 dumps and restarts – magnetic tape*	273
trace	routines *3.4.5 diagnostics*	144
4.2 analysing data collection	routing of source documents	223
basic	run structures *4.4.1 structure of magnetic tape systems*	238
job	running *3.6.2 command languages*	166
parallel	running *5.4.2 pilot scheme and parallel running*	336
pilot schemes and parallel	running *5.4 changeover*	335
3.5.6 program handling software	run-time amendment	161
peripheral limited and processor limited	runs *4.4.8 multiprogramming with magnetic tape*	270

Key word		Page No.
scheduling	runs *4.4.8 multiprogramming with magnetic tape*	271
mixing	run types *4.4.8 multiprogramming with magnetic tape*	270
time	scales *5.2.1 planning education*	326
5.2.2 types of courses	scheduling	329
4.5.1 magnetic tape files – structure	scheduling for multiprogramming	275
job	scheduling *3.6 operating systems*	171
principles of	scheduling *3.6.5 job scheduling*	171
5.4.1 changeover planning	scheduling of changeover	334
pilot	schemes *5.4.2 pilot schemes and parallel running*	336
5.4 changeover pilot	schemes and parallel running	335
3.4.4 languages	scientific languages	139
3.5.4 file control	scratching	157
3.1.6 timesharing	security	94
file	security *3.6.4 file storage*	171
4.6.1 direct access file storage the	seek area concept	298
3.4.3 coding	segmentation and overlays	137
4.6.3 indexing methods for sequential files	self indexing	303
mark	sensing *3.7 data preparation*	191
characteristics of mark	sensing systems *3.7.3 mark sensing*	192
4.6 direct access processing	sequential files	301
indexing methods for	sequential files *4.6 direct access processing*	303
overflow of	sequential files *4.6 direct access processing*	306
timing characteristics of	sequential files *4.6 direct access processing*	309
4.6.1 direct access file storage	sequential processing	298
4.6 direct access processing	serial files	301
4.6.1 direct access file storage	serial processing	297
job	set-up *3.6.2 command languages*	166
open and closed	shop *3.4.6 remote testing*	145
5.5.2 converting existing files	simulated updating	339
5.2.2 types of course	simulation	329
3.4.4 languages	simulators	141
3.1.3 core storage central processors	single-level storage	89
basic peripheral control	software *3.5.1 housekeeping packages*	152
data editing	software *3.5.1 housekeeping packages*	152
data management	software *3.5 utility software*	159
program handling	software *3.5 utility software*	160
use of	software *5.5.2 converting existing files*	339
utility	software *3. data processing equipment*	151
file processing	software controls *4.3 systems controls*	227
3.6.2 command languages	software utilisation	166
3.5.3 sorts	sort keys	155
3.5.2 generators	sort/merge generators	154
minimising	sort operations *4.4.3 sorting and merging*	248
3.5.3 sorts	sort techniques (magnetic tape)	156
3.5.3 sorts	sort techniques (other devices)	157
3.8 punched card equipment	sorter	194
principles of	sorting *3.5.3 sorts*	154

INDEX

	Key word	Page No.
4.4 processing methods for magnetic tape	sorting and merging	244
4.4.3 sorting and merging	sorting multi-reel file	250
3.5 utility software	sorts	154
3.7 data preparation	source document direct to computer	193
routing of	source documents *4.2 data collection requirements*	223
4.2 analysing data collection	source documents and data preparation	222
5.5.1 file creation: data collection	special exercise	338
program	specification *3.6.8 systems design for operating systems*	175
3.7.1 punched cards	speeds of card punching	186
5.2.1 planning education	staff categories	325
1. the role of the systems analyst	stages of systems analysis	27
implementation	stages *5.1.1 user participation*	323
1.2 the data processing function	steering committee	24
3. data processing equipment	storage	115
core	storage *3.1 processors*	88
core	storage *3.3 storage*	116
file	storage *3.6 operating systems*	168
single-level	storage *3.1.3 core storage central processors*	89
3.5.1 housekeeping packages	storage device control	153
direct access file	storage principles *4.6.1 direct access processing*	296
3.6.4 file storage file	storage utilisation	170
3.4.5 diagnostics	store prints	145
data processing department	structure *1.2 the data processing function*	25
file	structures *3.6.8 systems design for operating systems*	175
record	structures *5.5.2 converting existing files*	339
3.4.7 programming techniques	subroutines	149
2.2.4 methods of fact recording	suggestions section	70
tabulator and	summary punch *3.8 punched card equipment*	200
3.4.7 programming techniques	switches	149
computer procedures flowchart	symbols *2.2.4 methods of fact recording*	69
message	switching *3.2.11 communications equipment*	114
stages of	system approval *2.1.3 identifying management requirements*	40
operating	system control *3.6.2 command languages*	166
4.3 systems controls	system controls and auditors	234
2.3.3 analysis	system review checklist	75
5.3.3 procedure acceptance	system timings	333
line	systems *3.2.11 communications equipment*	113
operating	systems *3 data processing equipment*	164
characteristics of mark sensing	systems *3.7.3 mark sensing*	192
systems design for operating	systems *3.6 operating systems*	174
3.5.5 data management software	systems applications	160
1. the role of the systems analyst stages of	systems analysis	27
3.7.2 paper tape	systems considerations	191
6. documentation standards contents of	systems definition	354
3.6 operating systems	systems design for operating systems	174
2.2.4 methods of fact recording	systems file	54

386

	Key word	Page No.
Part 5	systems implementation	321
Part 7	systems maintenance	363
7. systems maintenance	systems review	369
identifying areas for	systems study *2.1 establishing systems objectives*	38
3.4.7 programming techniques	table look-up	151
decision	tables *3.4 programming*	129
3.8 punched card equipment	tabulator and summary punch	200
4.6.4 overflow of sequential files	tagging	307
keyboard to magnetic	tape *3.7 data preparation*	192
magnetic	tape *3.3 storage*	117
sort techniques (magnetic	tape) *3.5.3 sorts*	156
paper	tape *3.7 data preparation*	188
paper	tape codes *3.2.5 paper tape punch*	102
paper	tape codes *3.2.4 paper tape reader*	102
paper	tape punch *3.2 peripheral units*	102
paper	tape reader *3.2 peripheral units*	100
block size and	tape utilisation *4.5.3 block organisation, magnetic tape*	285
project	teams *5.1 involvement in implementation*	323
programming	techniques *3.4 programming*	146
3.5.3 sorts sort	techniques (magnetic tape)	156
3.5.3 sorts sort	techniques (other devices)	157
5.3 testing and acceptance	test data	330
procedure	test data *5.3.1 test data*	331
program	test data *5.3.1 test data*	330
on-line	testing *3.4.6 remote testing*	146
program	testing *3.4.5 diagnostics*	143
remote	testing *3.4 programming*	145
5.3.2 program acceptance	testing against expected results	332
5. systems implementation	testing and acceptance	330
processor	time *4.5.4. timing of magnetic tape*	289
5.2.1 planning education	time scales	326
peripheral and operating	times *4.5.4 timing of magnetic tape*	290
3.1 processors	timesharing and multiprocessing	93
5.4.1 changeover planning	timing	335
system	timings *5.3.3 procedure acceptance*	333
4.5 the design of a magnetic tape system	timing and evaluation of magnetic tape procedures	287
4.6 direct access processing	timing characteristics of sequential files	309
4.6.5 timing sequential files	timing estimates	310
4.5.4 timing magnetic tape	timing estimates	287
general	timing method *4.5.4 timing of magnetic tape*	287
batch	totals *4.2.3 source documents and data preparation*	225
batch	totals *4.4.2 input transcription programs*	240
3.4.5 diagnostics	trace routines	144
types of	training *5.2 education for implementation*	326
4.4 processing magnetic tape input	transcription programs	240
3.2.1 interface	transfer modes	95
peripheral	transfers *3.1.6 timesharing*	93
3.2.11 communications equipment GPO	transmission facilities	115

	Key word	Page No.
4.1.4 choice of output medium	turn-around documents	214
error	types *3.4.5 diagnostics*	141
7. systems maintenance	types of change	365
3.6.4 file storage	types of file	168
7. systems maintenance	types of flexibility	368
5.1.2 project teams	types of project	324
5.2 education for implementation	types of training	326
3.2 peripheral units	typewriters	105
arithmetic	unit *3.1.2 control logic*	87
control	unit *3.1.2 control logic*	87
interrogation and display	unit control *3.5.1 housekeeping packages*	153
peripheral	units *3. data processing equipment*	94
retrieval	unit *3.5.5 magnetic card file*	124
visual display	unit *3.2 peripheral units*	106
simulated	updating *5.5.2 converting existing files*	339
distinctions between	updating and maintenace *4.4.4 file updating methods*	254
4.4 processing magnetic tape file	updating methods	250
the nature of basic	updating runs *4.4.4 file updating methods*	254
7.1 types of change	user attitudes	367
5.1 involvement in implementation	user participation	323
block size and tape	utilization *4.5.3 block organisation for magnetic tape*	285
file storage	utilization *3.6.4 file storage*	170
main memory	utilization *4.5.3 block organisation for magnetic tape*	282
software	utilization *3.6.2 command languages*	166
3. data processing equipment	utility software	151
5.3.3 procedure acceptance	validation procedures	333
4.5.3 block organisation for magnetic tape	variable length blocks	285
4.5.2 record design for magnetic tape	variable length records	279
3.7.1 punched cards	verifying punched cards	187
2.2.4 methods of fact recording	vertical section chart	67
3.2 peripheral units	visual display unit	106
7.1 types of change	volumes	367
5.2.2 types of course	week-end courses	328
foreground and background	working *3.6.1 general principles (operating systems)*	164
on-line and off-line	working *3.6 operating systems*	167
6. documentation standards	working files	346
4.3.1 standard hardware/software controls	write permit or inhibit rings	227